THINKING COMPLEXITY

Complexity & Philosophy

Volume 1

ISBN: 978-0-9791688-7-1

Edited by

Paul Cilliers

Published and distributed by:

ISCE Publishing

395 Central Street
Mansfield, MA 02048

Thinking Complexity
Complexity & Philosophy - Volume 1
Library of Congress Control Number: 2007936999
ISBN: 978-0-9791688-7-1

Cover image provided courtesy of *pixelio.de*

Printed in the United States of America

Contents

Detailed Contents

Chapter 4
Archetypal Dynamical Systems and Semantic Frames in Vertical and Horizontal Emergence

Chapter 5
Hopeful Monsters: Literary Teleology and Emergence

Chapter 6
The Challenge of Complexity: Unfolding the Ethics of Science In Memoriam Ilya Prigogine

Chapter 7
Ethical Implications of the Laws of Pattern Abundance Distribution

Chapter 8
What is Complexity Science? Towards the End of Ethics and Law Parading as Justice

Chapter 9
Complexity and TINA

Chapter 10
Using Complexity Science in Organization Studies: A Case for Loose Application

Chapter 11
Knowledge, Complexity and Understanding

Chapter 12
Complexity Theory and Continental Philosophy Part 1: A Review of Letiche's Phenomenological Complexity Theory

Chapter 13
Complexity Theory and Continental Philosophy Part 2: A Hermeneutical Theory of Complexity

Chapter 14
Complexity in a Complex Europe: Reflections on the Cultural Genesis of a New Science

Chapter 15
Human Knowing and Perceived Complexity: Implications for Systems Practice

Chapter 16
On the Importance of a Certain Slowness

Complexity and Philosophy

Albert Einstein once pointed out, "The results of scientific research very often force a change in the philosophical view of problems which extend far beyond the restricted domain of science itself" (Einstein writing with Leopold Infeld). This is certainly true with regard to contemporary research into complex systems since, by stimulating a re-thinking of many of our fundamental assumptions about the nature of our organizations as systems, complexity theory is radically altering the sundry folklore and outmoded conceptual freight underlying traditional models of organizations. By re-thinking assumptions, the varied sciences of complex systems are inspiring a re-look at age old philosophical and metaphysical issues.

The new insights offered by complexity theory, therefore hold the promise that Schopenhauer once gave voice to, "...the corrected, extended, and more thorough knowledge of nature is the very knowledge that always undermines and finally overthrows the metaphysical assumptions that until then have prevailed." That is why *E:CO* includes a section committed to offering fresh and insightful papers on the interface of complexity and philosophy. In this section, researchers and scholars grapple with foundational questions emerging from the debate between modernist and post-modernist perspectives, a debate which has led to intriguing and fresh speculations in ontology, epistemology, cognition, semiotics, ethics, and aesthetics. Again, these articles have been selected through their recognition that new constructs are not merely ends in themselves, but can offer new methods with pragmatic benefits.

Jeffrey Goldstein

The Philosophical Importance of Thinking Complexity

Paul Cilliers
University of Stellenbosch, ZAF

Complexity theory has now been part of a wider intellectual world for the best part of two decades. One could probably mark the start of this interest with the publication of the two popular books by Lewin and Waldrop (both in 1993). In this time, thinking about complexity has had a marked influence in a number of disciplines, including but not limited to: Sociology, Health Care, Political Science, Anthropology, Management and Organizational Science, Sustainability Studies, Economics, Literary Studies and the Arts, in the Social Sciences; Physics, Biochemistry, Biology, Genetics, Ecology, Mathematics and Computational Theory in the Natural Sciences. Perhaps even more importantly, talk about the nature of complexity has generated a new and vibrant interaction between these various disciplines.

What is a little strange, however, is the relatively small impact Complexity has had on professional Philosophy. Two aspects of Philosophy in general may explain something of this strange state of affairs. In the first place, much of complexity theory has resulted from developments in mathematics and computational theory. This is not the normal domain of most philosophers. Complexity has been discussed in philosophy of science, mathematics and computation to a certain extent, but not really in philosophy of culture and social philosophy. A second reason may be that philosophy has somehow always been engaged with complex issues, even if it has not been done in the language used by contemporary complexity theorists. Many philosophers may feel, incorrectly in my view, that the language of Complexity is just a banal form of talking about things they have dealt with in a more subtle way for a century or two.

There is a certain irony in this state of affairs since I am convinced that the most important contribution made by Complexity is on a philosophical level. It has radically changed the way we think about science and society, with specific reference to notions like fundamental truth, objective knowledge, reductionism and causality. None of these notions have been dismissed, but the perspective from Complexity has necessitated a re-evaluation of their role and status. Thus, to my mind, Complexity has had less of an influence on the actual scientific methods we use to gather knowledge about the world, but a huge influence on how we interpret the results we obtain using these methods.

It is for this reason that one may distinguish between Complexity Theory, and what one could call Complexity Thinking. This second category, I would argue, comprises the *acknowledgement* of complexity as a vital first step. Once we realize that we are dealing with complex things, and we accept the consequences of this, our approach to what we are doing, irrespective of how we are actually doing it, will change fundamentally. The "complexity attitude" may be thoroughly informed by the findings of complexity theory, but not fully determined by it, for if that were the case, complexity theory would have become the new source of final truth and in the process contradict some of its own premises!

This position is, of course, a contested one. There are many who believe that complexity theory is uncovering fundamental truths about the nature of nature and that we are making progress towards a more comprehensive science of everything. This way of understanding Complexity will see less of a break with traditional science and continues the search for regularities and laws which have escaped us thus far. Those focussing on chaos, universality, power laws and fractal mathematics generally, but not exclusively, fall in this category.

There is absolutely no way in which I want to dismiss the importance of mathematical and computational complexity theory and the insights it generates. I just think that it is important to recognise the difference between these views and reflect upon the implications thereof. The distinction is sometimes marked by referring to "hard" complexity and "metaphorical" complexity. In this understanding, the first category would imply true scientific activity whereas the second refers to the softer, more interpretative strategies of the social sciences. This may be a useful way of thinking about the difference, but it does mask some of the underlying philosophical problems.

In order to make these problems more explicit, Morin (2007) distinguishes between what he calls "restricted" and "general" complexity. He argues that much of "hard" complexity science has not escaped the positivist and reductionist paradigms of traditional science. Although it produces significant shifts in the boundaries of a modernist worldview, it does not transcend it. What we need, he insists, is an acknowledgement of the real difficulties complexity brings with it. To a large extent we still lack the language in which to explicate a "general" understanding of complexity, but we should resist an unreflective use of a scientific discourse which re-instates an essential, if disguised, reductionism. If we do this, we deny some of the essential insights we have gained from complexity theory.

I think Morin is largely correct. One should make clear, nevertheless, that this position is not an excuse for relativism. It constitutes a challenge to develop a new kind of scientific understanding, it does not want to argue that sloppy work is okay. This position does not make things easier, it makes it more difficult, and we should acknowledge these difficulties (see Cilliers, 2005).

This volume of essays should contribute to the ongoing discussion about the nature of complex things. No attempt has been made to select only articles conforming to a specific understanding of complexity, nor does the editor want to impose an interpretation of these articles that would try to reconcile the many differences between them. The idea is exactly to present them *in* their difference and thereby to stimulate the discussion between the many viewpoints represented here. The only criterion used for selection was that all of them appeared as refereed papers in the journals *Emergence: Complexity and Organization* and its predecessor, *Emergence*.

Reading through these papers, it will soon become apparent that all the central philosophical themes affected by complexity thinking are addressed: Pluralism, Reductionism, Emergence, Political and Social Science, Organizational Science, Ethics and Aesthetics. The works in this volume should be approached with as little prejudice as possible, and also with a certain charity. They are dealing with difficult issues and it is not easy to talk about difficult things. Even so, I am convinced that the reader will come away with a general feeling of excitement. In each article there is an engagement with the world in which we live, coupled with a desire to understand it better. Ultimately, that is what Philosophy is all about.

Perhaps Complexity has not had such a great impact on hard core Philosophy, but Philosophy does owe Complexity a certain debt. I think that the language of Complexity has provided a way for a much larger audience to discuss a number of crucial philosophical problems. In the long run this may help the generate philosophies which are more accessible outside the lofty halls of Philosophy departments. It will also help to remind us that we all have to be philosophers at least some of the time.

This volume would not have been possible without the existence of *E:CO* and the efforts of its managing editor, Kurt Richardson. Competent and enthusiastic editorial assistance was supplied by Rika Allen.

References

Cilliers P. (2005). "Complexity, deconstruction and relativism," *Theory Culture & Society*, 22(5): 255-267.

Lewin, R. (1993). *Complexity: Life on the Edge of Chaos*, London, UK: Phoenix.

Morin, E. (2007). "Restricted complexity, general complexity," in C. Gershenson, D. Aerts, B. Edmonds (eds.), *Worldviews, Science and Us: Philosophy and Complexity*, Singapore, New Jersey, London: World Scientific Publishing Co., pp. 5-29

Waldrop, M. (1993). *Complexity: The Emerging Science at the Edge of Order and Chaos*, New York, NY: Simon & Schuster.

1. Why Integrative Pluralism?*

Sandra D. Mitchell
Department of History and Philosophy of Science, University of Pittsburgh, US

Introduction

This article is an exposition and defense of a perspective I call 'integrative pluralism'. I will argue that integrative pluralism is the best description of the relationship of scientific theories, models, and explanations of complex biological phenomena. Complexity is endemic in biology, and various features of multicomponent, multilevel, evolved systems constitute it. The types of scientific representations and the very methods we use to study biological systems must reflect both that complexity and variety. Developing models of single causal components, such as the effects of genetic variation, or of single-level interactions, such as the operation of selection on individuals, give valuable, if partial, accounts. These explanations need to be integrated in order to understand what historical, proximal, and interactive processes generate the array of biological phenomena we observe. Both the ontology and the representation of complex systems recommend adopting a stance of integrative pluralism, not only in biology, but in general.

Clearly, the way the world is dictates what we can say about it. The way our representations are structured also plays a significant role in the scientific accounts we develop. Theories and models are idealized, partial descriptions, couched in the conceptual frameworks of the day, framed in a language that carries meanings from the broader social context. The suggestion that our current best theories of the nature of nature exactly capture the world in all its details is hubris. The idealized and partial character of our representations suggests that there will never be a single account that can do all the work of describing and explaining complex phenomena. Different degrees of abstraction, attention to different components of a system, are appropriate to our varying pragmatic goals and conceptual and computational abilities.

If different models are perceived as partial solutions to a question, then one might argue that a theory of division of labor in social insects, for example, would be one that correctly unified the partial accounts. However, while integration of the partial accounts is indeed required for explaining a concrete particular, unification at the theoretical level is unlikely to be very robust. This is due to the nature of the complexity characterizing the domain of phenomena studied. It is the diversity of the 'solutions' to adaptive problems and the historical contingencies influencing those variable paths that preclude global, theoretical unification. Ants, for example, exhibit divi-

---

* Excerpted and revised from *Biological Complexity and Integrative Pluralism*. Copyright 2003 Cambridge University Press. Reprinted with the permission of Cambridge University Press.

sion of labor behavior similar to that of bees. Individual task performance varies with age, learning, and in response to internal and external contexts. Ant colonies, however, contain negligible genetic diversity compared with bee colonies. Thus, the theoretical constituents that would be integrated in the explanation of division of labor in ants would not be the same set as those required for the explanation of the 'same' phenomenon in bees. For ants, a self-organization model based on genetic diversity would often not apply. Even that claim is contingent, since Boomsma, *et al.* (1999) has detected in one species of ant nearly as much genetic diversity as in honeybees. Thus, what in fact explains the existence and characteristics of division of labor in social insects, what appeared at first sight to be the 'same' phenomenon requiring a single explanation, will be itself contingent on the particular features and pathways that occur in each case. At the same time, there is only one 'true' integrated explanation for honeybees and one 'true' integrated explanation for leaf-cutting ants, and hence competition among explanations in specific cases will, and does, occur. Nevertheless, compatible pluralism will remain for the models of potentially contributing causes, if not in the application of the models to explain a specific, concrete event.

A promising model of pluralism can be forged from understanding that causal models are abstractions that will always remain idealizations. By making simplifying assumptions regarding the noninterference of other potential causes, causal models describe only what would be expected in idealized circumstances (Levins, 1968). This conception of theories helps to explain how it can be that models of different causal factors *qua* models do not directly conflict. However, even if the models may be jointly consistent, in the application of models to the explanation of a concrete case, conflict can arise. In actual cases, multiple causes are likely to be present and interact, and other local elements may also contribute to a specific causal history. Thus in explanation, models of variant possible contributing factors must be integrated to yield the correct description of the actual constellation of causes and conditions that brought about the event to be explained.

"*Our choice of models, and to some extent our choice of words to describe them, is important because it affects how we think about the world... [O]ur choice of model decides what phenomena we regard as readily explicable, and which need further investigation*" (Maynard Smith, 1987: 120).

Pluralism in scientific theory and practice has certainly been studied and advocated before. As a result of historical investigations of scientific methods, Feyerabend was led to defend *epistemological anarchism* (1975, 1981). He endorsed any method as being as good as every other in generating acceptable scientific results. Like Feyerabend's, the most recent defenses of pluralism in science have launched their accounts from an epistemological perspective. Feminist stance theory, for example, grounds pluralism in a *perspectivalism* based on individual and group social experience (Harding, 1986; Longino, 1990). Other disunity supporters argue from the partial character of descriptions or diverging areas of interest of the researchers

(Cartwright, 1980, 1982, 1989, 1994; Dupré, 1983, 1996). In contrast, a few have linked pluralism with ontological views about natural kinds (Dupré, 1983, 1993; Hacking, 1996). While my approach draws on many of the insights of these philosophers, it differs from them in grounding pluralism jointly on metaphysical and epistemological arguments, and appealing to complexity as a critical tool for understanding the nature and limits of diversity in scientific methods and representations.

It has long been argued that biology has no laws (Beatty, 1995; Smart, 1968). Yet biologists speak of 'laws' in their writings. One of Mendel's 'laws' claims that with respect to each pair of alleles at a locus on the chromosome of a sexual organism, 50 percent of the organism's gametes will carry one representative of that pair, and 50 percent will carry the other representative of the pair. Recently, a number of biological scaling 'laws' have been discovered. These include Kleiber's law that metabolism increases in proportion to body mass raised to the 3/4 power, and the scaling law that respiratory rate is inversely proportional to body mass raised to the 1/4 power (Pool, 1997; West, *et al.* 1997).

Why do some philosophers fail to count these results of biological investigation as laws? How are they different from Proust's law of definite proportion, or Galileo's law of free fall or the conservation of mass-energy law? Those who argue that there are no laws in biology point to the historical contingency of biological structures and the particularity of the referents in biological generalizations as grounds for excluding the law designation. In considering the problem of the existence of biological laws I was led to a general reflection on laws in science. My conclusion is that we need to think about scientific laws in a very different way: to recognize a multidimensional framework in which knowledge claims may be located and to use this more complex framework to explore the variety of epistemic practices that constitute science.

Why Reductionism is Compelling

A compelling argument can be made for reductionism. We live in one world, not many worlds. Further, the material from which all the entities in the world are built is ultimately one kind of 'stuff', that is, matter. It is the job of scientists to represent the features of that one world in a way that allows us to explain the patterns of phenomena we observe, to predict what will occur in future situations, and to permit us to intervene in ways that allow us to accomplish our practical goals. The next step in this argument is to add the assumption that the representations scientists come to accept as true stand in some sort of strong mapping relation with the actual features or structures of the world. The strongest form of this argument would assume that the representations most widely accepted by the scientific community are direct mirrors of the world's structure.

If this were the case, then we would expect generally accepted, but diverse, representations of a given domain - for example, the different accounts of division of labor in social insects - to stand in some strong mapping relation to each other. What is that relation? Minimally, we would expect the representations to be consistent with each other. If the world is one, and scientific claims accurately describe it, then two contradictory statements cannot both be true. But reductionists argue for a stronger relation than this. They argue for intertranslatability, or derivability. Ultimately, there is one fundamental, maximally accurate description from which the others can be derived. These are the earmarks of a reductionist view of science. While intertranslatability is symmetric, derivation is not. There is a preferred ordering to the direction of such a reduction, and that ordering is generally provided by another metaphysical assumption, namely, compositional materialism.

The material composition assumption that every object is made up of one type of substance, namely, matter, suggests that there is some basic level of description of the material building blocks (Moser & Trout, 1995). The privileged level of description is taken to be the most fundamental, and one can come to understand the more complex objects by knowing the properties of the simple components and the composition functions. So if atoms make up molecules, and molecules constitute chemical elements, and elements make up different types of material objects, and material objects are the parts of cells, and cells make up organisms, and organisms make up societies, then if we understand atoms (or quarks or whatever we take as fundamental) and how these combine to form less fundamental objects, that should suffice. Scientific truths of biology then could be restated as truths about chemistry, those in chemistry in terms of physics. So the reasoning goes.

Indeed, if we believe that a description at the most fundamental level - physical or material description - is sufficient to describe the causal interactions responsible for all changes of state, that is, if we endorse the doctrine of causal completeness, then while descriptions, explanations, and predictions in a language of biology might be convenient (and true, if translatable or reducible to the fundamental level of representations), they are not necessary. We could, in principle, reduce all the diversity of the representations of contemporary science to statements describing the fundamental elements (see Causey, 1977).

Yet Reductionism Doesn't Capture the Realities of Scientific Inquiry

Grounds for rejecting reductionism for all cases are found in a more comprehensive analysis of the nature of scientific representations. The required simple mirroring relationship between theory and world does not hold. Since Kant, most philosophers accept that every representation will be shaped, in part, by the concepts that humans bring to the task of describing the world. While compositional materialism may be correct that all entities are made up of matter, the inference to a logical 'composition' relationship between the 'entities' in our scientific theories is not immediate.

Scientific representations are abstractions or idealizations (Cartwright, 1980, 1982, 1989, 1994; Dupré, 1983, 1993, 1996; Wimsatt, 1987). They can represent only partial features of individuals rather than the individuals themselves as complex causal agents. For example, parasite and host theories explain population-level interactions and identify abstract individuals in terms of their roles in those interactions. The theories are then used to explain actual interactions between concrete individuals such as particular malaria viruses and humans. Yet the abstract individual identified by its functional role in the population models clearly does not exhaustively describe the individual as a causal agent (Dupré, 1993). An individual human being is truly described in different theories at the same time as a host to a parasite, a consumer in an ecosystem, and a phenotypic expression of a set of genotypes, as well as a mammalian organism, a homeostatic endotherm, an organization of multiple cell types, and so on.

Actual, complex events or concrete individuals - the constituents of the one-world ontology - are, at the same time, instances of objects of multiple abstract theories concerned with different compositional levels. Reductionism requires replacing the higher-level abstractions by lower-level ones. Yet the abstractions, which constitute theoretical objects at the different levels, do not constitute identical representations across levels. That is, even if the descriptions at the various levels are all accurate, by being partial they may not be representing the same features of nature and hence would not stand in any straightforward derivability or intertranslatability relation, nor form a neat, nested hierarchy. Dupré defends this type of picture. He is a materialist but denies that "there is any interesting sense in which ontological priority must be accorded to the allegedly homogenous stuff out of which bigger things are made" (1993: 89). Thus, Dupré endorses a weak, compositional materialism, that is, the view that "whatever kinds of things there may be, they are all made of physical entities" (p. 92) that permits anti-reductionism. The material from which entities are composed does not carry all the explanatory weight. However, it is the association of matter with cause and the further alignment of cause with explanation that makes materialism and antireduction difficult bedfellows. In considering these issues in general, Dupré addresses the argument that runs from the requirement that explanations at different structural levels must at least be consistent with one another. That such consistency is ensured by reduction is obvious, but Dupré points out that the only way to reach such consistency entails that reduction rests on a particular view about causality, namely, the assumption of *causal completeness.*

Thus, reductionists present an even stronger argument that higher-level descriptions are dispensable. Causal closure is a doctrine about sufficient cause. Dupré considers this argument and rejects it. Dupré's reconstruction of the reductionist's argument is as follows: Consider an event at the microphysical level. Dupré uses the example of the motion of an electron in one's index finger. As a person moves one's finger to type the letter 'b' on a keyboard, the electron has to move. If the microlevel is causally complete, then there is a set of events at that level, prior to the motion of the electron, that is sufficient to cause it to so move. Nevertheless, an explanation might be given at a macro level, that is, by appealing to one's intention to type the letter 'b' to explain the motion of the finger that, of course, entails the movement of the electron. Since what is going on at the micro level, given the assumption of causal completeness, is sufficient to bring about the motion of the electron, the macro story must be at least consistent with the micro causal story. Furthermore, since causal completeness presumes sufficiency at the micro level, nothing more at the macro level is necessary to bring about the movement of the electron. Hence, macro-level events appear to be causally inert. Since on this account, explanation is in terms of causes and all one needs for the causal story is microlevel events, all explanation could be run at the micro level. But this is just to restate the doctrine of in-principle reductionism.

This is a powerful argument, and one that sits behind many contemporary defenses of some form of reduction. In the face of it, many have promoted what Dupré correctly identifies as a weak form of reductionism, namely, *supervenience*. Others have bitten the bullet on causal explanations being the domain of the micro level and have opted for a defense of some other kind of explanation as appropriate to macro-level science. But Dupré wants to defend a much more robust pluralism and so claims that "... a central purpose of the ontological pluralism [he has] been defending is to imply that there are genuinely causal entities at many different levels of organization. And this is enough to show that causal completeness at one particular level is wholly incredible" (1993: 101). That may well be his purpose, but we have to look at whether or not he can get there with the arguments he provides. Dupré attempts to invert the reductionist *modus ponens* (causal completeness requires reductionism) into an antireductionst *modus tollens* (the failure of reductionism implies the falsity of causal completeness). While Dupré's account of the relationship between the two theses of causal completeness and reduction is correct on a purely ontological interpretation, his arguments do not accomplish the inversion he desires.

What Dupré uses as evidence that reductionism fails is the implausibility of it succeeding in any of a variety of the cases in biology and psychology that he considers. But all this shows is that reduction is unavailable in fact in these cases. It does not show that it is impossible in principle. It is the second, stronger claim that is needed to overturn causal completeness, for it is in principle reduction that figures as the conclusion of the reductionist argument.

I suggest a different argument that can be launched against reductionism. The reductionist argues that if the world is composed of physical matter and if all composite entities are made up of just physical matter, then if there is an account of the complete cause of events at the physical level, nothing more causally, and hence explanatorily, is required by higher-level descriptions. But how do we understand that? There are two ways.

First, metaphysically, it is the view that there is at the physical level a single, unique process of physical interactions that bring about a physical result. Unless one adopts a position of uncaused events, this does seem inescapable. But what about representing this process in physics? One can believe that all events are caused and at the same time argue convincingly that all the factors contributing to the complete cause cannot be represented by any single theory (and the representational mechanism that instantiates it) in physics. The local, contingent constituents of every causal process just, are not included in the scope of physical theory, as Cartwright has argued (1994), and these will always be part of the complete cause. So there may well be causal closure at the level of physical entities, while there is always incompleteness, or causal openness, in the representation or theorizing about those processes, the representations that make up the physics entities.

There is a further source for disarming causal completeness as the ground for reduction. That is the view that there are two aspects to composite, complex objects or events: the material and the manner in which the material is arranged, that is, matter and form, or material and structure. Some patterns in the world that we identify as causal processes may depend as much on the structural characteristics of complex objects as on the material characteristics. In fact, as Bechtel and Richardson (1993) have suggested, there may well be a continuum of contributions from matter and structure such that actual causal processes occupy many different locations.

How does this bear on causal closure? Well, if the physical level is construed only materially, then structure is a level up and causally significant. Hence causal closure is false. If, however, structure is included in the physical level, then macro objects are physical and closure applies to a collection of micro and macro objects and events. Either way, reduction to a purely material physical level is thwarted.

By breaking the connection between physical material and physics representations, the arguments for causal completeness and theory reduction are no longer as closely linked, and Dupré's reverse inference from nonreduction to noncompleteness does not go through. Nevertheless, there are conceptual and empirical arguments against the inference from causal closure to reduction that I believe make a strong case for the rejection of reductionism.

There are recent, seemingly new, arguments for reductionism, especially in the 'new wave' reduction of psychology to neurophysiology (Bickle, 1997; Churchland, 1986). These permit appropriate modification of terms and claims of the two sciences to accommodate the elimination of false claims by reduction to lower-level true claims, thus recognizing that the mapping between terms need not be direct. There are specific objections, concerning the translatability, eliminability, and causal efficacy of functional properties that challenge the newer versions of reduction (Schouten & De Jong, 1999). Each new wrinkle in the reductionist position elicits new responses from anti-reductionists. Rather than counter every new twist in the reductionist argument, I will instead attempt to articulate and explore an alternative view of the relations among scientific claims.

Alternatives to Reductionism

'Anything goes' pluralism is associated with Paul Feyerabend's epistemological anarchism (Feyerabend, 1975, 1981). In the absence of foundational standards of justification, Feyerabend argues that any and all forms of argument are acceptable in science. Some versions of social constructivism (see Collins, 1982, 1985; Latour, 1988; Woolgar, 1988) can be taken to promote 'anything goes' pluralism, at least as far as truth is concerned. Perhaps it would be more aptly identified as 'anything that scientists accept goes'. This shifts the criteria by which different scientific claims are granted authority from their representational adequacy to their social authority. Thus, one might argue that as long as some group of self-identified scientists agree to accept a claim about the world, then it 'goes' and the constraints on the ways in which scientific groups are formed may be loose enough to condone an interpretation of 'anything goes'.

I find both the advocacy of retaining all, possibly inconsistent, theories that emerge from a community of investigators and the insistence that any collection of analyses of the same phenomena must be reduced to a single theory equally unacceptable. The challenge is to define clearly the middle ground: How can a diverse, well-confirmed, but irreducible set of theories be used collectively to achieve a more complete understanding than any of the theories taken in isolation?

A *prima facie* problem for reductionism is the apparent diversity of theories in a science such as ecology. It has been argued that some of the compatibility of diverse accounts can be explained by the divergence of questions and interests that the scientist brings to the table. Mayr, Tinbergen, and Sherman have all defended a 'levels of analysis' account of the plurality of theories and explanations that is almost dogma within the biological community (Mayr, 1961, 1982; Sherman, 1988; Tinbergen, 1963). Mayr distinguished between proximate 'how' and ultimate 'why' questions in the face of the possible encroachment of genetics into whole organism biology (Beatty, 1994). Tinbergen elaborated four different kinds of questions posed in the study of ethology. A recent revival of this model by Sherman further subdivides the four 'levels' of questions that partition biological research. These are: evolutionary origins, functional consequences (the two 'why' questions), ontogenetic processes, and mechanisms (the two 'how' questions). Sherman's addendum divides questions of mechanisms into those that target physiology and those that target cognition (Sherman, 1988).

For example, biologists explaining division of labor in social insects might approach this phenomenon with four different questions. Those concerned with evolutionary origins would investigate why this behavior arose when it did multiple times in the lineages of social insects; those with functional concerns would ask what current consequences the trait has on colony ergonomic efficiency and hence fitness; those with queries about ontogenetic processes would aim to explain how the various behaviors come to be expressed in the individual insects over time; and those interested in mechanisms would detail the environmental triggers and hormonal or cognitive mechanisms that then issue in the behaviors. Different questions invoke different explanatory schemata. Sherman claims, "Every hypothesis in biology is subsumed within this framework; competition between alternatives appropriately occurs within and not among levels" (Sherman, 1988: 616). Thus, answers to questions at the different levels represent compatible components of a pluralistic, multidimensional body of knowledge. On this view, there is no need to account for intertheory relations among the levels.

Even if we granted that this describes current scientific practice, the question remains as to what relations should connect these autonomous enterprises. One reading of this type of compatibilism leads to an isolationist stance with respect to the separate analyses. If there is no competition between levels, there need be no interaction among scientists working at different levels either. The problem with the isolationist picture of compatible pluralism is that it presupposes explanatory closure within each 'level of analysis' and a narrowness in scope of scientific investigation that precludes the type of fruitful interactions between disciplines and subdisciplines that has characterized much of the history of science (see Darden & Maull, 1977). Even when the unlikely situation occurs that a scientist is narrowly concerned with only one level of analysis, it is a mistake to think that the answers to other questions have no bearing on the investigation at that level. There may be causal dependence or causal interaction between processes described by the different analyses. If so, remaining within a single level will fail to provide understanding for the questions addressed at that very level (Bechtel & Richardson, 1993). Thus, the division of levels of analyses, different questions with different answers, could be mistakenly interpreted as a justification for unproductive heuristics for scientific investigation.

The answers to different questions appeal to distinct abstract models of causal processes as well as specify their application to concrete instances. Fundamental scientific laws used in explanations of concrete phenomena have *ceteris paribus* clauses (Cartwright, 1980). In other words, they describe what is to be expected in idealized situations, when only one of a set of potential causal factors is operating, that is, when nothing else is interfering. All simple models also suffer by design from this type of unrealism. If we accept that multiple causal factors can, and often do, interact in the production and maintenance of the phenomena that the real cases are complex, without the *ceteris paribus* proviso, the laws would be literally false. With the proviso, however, such models do not themselves directly account for many, if any, real cases. That is, while abstract models describe the effects of the operation of a single causal process, our world does not normally approximate the ideal world that the models directly represent. The concrete explanatory situations on which we bring the abstract models to bear are messy, perhaps unique products of historical contingencies and interacting, multiple causal factors. In this sense, the simplified, abstract, theoretical models developed in science are literally false (cf. Wimsatt, 1987). This does not make them undesirable or useless (cf. Richerson & Boyd, 1987). Indeed, simplification allows models to be mathematically and empirically tractable for increasing precision of assumptions, crisper testability, and allowing counterintuitive results to be generated. The robust convergence of results of a variety of simple models is evidence that the result does capture a feature found in the complex world.

Nevertheless, the pragmatic virtue of simplicity is most frequently bought at the cost of realism in explanation (Levins, 1968). Though, of course, it is possible that a single causal process may completely determine a particular event, given the complexity and diversity of the phenomena, simple models are more likely to capture partial causes. In this situation, individually more complete models may be developed, if there are means for replacing the false assumptions in the model by more realistic descriptions (say, in models of language acquisition or measures of genetic fitness). At the most concrete level in generating an explanation, a model may introduce all of the relevant features that uniquely characterize a given event (say, the ecological conditions of a particular lake). However, the cost here is that of generality. A full-scale map of a town would express the greatest realism; however, it would be as useless for finding city hall on a map as one that represented the town as a single point. So, too, all that is true in a concrete model of Clear Lake will likely not apply directly to a neighboring lake. It is by distinguishing the idealized models from their applications that we can identify the location and scope of integration.

In short, *isolationist pluralism* employs a levels-of-analysis framework to endorse a strategy of limiting interactions between various theories offering explanations in a given domain. While some scientists may restrict their interests to a specific level only, this is not necessarily the case. Pluralism better describes the causal models that, by modeling the contribution of individual causes, necessarily abstract away the operation of other compounding factors. By so doing, they can make no incompatible claim about the operation of the ignored causes. Once this structure of causal models is recognized, one may understand why competitive interactions arise within a level (e.g., how best to measure fitness in determining the current function of a cultural trait). However, the model must be distinguished from its application. In application, one can immediately see that causal models that provide answers at different levels are indeed related. Thus, although pluralism is to be defended, it is not the pluralism of questions and the consequent independence of answers, but rather a pluralism of models of causal processes that may describe contributing factors in a given explanatory situation. This is not to recommend an 'anything goes' pluralism. Not all explanations are equally good. Hence, to defend a strategy of pluralism for causal models and criticism of explanatory applications of those models requires a further account of how idealized models are to be integrated into explaining concrete, nonideal cases.

Integration

I have argued elsewhere that the dual complexity of the phenomena studied by scientists and the diverse interests and pragmatic constraints on the representations scientists devise to explain the phenomena conspire against simple pictures of scientific knowledge (Mitchell, *et al.*, 1997; Mitchell, 2000). Correspondingly, the strategy for integrating diverse theories and explanations will not be algorithmic. This is evident from even a superficial investigation of how genetics and population biology are jointly modeled, or of current models of how the biochemistry of hormone production in a developing organism affects and is affected by the external environmental conditions in which the organism finds itself. The genetics of a population will constrain the variation on which natural selection can operate, and the operation of natural selection can change the genetic constitution of the population. Complex systems, like those studied by biology, are going to harbor multiple, interacting forces at different scales, with variable temporal orders operating in diverse combinations in different particular situations. Integration of theories and models in such cases will not be as simple or global as in the case of, say, vector addition of electromagnetic and gravitational forces in physics.

The work of Michael Friedman (1974) and Philip Kitcher (1981) seems to acknowledge the persistence of multiple potential descriptions of individual phenomena - the pluralism side of the story - but they marry it to a unificationist goal rather than an integrationist goal. Indeed, they argue that the unification of ever more phenomena under one theoretical schema is central to scientific explanation. This can be seen in Kitcher's criteria for identifying scientific progress with the acceptance of theories embodying the fewest explanatory schemata for the widest phenomenal coverage. This type of explanatory unification is certainly evident in some of the best-known cases in the history of science, including, famously, Newton's 'unification' of terrestrial and celestial mechanics, Lyell's 'unification' of the historically distant and current geological forces of change, and Darwin's 'unification' of explanations of diversity and distribution of phenotypic features of humans and nonhumans by appeal to a single schematic principle of natural selection on heritable variation. Subsumption of diverse phenomena by appeal to increased generality and abstraction is indeed one way in which a plurality of accounts can be related. This constitutes a type of theoretical integration (see also Morrison, 2000). However, I do not think it is the only mode of relationship. Therefore, identifying theoretical unification as the only means of doing good science is a mistake that removes the impetus to understand the value of diverse integrative strategies in scientific inquiry.

Darwin's great unifying insight in *The Origin of Species* (1859) was to see that the very different situations of the diversity of morphology of the thirteen species of finches he observed in the Galapagos Islands, the variety of domesticated pigeons in England, and the precise construction of the orchids he investigated were all explained by a single schema: natural selection operating gradually over slight heritable variation in individual members of a population.

While selection characterized at this level of abstraction unifies quite a significant number of phenomena, there are reasons to go both more abstract and less abstract with corresponding increases and decreases in generality. Within biology, hierarchical selection theory expands the biological targets from Darwin's individual organisms in a population both up to kin groups, trait groups, unrelated groups, and possibly species levels and down to gametes and genes. Darden and Cain (1989) move up the scale and beyond biology. "Natural selection, clonal selection for antibody production, and selective theories of higher brain function are examples of selection type theories. Selection theories solve adaptation problems by specifying a process through which one thing comes to be adapted to another thing" (p. 106). (See also Skipper, 1999.)

Within biology, even individual or organismic selection is broken down into *r* and *k* selection. *r* selection characterizes the processes occurring in populations that are not near the carrying capacity of their environment, and hence selection occurs by increasing fecundity, while *k* selection operates in populations at their quantitative limit, and hence the mechanism is instantiated by increased survival of offspring rather than number of offspring. The consequences of these different kinds of organismic selection are not always in concert. A number of studies both in the lab and in the field show mixed results of the relationship between fecundity and survival: out of 22 lab studies, a negative relationship was found in seven (and positive in five); out of 41 field studies, a negative relationship was found in 23 (and positive in four); the remaining cases were nonsignificant (Stearns, 1992). Indeed, the more concretely described individual processes that constitute organismic selection may operate antagonistically, additively, or synergistically. Hence, collapsing the different processes constituent of natural selection operating on organismic variation in a population into a single representation of the process of selection can obscure what may be important differences. The point is that different levels of abstraction are required for different tasks (see also Mitchell, 2000 for a version of this argument applied to chemistry).

Specific theoretical unifications, while being one form of integrating diverse causal models, must be justified by appeal to more than the fewer-the-better argument. That is, one must at least specify what it is 'better' for.

Integrative Pluralism

Integration, the alternative to both reduction and isolation, occurs at many levels of abstraction and is driven by a variety of pragmatic interests. Establishing the philosophical arguments for the need for some form of integrative pluralism is clearly only the first step toward a better understanding of science. Indeed, the arguments I have given for expecting pluralism imply that the types of integration within science will also be varied and diverse. No single theoretical framework, no simple algorithm, will suffice. This is also evident from the type of case study-driven work that has already considered questions of integration. Darden and Maull (1977) outlined several kinds of 'interfield' relations that might characterize integration across two theoretical boundaries: physical localization and part-whole relation, physical description of an entity described in another theory, identifying a structure underlying a function, and identifying a cause in one field of an effect described in another (see also Darden, 1986, 1991). Bechtel (1986) identified other 'cross-disciplinary' patterns that are nonreductive: conceptual links between disciplines to induce modification of perspective in one or the other, recognition of new levels of organization to solve unsolved problems in existing fields, using research techniques from one field to develop theoretical models in another, extending and applying a theoretical framework from one field to another, and "developing a new theoretical framework that will reconceptualize research in now separate domains as it tries to integrate them" (p. 47). Bechtel and Richardson (1993) have further explored integrative strategies in their more recent work, especially with respect to crossing different levels of organization. These studies call for further cases in order to catalog what appears to be a wide variety of ways in which integration proceeds.

My own investigations of interdisciplinary work between developmental and evolutionary models (Mitchell, *et al.*, 1992) and also between biology, writ large, and cultural models (Mitchell, *et al.*, 1997) have suggested a set of working hypotheses. In that work I proposed three types of integration: (1) mechanical rules, (2) local theoretical unification, and (3) explanatory, concrete integration.

Mechanical rules can be used to quantitatively determine the joint effects of independent additive causal processes explained by different theories. Vector addition on the contributions of electromagnetic and gravitational forces to resultant motion is an example. The integration of theories is simply a demonstration that they are simultaneously applicable in a linear way. Sewell-Wright attempted to do the same for the effects of mutation and selection on gene frequencies (see Sober, 1987). *Prima facie*, this type of integration seems appropriate for causes that are additive and operate on the same entities for comparable time periods. However, some biological phenomena do not seem to be amenable to mechanical rule integration. Think of the slime mold. When there is sufficient food in the environment, slime mold exists as independent amoebae. They move, feed on bacteria, and reproduce by cell division. When food becomes scarce, a chemical signal stops cell division and they move toward each other to form a multicellular 'pseudoplasmodium' of tens of thousands of cells. This new association of cells then differentiates into a stalk supporting a fruiting body that produces spores. The spores are launched to spread to a new environment (with better chances of food), and life as individual amoebae begins again (Keller & Segel, 1970). This type of 'emergent' effect of the interaction of individual components of a complex system is typically nonlinear, and thus the individual component causes are nonadditive. Therefore, theories explaining the causes of emergent phenomena requires consideration of interactions.

The second model of integration, *local theoretical unification*, aims to develop models in which a number of features of a complex process are jointly modeled. This is similar to the explanatory unification counseled by Friedman and Kitcher. However, as I argued above, the appropriate scope of the unity and corresponding degree of abstraction will be settled by a combination of pragmatic and ontological constraints. The problem of scale in ecology illustrates this. In discussing the trade-off of detail for generality in modeling evolution on different classes of entity and ecological relationships among populations, Levin comments, "Here ... the problem is not to choose the correct scale of description, but rather to recognize that change is taking place on many scales at the same time, and that it is the interaction among phenomena on different scales that must occupy our attention" (1992: 1947). Furthermore, "In general, one must recognize that different processes are likely to be important on different scales, and find ways to achieve their integration" (p. 1950). A more concrete example of this is found in the discussion of 'top-down' and 'bottom-up' theories of the regulation of trophic structure and species composition in an ecosystem. Liebold, *et al.* (1997) argue that these "two artificially distinct perspectives" (p. 468) that model the effects on trophic structure of predation and resources, respectively, are both supported by empirical evidence. They propose a 'synthesized' model that takes both forces into account and thereby links community and ecosystem approaches. At the same time, they recognize that their model may still represent only "one subsystem in the more complex array of food web interactions," a local unification in my terminology, or, alternatively, may predict "the cumulative behavior of many subsystems that act in parallel but roughly additive ways" (p. 483), thus constituting something like a mechanical rule.

The third type of integration, *explanatory, concrete integration*, appears to occur in cases of high complexity and pressing pragmatic goals (see Oreskes, *et al.*, 1994). That is, when a large number of at most partially independent factors participate in structuring a biological process, and where those factors span time and dimension scales as well as standard scientific disciplines, even modest theoretical unification will be elusive. Think of the changes of state of a complex ecosystem such as that of Lake Erie. There is an ongoing modeling project to consider the lake-wide effects of the invasion of zebra mussels, declining phosphorus loading, continuing toxic contamination, and fish harvesting on the structure of the fish community of that lake (see Culver, 1999; Koonce & Locci, 1999). The different factors contributing to these effects are large and diverse, including the chemicals silica, ammonia, nitrate, phosphate, total phosphorus, and phosphorus in sediments; five taxa of phytoplankton, six taxa of herbivorous zooplankton (including zebra mussel veligers), and three taxa of predatory crustacean zooplankton; zebra mussels and four taxa of other macrobenthos; and eleven taxa of planktivorous fish and six taxa of piscivorous fish. The model also represents seasonal and spatial variation in solar radiation. "Phytoplankton, zooplankton, and planktivorous fish are sensitive to light, temperature, and nutrient concentrations they experience in nonlinear ways.

Hence one cannot adequately model the functions of the pelagic zone using the basin-wide averages of state variables" (Culver, 1999). Features of the method of integration of these multiple factors for a single lake may be local to Lake Erie or may be symptomatic of a class of situations, but are unlikely to be global and algorithmic. As such, I suggest, it may be better described as explanatory, concrete integration.

Conclusion

I take this project to be similar in spirit to Carnap's analysis of the acceptance of different linguistic forms within science. He concludes his investigation of the relative worth of using thing language, abstract language, or not speaking at all:

"*The acceptance or rejection of abstract linguistic forms, just as the acceptance or rejection of any other linguistic forms in any branch of science, will finally be decided by their efficiency as instruments, the ratio of the results achieved to the amount and complexity of the efforts required. To decree dogmatic prohibitions of certain linguistic forms instead of testing them by their success or failure in practical use is worse than futile; it is positively harmful because it may obstruct scientific progress. The history of science shows examples of such prohibitions based on prejudices deriving from religious, mythological, metaphysical, or other irrational sources, which slowed up the developments for shorter or longer periods of time. Let us learn from the lessons of history. Let us grant to those who work in any special field of investigation the freedom to use any form of expression which seems useful to them; the work in the field will sooner or later lead to the elimination of those forms which have no useful function. Let us be cautious in making assertions and critical in examining them, but tolerant in permitting linguistic forms*" (Carnap, 1950).

I endorse both Carnap's pragmatic standard and his plea for toleration. However. I believe it should be applied not only to linguistic expressions within science, but also to philosophical expressions about science.

In defending integrative pluralism, an image of science that makes room for compatible pluralism, I have attempted to steer clear of two undesirable methodological pitfalls. The first is an isolationist stance that partitions scientific investigations into discrete levels of questions and their corresponding answers in a way that precludes the satisfactory investigation of any of the levels. The second is an uncritical anarchism that endorses all and any propositions. Neither of these positions correctly locates where and when competition in fact occurs between theories and explanations in biology. I have appealed to the idealized structure of scientific models and emphasized the distinction between the model and its application to a concrete situation. While the idealized and abstract character of models allows compatibility at the theoretical level, the realistic and concrete nature of explanation entails integration and resolution. Given the multiplicity of causal paths and historical contingency of biological phenomena, the type of integration that can occur in the application of models, that is, their use in explanations, will itself be piecemeal and local. The result is that pluralism with respect to models can and should coexist with integration in the generation of explanations of complex and varied biological phenomena. By extension, integrative pluralism is a fruitful approach to the generation of explanations of most complex phenomena.

References

Beatty, J. (1994). "Ernst Mayr and the proximate/ultimate distinction," *Biology and Philosophy,* 9: 333-356.

Beatty, J. (1995). "The evolutionary contingency thesis," in G. Wolters, and J. G. Lennox (eds.), *Concepts, theories, and rationality in the biological sciences*, Pittsburgh: University of Pittsburgh Press, pp. 45- 81.

Bechtel, W. (1986). *Integrating scientific disciplines*, Dordrecht: Martinus Nijhoff.

Bechtel, W. and Richardson, R. C. (1993). *Discovering complexity: Decomposition and localization as strategies in scientific research*, Princeton: Princeton University Press.

Bickle, J. (1997). *Psychoneural reductionism: The new wave*, Cambridge, MA: MIT Press.

Boomsma, J. J., Fjerdingstad, E. J. and Frydenberg J. (1999). "Multiple paternity, relatedness and genetic diversity in Acromyrmex leaf-cutter ants," *Proceedings of the Royal Society of London Series*, B(266): 249-254.

Carnap, R. (1950). "Empiricism, semantics, and ontology," *Revue Internationale de Philosophie*, 4: 20- 40.

Cartwright, N. D. (1980). "The truth doesn't explain much," *American Philosophical Quarterly*, 17: 159-163.

Cartwright, N. D. (1982). *How the laws of physics lie*, Oxford, UK: Oxford University Press.

Cartwright, N. D. (1989). *Nature's capacities and their measurement*, Oxford, UK: Oxford University Press.

Cartwright, N. D. (1994). "Fundamentalism vs. the patchwork of laws," *Proceedings of the Aristotelian Society*, XCIV: 279-292.

Causey, R. L. (1977). *Unity of science*, Dordrecht: Reidel.

Churchland, P. S. (1986). *Neurophilosophy: Toward a unified science of the mind-brain*, Cambridge, MA: The MIT Press.

Collins, H. M. (ed.) (1982). *The sociology of scientific knowledge: A sourcebook*, Bath, UK: Bath University Press.

Collins, H. M. (1985). *Changing order: Replication and induction in scientific practice*, London, UK: Sage.

Culver, D. A. (1999). "Ecological modeling of Lake Erie trophic dynamics." http://zeus.uwindsor.ca/erie2001/Modelling_Summit/culvermod.htm.

Darden, L. (1986). "Relations among fields in the evolutionary synthesis," in W. Bechtel (ed.), *Integrating scientific disciplines*, Dordrecht: Nijhoff, pp. 113-123.

Darden, L. (1991). *Theory change in science: Strategies from Mendelian genetics*, New York, NY: Oxford University Press.

Darden, L. and Cain, J. (1989). "Selection type theories," *Philosophy of Science*, 56: 106-129.

Darden, L. and Maull, N. (1977). "Interfield theories," *Philosophy of Science*, 44: 43-64.

Darwin, C. (1859). *On the origin of species*, Harvard, MA: Harvard University

Press (1964).
Dupré, J. (1983). "The disunity of science," *Mind*, 92: 321 346.
Dupré, J. (1993). *The disorder of things: Metaphysical foundations of the disunity of science*, Cambridge, MA: Harvard University Press.
Dupré, J. (1996). "Metaphysical disorder and scientific disunity," in P. Galison, and D. Stump (eds.), *The disunity of science*, Stanford: Stanford University Press, pp. 101-117.
Feyerabend, P. K. (1975). *Against method*, London, UK: New Left Books.
Feyerabend, P. K. (1981). *Philosophical papers*, Cambridge, UK: Cambridge University Press.
Friedman, M. (1974). "Explanation and scientific understanding," *Journal of Philosophy*, 71: 5-19.
Hacking, I. (1996). "The disunities of the sciences," in P. Galison, and D. Stump (eds.), *The disunity of science*, Stanford: Stanford University Press, pp. 37-74.
Harding, S. G. (1986). *The science question in feminism*, Ithaca: Cornell University Press.
Keller, E. F. and Segel, L. A. (1970). "Initiation of slime-mold aggregation viewed as an instability," *Journal of Theoretical Biology*, 26: 399-415.
Kitcher, P. (1981). "Explanatory unification," *Philosophy of Science*, 48: 507-531.
Koonce, J. F. and Locci, A. B. (1999). "Resoluton of issues of scope and detail in the development of the Lake Erie ecological model," http://zeus.uwindsor.ca/erie2001/Modelling_Summit/Kooncemod.htm.
Latour, B. (1988). "The politics of explanation," in S. Woolgar (ed.), *Knowledge and reflexivity: New frontiers in the sociology of knowledge*, London, UK: Sage, pp. 155-176.
Levin, S. A. (1992). "The problem of pattern and scale in ecology: The Robert H. MacArthur Award Lecture," *Ecology*, 73: 1943-1967.
Levins, R. (1968). *Evolution in changing environments*, Princeton: Princeton University Press.
Liebold, M. A., Chase, J. M.,Shurin, J. B. and Downing, A. L (1997). "Species turnover and the regulation of trophic structure," *Annual Revue of Ecology and Systematics*, 28: 467-494.
Longino, H. (1990). *Science as social knowledge*, Princeton: Princeton University Press.
Maynard Smith, J. (1987). "How to model evolution," in J. Dupré (ed.), *The latest on the best*, Cambridge, MA: MIT press, pp. 119-31.
Mayr, E. (1961). "Cause and effect in biology," *Science*, 134: 1501-1506.
Mitchell, S. D. (2000). "Dimensions of scientific law," *Philosophy of Science*, 67: 242-265.
Mitchell, S. D. and Page, Jr., R. E. (1992). *Idiosyncratic paradigms and the revival of the superorganism*, Bielefeld: Center For Interdisciplinary Studies.
Mitchell, S. D., Daston, L., Gigerenzer, G., Sesardic, N. and Sloep, P. (1997). "The how's and why's of interdisciplinarity," in P. Weingart, S. D. Mitchell, P. Richerson, and S. Maasen (eds.), *Human by nature: Betwee biology and the social sciences*, Mahwah: Erlbaum Press, pp. 103-150.
Morrison, M. (2000). *Unifying scientific theories: Physical concepts & mathematical structures*, Cambridge, UK: Cambridge University Press.
Moser, P. K. and Trout, J. D. (eds.) (1995). *Contemporary materialism*, London, UK: Routledge.
Oreskes, N., Shrader-Frechette, K. and Belitz, K. (1994). "Verification, validation, and confirmation of numerical models in the earth sciences," *Science*, 263(5147): 641-646.

Pool, R. (1997). "Why nature loves economies of scale," *New Scientist*, 154(2077): 16.
Richerson, P. J. and Boyd, R. (1987). "Simple models of complex phenomena: The case of cultural evolution," in J. Dupré (ed.), *The latest on the best: Essays on evolution and optimality*, Cambridge, MA: MIT Press, pp. 27- 54.
Schouten, M. K. D. and De Jong, H. L. (1999). "Reduction, elimination, and levels: The case of the LTP-learning link," *Philosophical Psychology*, 12: 237-262.
Sherman, P. W. (1988). "The levels of analysis," *Animal Behavior*, 36: 616-619.
Skipper, R. (1999). "Selection and the extent of explanatory unification," *Philosophy of Science: Supplement*, 66: S196-209.
Smart, J. J. C. (1968). *Between science and philosophy: An introduction to the philosophy of science*, New York, NY: Random House.
Sober, E. (1987). "What is adaptationism," in J. Dupré (ed.), *The latest on the best*, Cambridge, MA: MIT Press, pp. 105-118.
Stearns, S. C. (1992). *The evolution of life histories*, Oxford, UK: Oxford University Press.
Tinbergen, N. (1963). "On the aims and methods of ethology," *Z. Tierpsychol*, 20: 410-433.
West, G. B., Brown, J. H. and Enquist, B. J. (1997). "A general model for the origin of allometric scaling laws in biology," *Science*, 276: 122- 130.
Wimsatt, W. C. (1987). "False models as means to truer theories," in M. H. Nitecki, and A. Hoffman (eds.), *Neutral models in biology*, Oxford, UK: Oxford University Press, pp. 23-55.
Woolgar, S. (ed.) (1988). *Knowledge and reflexivity: New frontiers in the sociology of scientific knowledge*, London, UK: Sage.

2. Complexity Science: A 'Gray' Science for the 'Stuff in Between'

Kurt A. Richardson[1], Paul Cilliers[2] & Michael Lissack[1]
1 Institute for the Study of Coherence and Emergence, US
2 University of Stellenbosch, ZAF

Introduction

It is becoming rather monotonous continually reading articles that tell us how the concept of and the requirements for the modern organization are changing, how these are more complex than ever, and how a paradigm shift is necessary in order to facilitate our continued analysis, and management, of such entities. We are told that we must distribute decision making, encourage individual autonomy, and strive to innovate in the rapidly changing environment that characterizes the apparent New World Order. The list is as far reaching as it is impressive. These concepts coincide with a new, or at least emerging, description of organizations. This "paradigm" appears, from particular presentations at least, wholly to reject the long-held prevailing paradigm of the mechanistic, efficiency-driven, hierarchical, command-and-control organization. (We would question the "whollyness" of this position.)

Complexity science has emerged from the field of possible candidates as a prime contender for the top spot in the next era of management science. The number of management trade books on the subject has exploded, with provocative titles such as *Leading at the Edge of Chaos* (Conner, 1998), *ReWiring the Corporate Brain* (Zohar, 1997), and *Adaptive Enterprise* (Haeckel, 1999), to name but a few. The majority of the popular writings seem to claim that the "old" thinking is dead and needs to be (wholly) replaced with "new" thinking, and that a new, all embracing perspective, sometimes referred to as "complexity thinking," is available that will solve all our apparent woes. Of course, much of this is the hype that accompanies any "New Science," and we should know by now that the inevitable disappointment is also not far away.

Despite the promise indicated by various authors within the field, complexity science has thus far failed to deliver tangible tools that might be utilized in the examination of complex systems. In an attempt to derive some *actionable knowledge* (Argyris, 1993) from the field, the aims of this article are briefly to familiarize the reader with the more popular aspects of complexity science, and then, by focusing on the issue of incompressibility, to provide a provisional outline epistemology that attempts to incorporate the lessons derived from computer-based observations of complex systems' behavior and mathematical analysis of simple nonlinear systems. It is difficult to provide a complete presentation within such a limited format and for the interested reader a much extended version of this article is available (Richardson *et al.*, 2000).

The Theory of Complexity Science

What is a Complex System?

The general message from the popular complexity science literature seems to be that, where we once focused on the parts of a system and how they functioned, we must now focus on the interactions between these parts, and how these relationships determine the identity not only of the parts but of the whole system. As everything is connected to everything else, the notion of a distinct system as an entity becomes very blurred – where are a system's boundaries and how do we define them?

A complex (adaptive) system can be simply described as a system comprised of a large number of entities that display a high level of interactivity. The nature of this interactivity is mostly nonlinear, containing manifest feedback loops. A direct result of this is that sometimes it can be very difficult to associate effect with cause: Is the concept of the "learning organization," so popular in current management streams, oxymoronic? At the single-loop level at least there is cause for concern. Nonlinear interconnectivity also places fundamental limitations on our ability to validate models of complex systems.

There are a number of basic observations that have been made through the examination of such systems, primarily through the use of computer simulation and the mathematics of nonlinearity. The following sections will discuss the nature and implication of these observations in turn. For a more complete list refer to Cilliers (1998).

- *System memory/history* (Cilliers, 1998: 4). A complex system has memory/history captured at both the micro- (e.g., personal experiences, personal opinions, worldview) and macroscopic (e.g., culture, ritual, value system) levels. Therefore, system history plays an important role in determining the current state of the system as well as affecting system evolution.
- *A diversity of behaviors* (e.g., Allen, 2001). A rich diversity of qualitatively different operating regimes exists that the system might adopt. This is a result of the nonlinear nature of the relationships that describe the interactivity between the different system constituents.
- *Chaos and self-organization* (e.g., Auyang, 1999). The system evolution is potentially incredibly sensitive to small disturbances (a phenomenon popularly referred to as deterministic chaos), as well as being potentially incredibly insensitive to large disturbances (as a result of self-organization or, alternatively, anti-chaos). All possibilities in between also exist. Complex systems are often quite robust.
- *The incompressibility of complex systems* (Cilliers, 1998: 4). Complex systems are incompressible, that is, it is impossible to have an account

of a complex system that is less complex than the system itself without losing some of its aspects. Incompressibility (resulting from nonlinearity) is probably the single most important aspect of complex systems when considering the development of any analytical methodology, or epistemology, for coping with such systems. The following section will explore the ramifications of incompressibility in greater detail.

The Incompressibility of Complex Systems

As mentioned above, in a complex system everything is connected to everything else, whether directly or indirectly. The concept of "boundary" therefore becomes problematic. Just because an obvious physical boundary is judged to exist doesn't mean that it should be immediately assumed that this is the correct boundary for sense-making. And, just because particular boundary judgments were used in the past doesn't mean that they are appropriate again, even if the two situations appear to be the same. In fact, it is asserted that the boundaries that decision makers infer around a system are more a feature of our need for a bounded description than a feature of the system itself (Cilliers, 1998: 4). Boundaries are often drawn where we want them, and this may not be the best for the job at hand. At a fundamental level, boundaries are inferred in order to allow us to begin to make sense of our surroundings. Hard, enduring boundaries do not exist in nature; all perceived boundaries are transient when viewed from a sufficiently broad time frame. This does not mean that making the assumption that such boundaries exist is an unacceptable approximation in a wide variety of situations. An important aspect of sense-making, which is beyond the scope of the current discussion, is how both implicit and explicit assumptions create, or force, the boundary for analysis (see Richardson, *et al.*, 2000). For now, we will limit our discussion to the paradox of incompressibility versus our human need for boundaries, or compressibility.

As already stated, complex systems are *incompressible*. What this means is that if a model of a complex system were to be constructed that captured all the possible behaviors exhibited (both current and subsequent) by the system being represented, then that model must be *at least* as complex as the system of interest. The reason for this is that there will always be something outside of the boundary (i.e., the boundary inferred by the model) that would affect the system's behavior in some way at some time.

Let us adopt a skeptical stance for a moment: Because complex systems are sensitive to small changes, or small errors in our assumptions – that is, a small misplacement of the model boundaries – the model might be wholly inappropriate for the decision that it supposedly supports. "Something is better than nothing" would therefore be a misleading guideline to bandy around the analytical community. ("Take nothing for granted" is possibly more suitable.) To model a complex system accurately, we would have to model life, the universe, and everything. As sensemakers, we would have to take the first proposition of Wittgenstein's *Tractatus*, "The world is all that is the case" (Wittgenstein, 1997), completely literally. Acknowledging that there is only one complex system is important, however, since it forces the analyst to recognize the narrow scope and provisionality of their representations. Given that no absolute, enduring boundaries exist in reality, the use of the term "system" can be misleading, as it suggests the existence of *completely* autonomous entities. Maybe we should rename complexity science the "science of partial complex systems." This usage would make explicit the fact that when considering any problem we are in fact investigating a *part* of a complex system. As such, all the hypotheses and concerns raised by a "science of partial complex systems" would be appropriate for all analyses, rather than just special cases.

Assuming the notion of incompressibility to be correct, what does this mean for analysis?

Incompressibility essentially negates the possibility of the existence of a globally and permanently valid perspective, or paradigm. Furthermore, it means that there cannot be a perspective, paradigm, framework, etc. that can be used to wholly describe any subsystem embedded within *the* complex system. (Note that how we define any "subsystem" will be dependent upon our perceptions and the use of our description rather than a permanent feature of the real world.) This observation may seem to deny the usefulness of sense-making altogether. What it means, however, is not that we should not attempt to make sense, but that we should be strongly aware of, and blatantly open about, the provisionality of any perspective that might be utilized in underpinning an analysis of any problem – we must demonstrate considerable humility. Without this scientific "humility," we will continue to believe that our current understanding is true and defines all that is possible (and desirable).

The skeptical interpretation of the implications of incompressibility does not offer much in the way of advice, or *actionable knowledge* for analysts. It essentially argues for a "paradigm-less" approach toward analysis in which categorization of any sort must be avoided: a plainly impractical and absurd argument, particularly given the category-based functioning of the human mind. Whether skeptics like it or not, we rely heavily on categorization to make sense of the world and to legitimate our decisions and resulting actions. The skeptical interpretation is best seen as a very important and profound health warning.

Local vs. Non-Local Knowledge

Consideration of incompressibility leads neatly into the debate over whether non-locally valid descriptions of systems, that is, descriptions that are valid over a broad range of different contexts, are possible or whether we must accept the overwhelming context dependence of any description. (The reader should note that "local" should not be understood in the purely spatial sense. A certain factor may be "far away," but if the *information* is available locally, it is local.) As has been indicated by most of the discussion thus far, there is no black-and-white answer, which in itself indicates that context is critical. Using the concepts of the phase portrait and the attractor basin helps us understand this question.

Let us assume that we have developed a model in which we have strong confidence, as it appears to account for much of the (partial complex) system's currently observed behavior. The question arises as to whether we can now take this model and make predictions about the future operation of the system. The answer, from a complex systems perspective, is that if the qualitative nature of the assumptions that describe the new context remain valid, then the model will be useful; that is, if we remain in the same attractor basin within the "assumption space," then the knowledge derived from such a model can be straightforwardly translated into the new context. Qualitative changes in context prevent such a translation from occurring. So, at first consideration it seems that knowledge is strongly context dependent, but this dependence does not necessarily devalue this knowledge in light of a new context. Playing the skeptic again, the recognition that a new context is qualitatively similar to another is strongly subjective, and so some feature might be overlooked, however small, that would lead to the two contexts not being reducible to each other. There is considerable background "noise" in making such a judgment and, according to the phase picture, the impact of this noise depends on whether a separatrix has been crossed.

The essential lessons from this discussion on incompressibility are again diffused across a continuum. At worst, knowledge is so incredibly context specific that the search for understanding that is valid in other contexts is utterly futile; this has led to calls announcing the death of the expert (e.g., Taket & White, 1994). Consequently, attempts to make use of such knowledge in different contexts would be completely irresponsible, leading to misguided advice and action. At best, knowledge based on a particular context is indeed valid for a bounded range of other contexts, but this validity should never be taken for granted; suggesting that the expert isn't quite dead yet (Richardson & Tait, 2008). The quest for frameworks that attempt to describe the many contexts of complexity is not futile, but any frameworks developed should be regarded with a healthy skepticism when it comes to making use of them in specific circumstances. As an example, to characterize "wholes" within the complexity field itself consider the following.

The London School of Economics has a "complexity project" that is developing a complexity lexicon. The project researchers are encouraging the use of this lexicon when considering complexity. This might be considered a worthy aim by definition-minded modernists, but it must be remembered that there are an infinite number of ways to talk about complexity, and that the words themselves have different associations when used in different contexts – the transference of meaning is strongly context dependent. The meanings of words should not be defined and enforced at the global level, but should be allowed to be intersubjectively negotiated (e.g., Mingers, 1995) at the local level, through, say, critical dialog (e.g., Robinson, 1993). The prescribed lexicon will undoubtedly provide a sound starting point, but we should be overtly aware of how language, which is based in a particular perspective, limits our "vision." Again, to some this awareness may seem to be a trivial matter, but we believe it to be crucial, assuming that the world is complex and that the need for quasi-"paradigmless," or multi-perspective, thinking follows naturally from this.

"*This much is certain: the quest for comprehensiveness ... is not realizable. If we assume that it is realizable, the critical idea underlying the quest will be perverted into its opposite, i.e., into a false pretension to superior knowledge and understanding*" (Ulrich, 1993).

In many ways, complexity science provides insights concerning analysis that might be seen as nothing more than common sense. The need for an awareness of the provisionality of all understanding may seem obvious but, as a community, we seem to have forced this, and other issues, into the background. Maybe this is because we feel so much securer with the prevailing modernist view of Absoluteness and Truth: It's so much neater.

Complexity Thinking as Epistemology

The aim of this article is not to question the basic observations made concerning "the complex system," but to understand how the implications of these observations affect analysts' abilities to discover "truths" (with a small *t*) concerning such systems. Elsewhere (Richardson, *et al.*, 2000) we have also suggested, in attempting to follow through these implications, that chunking the area of analysis into distinct "paradigms" is misleading, our insinuation being that complexity thinking leads to a break from traditional paradigm-based thinking, and the necessary removal, or at least "fuzzification," of the boundaries that allow us to recognize a paradigm as a paradigm. As a result, attempts to rigidly define the boundaries of the complexity paradigm are, we argue, contradictory to the fundamental complexity message. This does not mean that attempts to do so cannot be useful, but it does mean that the boundaries should be seen as provisional and definitely local, i.e., context specific. We must each play the role of a skeptic until such a time that we need to "fake" being "positivist" so that action can be justified and initiated. But, in so doing, and in recognition also of the fact that the system evolves, we need to critically review and possibly change the sense-making boundaries on an ongoing basis.

In the previous section the implications of incompressibility on analysis were explicated. In this section, we offer a high-level conceptual approach to analysis that acknowledges the difficulties previously discussed.

Given that no one perspective can capture the inherent intricacies of complex systems, the analysis of complex systems requires us to consider a number of perspectives. The underlying premise for this is that by exploring a number of perspectives, a richer appreciation of the "state of affairs" or "problematic situation" of interest will be developed, resulting in more informed decision making. In considering a variety of perspectives, a negotiation between these perspectives is encouraged that fuels and guides the exploration process. The pros and cons of each perspective would be examined in light of both the supporting and contradictory evidence offered by the other perspectives. This evidence may be in the form of individuals' experiences, the numerical output of a particular computer model, etc. As the different perspectives are played against and with each other, new perspectives emerge that are, at least, an eclectic mixture of the parts of the constituent perspectives that seem most relevant to the state of affairs under consideration. This intra- and interperspective exploration, or boundary critique, will identify other perspectives that might be worthy of inclusion, further fueling the exploration process. After a number of exploration cycles, perspectives that are deemed acceptable are left. This end point might come about in a number of ways: the creativity, fueled by the differences between the various perspectives or a variety of creative thinking exercises, may die out; the perspectives might naturally converge in a way that satisfies the basic needs of each perspective; or, a particular perspective becomes dominant and forces an end to the exploration process. Remember that in using the term "perspective" we are not distinguishing between formal and informal models, and so the perspective of the person who controls the budgetary strings also vies for a position in this interperspective exploration. The end point of an analysis then becomes the point at which a perspective, which may have emerged during the analysis or was present at the beginning, becomes overwhelmingly dominant.

In short, a principal requirement of a complexity-based epistemology is the exploration of perspectives. It may be useful to associate the terms *weak* and *strong* exploration, where weak refers to intraperspective exploration and strong refers to interperspective exploration. Weak exploration encourages the critical examination of a particular perspective, which is undoubtedly driven by its differences with other perspectives. Strong exploration encourages the sucking in of all available perspectives in the considered development or synthesis of a situation-specific perspective. These two types of exploration are not orthogonal, and cannot operate in isolation from one another. The greater the number of perspectives available, the more in depth the scrutiny of each individual perspective will be. The deeper or broader the scrutiny, the higher the possibilities are of recognizing the value, or not, of other perspectives. Essentially, complexity-based analysis is a move from the contemporary authoritarian or imperialist (Flood, 1989) style, in which a dominant perspective bounds the analysis, to a more democratic style that acknowledges the "rights" and value of a range of perspectives, whether they be formal modeling methods or informal and (inter)subjective personal viewpoints. The decision as to what perspective to use is also deferred until after the exploration process (it is actually an emergent property of the process). While skepticism plays a central role in the exploration process, it plays a lesser role during implementation, at least initially. In order to implement a decision confidently, we must learn to fake positivism (something that comes quite naturally to most people), but always be aware that conditions will change and might require substantial rethinking of the implementation design itself.

The concepts of strong and weak exploration are all well and good, but decision makers would be frozen by the plethora of possibilities that such paradigmatic freedom offers – the familiar *paralysis by analysis*. How would such an approach be operationalized? It is clear that the manager must, in addition to other activities, be concerned with the management of the variety of perspectives; an activity that falls under the umbrella term of *facilitation*. What other frameworks, however limiting, might support such a perspective-based dialog and negotiation?

Operationalizations of a Complexity-Based Epistemology

Thus far, the article has discussed the epistemological implications of assuming that the world is best described as a complex system. Exploration both within and without different perspectives is encouraged, supporting the need for criticism, creativity, and pluralism. From a skeptical point of view, any attempt to operationalize such a complexity based epistemology, via a well-defined framework, would be in contradiction to the underlying tenets. From a pragmatic point of view, however, we must accept that frameworks are essential in providing at least a focus or starting point for analysis. What we must be strongly aware of is that the theoretical insights offered by any framework should not be used to *determine* our explorations, but considered as an offering of *direction*, or simply as a source of creativity to fuel the exploration process.

A number of well thought-out attempts have been made in the development of "meta-frameworks" that recognize the problematic nature of sense-making, offering guidelines as to how to manage the exploration process. These meta-methodologies have not been developed within the "official" complex systems research community, but within the management science community; more specifically, the operational research community. Examples of these developments include: the system of systems methodologies (Jackson, 1987); total systems intervention (Flood, 1995; Flood & Jackson, 1991); creative design of methods (Midgley, 1990); and critical appreciation (Gregory, 1992). In order to legitimate the various methodologies a variety of philosophies are drawn on, such as Habermas's early work on knowledge-constitutive interests as well as his later work on truth statements, rightness statements, and individuals' subjectivity, Foucault's theory of power, etc. For a good survey of the different methodologies and their associated philosophical underpinnings, see Midgley (1997).

On examining these different approaches, the reader may notice that generally each subsequent methodology attempts to make more explicit the role of ongoing critical reflection and the categorization process – partly driven by the ongoing critique of the different methodologies. In an extended version of this article (Richardson *et al.*, 2000), while acknowledging the coercive institutional forces (from regulative, normative, and mimetic pressures) acting to shape the form of any intervention, we discuss the culture in which any analysis is performed. We believe that if managers and scientists alike were to acknowledge the central role that critical thinking plays, there would be little need explicitly to design-in the activity. After all, hasn't examination of the underlying assumptions of any perspective always been associated with "good" analysis? If anything should be taken for granted it is the centrality of critical reflection, or boundary exploration and critique, to all forms of analysis. It is perhaps a poor reflection on the current analytical culture that critical thinking as an activity has to be made explicit.

Summary and Conclusions

A Modernist Argument for Affirmative Postmodernism?

By assuming the universe to be a complex system, complexity science offers an alternative perspective that supports the need for criticism, creativity, and pluralism through the notion of strong and weak exploration. It can be interpreted in such a way as to highlight the dangers of *any* categorization, via the concept of chaos, but also via incompressibility (by its acceptance of the need for categorization to "see" in the absence of a complete representation of everything). By illustrating the inherently problematic nature of boundary selection, complexity science warns of the risks of employing off-the-shelf perspectives, and the need to partake in an intra- and interparadigmatic negotiation to facilitate the development of shared (by those having a vested interest in the outcome), context-specific representations of perceived reality. In a way, dare we suggest it, complexity science provides a modernist argument for affirmative postmodernism. Boundaries are constructed for convenience. Quasi-paradigmless thinking should prevail until we are forced to take a position, that is, fake positivism and temporarily invoke an imperialist stance.

All contexts are unique. If they were not, then past experience would always be sufficient when confronting any situation. This uniqueness means that attempts to associate existing understanding with particular contexts is problematic. However, we need also to acknowledge that the recognition of contexts is not a black-and-white – i.e., trivial – exercise. Complexity science suggests that all our contexts should be considered "gray." As such, new perspectives must be tailored to "fit" the new context (definition of which is problematic in itself) through the synthesis of a variety of formal and informal paradigms (used in the term's broadest sense) via strong and weak exploration. Furthermore, complexity science (or "the science of partial complex systems") warns us of attempts to systematize the exploration process, but, at the same, acknowledges such a requirement. A healthy skepticism must prevail to prevent us from slipping into potentially "bad" habits.

Taking Responsibility

One more point before we conclude. Complexity science raises some ethical concerns that refer to the inevitability of choices that cannot be backed up scientifically or objectively (Cilliers, 2000). Why associate these concerns with ethics? First, because the nature of the system or organization in question is determined by the collection of choices made in it. There are, of course, choices to be made on all scales, major ones as well as all the seemingly insignificant small ones made all the time – and remember that the scale of the effect is not necessarily related to the scale of the cause. In a way, the history of the organization is nothing else but the collection of all these decisions. Secondly, since there appears to be no final objective or calculable ground for our decisions, we cannot shift the responsibility for the decision on to something else: "Don't blame me, the genetic algorithm said we should sell!" We *know* that all our choices to some extent incorporate a step in the dark, and therefore we cannot but be responsible for them. This may have a pessimistic ring to it, but that need not be the case. An awareness of the contingency and provisionality of things is far better than a false sense of security.

In conclusion, we find that complexity science offers an alternative way of legitimizing the current interest in boundary critique, creativity, and pluralism. Furthermore, in acknowledging the partiality and provisionality of any attempt to describe a particular "problematic situation," complexity science also raises concerns for how we recognize ethical behavior.

References

Allen, P. M. (2001). "Economics and the science of evolutionary complex systems," *SysteMexico*, special edn, "The autopoietic turn: Luhman's re-conceptualization of the social": 29-71.

Argyris, C. (1993). *Knowledge for Action: A Guide to Overcoming Barriers to Organizational Learning*, San Francisco: Jossey-Bass.

Auyang, S. Y. (1999). *Foundations of Complex-System Theories in Economics, Evolutionary Biology, and Statistical Physics*, Cambridge, UK: Cambridge University Press.

Cilliers, P. (1998). *Complexity and Postmodernism: Understanding Complex Systems,* London: Routledge.

Cilliers, P. (2000). "What can we learn from a theory of complexity?," *Emergence*, 2(1): 23–33.

Conner, D. R. (1998). *Leading at the Edge of Chaos*, Chichester, UK: Wiley.

Flood, R. L. (1989) "Six scenarios for the future of systems 'problem solving'," *Systems Practice*, 2(1): 75-99.

Flood, R. L. (1995). *Solving Problem Solving*, Chichester, UK: Wiley.

Flood, R. L. & Jackson, M. C. (1991). *Creative Problem Solving: Total Systems Intervention*, Chichester, UK: Wiley.

Gregory, W. J. (1992). "Critical systems thinking and pluralism: A new constellation," Ph.D. thesis, City University, London.

Haeckel, S. H. (1999). *Adaptive Enterprise: Creating and Leading Sense-and-Respond Organizations*, Boston: Harvard Business School Press.

Jackson, M. C. (1987) "New directions in management science," in M. C. Jackson & P. Keys (eds), *New Directions in Management Science*, Aldershot, UK: Gower.

Midgley, G. (1990). "Creative methodology design," *Systemist*, 12: 108-13.

Midgley, G. (1997). "Mixing methods: Developing systemic intervention," in J. Mingers & A. Gill (eds), *Multi-Methodology: The Theory and Practice of Combining Management Science Methodologies*, Chichester, UK: Wiley.

Mingers, J. C. (1995). "Information and meaning: Foundations for an intersubjective account," *Information Systems Journal*, 5: 285-306.

Richardson, K. A., Mathieson, G., & Cilliers, P. (2000). "The theory and practice of complexity science: Epistemological considerations for military operational analysis," *SysteMexico*, 1(1): 25-66.

Richardson, K. A. & Tait, A. (2001) "The death of the expert?," forthcoming.

Robinson, V. M. J. (1993). *Problem Based Methodology: Research for the Improvement of Practice*, Oxford, UK: Pergamon Press.

Taket, A. & White, L. (1994). "The death of the expert," *Journal of the Operational Research Society*, 45(7): 733-48.

Ulrich, W. (1993). "Some difficulties of ecological thinking, considered from a critical systems perspective: A plea for critical holism," *Systems Practice*, 6(6): 583-611.

Wittgenstein, L. (1997). *Tractatus Logico-Philosophicus* (originally published 1921), trans. D. F. Pears & B. F. McGuiness, London: Routledge.

Zohar, D. (1997). *ReWiring the Corporate Brain*, San Francisco: Berrett-Koehler.

3. The Concept of Emergence in Social Science: Its History and Importance

Geoffrey M. Hodgson
Affiliation

The terms "emergence" and "emergent property" date from the last quarter of the nineteenth century. However, the general idea behind these terms is older. It is redolent, for example, of the "law of the transformation of quantity into quality" laid down by G.W.F. Hegel in his *Logic* and subsequently taken up by Karl Marx and Frederick Engels. The philosopher Auguste Comte (1853, Vol. 2: 181) wrote of irreducible properties: "Society is no more decomposable into individuals than a geometrical surface is into lines, or a line into points." The idea of emergence was also hinted at by John Stuart Mill (1843, Bk 3, Ch. 6, Para. 2) with his idea of "heteropathetic" causation.

The word "emergent" in this context was first suggested by the philosopher George Lewes (1875, Ch. 3: 412). Subsequently, the philosopher of biology Conwy Lloyd Morgan (1927, 1933) wrote extensively on the topic. Following Mill and Lewes, Morgan (1927: 3–4) defined emergent properties as "unpredictable" and "non-additive" results of complex processes. In more detail, Morgan (1932: 253) explained:

"*...the hypothesis is that when certain items of "stuff," say o p q, enter into some relational organization R in unity of "substance," the whole R(o p q) has some "properties" which could not be deduced from prior knowledge of the properties of o, p, and q taken severally.*"

Morgan saw such properties as crucial to evolution in its most meaningful and creative sense, where (Morgan, 1927: 112):

"*...the emphasis is not on the unfolding of something already in being but on the outspringing of something that has hitherto not been in being. It is in this sense only that the noun may carry the adjective 'emergent'.*"

For Morgan, evolution creates a hierarchy of increasing richness and complexity in integral systems "as new kinds of relatedness" successively emerge (Morgan, 1927: 203). Also for Morgan, the "non-additive" character of complex systems must involve a shift from mechanistic to organic metaphors: "precedence should now be given to organism rather than to mechanism – to organization rather than aggregation" (Morgan, 1933: 58).

Morgan's formulation of the concept was explicitly acknowledged by a group of philosophers in the 1920s. Prominent among these was Samuel Alexander (1920) at the University of Manchester in the UK and Alfred Whitehead (1926) at Harvard University in the USA. The psychologist and philosopher William McDougall (1929) was conspicuous in the presentation and development of the concept. Similar ideas also appeared independently at the time. The American philosopher Roy Sellars (1922) developed the notion of "creative synthesis" to explain how new properties could emerge in complex systems. Accordingly, a set of ideas were established in philosophy in Britain and America in the 1920s that are remarkably redolent of the ideas that emerged later in complexity theory in the 1990s.

However, this earlier wave of complexity thinking did not last very long. Within philosophy, ontological speculation about the nature and properties of reality became highly unpopular, as positivism grew in influence in the interwar period. Positivists believe that "metaphysical" propositions that are not directly grounded on experience are "unscientific." The ontological concerns of Morgan, Alexander, Whitehead, McDougall, and Sellars were dismissed – along with the concept of emergence – as metaphysical and useless speculation.

Nevertheless, despite this setback, the concept of emergence did have an enduring impact. It survived at the fringes of biology and social science until its rediscovery later in the twentieth century.

Past Impacts of Emergence on the Social Sciences

One of the first social scientists to be influenced by Morgan was the institutional economist Thorstein Veblen. Morgan visited Chicago in 1896 and Veblen was crucially influenced by his ideas (Dorfman, 1934; Hodgson, 1998; Tilman, 1996). The influence of Morgan is evident in Veblen's treatment of institutions as phenomena that are dependent on individuals but are not reducible to them.

Prior to Veblen, many social scientists believed that social phenomena could be understood in terms of the biological characteristics of the populations involved. Human society, it was thought, could evolve no more rapidly than the individuals themselves. From such a standpoint, the rapid evolution of human civilization could only be explained if there were some Lamarckian process by which acquired characteristics could be inherited. Otherwise, there would be no explanation of the rapid evolution of the human genetic material that supposedly correlated with the evolution of human civilization in a few thousand years. Like other anti- Lamarckians, Morgan challenged this, arguing that the inheritance of acquired characteristics was not plausible. But this anti-Lamarckian standpoint created a problem: if there was no inheritance of acquired characteristics, then how could the relatively rapid evolution of civilization be explained?

Morgan (1896: 340) resolved this problem by suggesting that evolution occurs both at the level of human genes and at the level of human social institutions. These social institutions act as a storehouse of accumulating and evolving social customs, technology, and knowledge. Furthermore, as this cultural heritage itself evolves, it provides a new cultural and institutional environment for the development of each human individual. This evolving social environment unleashes new possibilities for each person, even if human nature and the human genetic endowment remained more or less the same. As a Darwinian opponent of the Lamarckian theory of biological inheritance, Morgan argued that it was not the human genetic endowment that had evolved significantly in the last few centuries, but the human social environment.

Morgan's Darwinian understanding of evolution led him to promote the idea of an emergent level of socioeconomic evolution that was not explicable exclusively in terms of the biological characteristics of the individuals involved. Evolution occurs at this emergent level as well, and without any necessary change in human biotic characteristics. Accordingly, the crucial concepts of emergence and emergent properties were liberated by the insistence of a barrier between acquired habit and biotic inheritance. The biological and the social spheres became partially autonomous, but linked, levels of analysis. On this basis, in later works, Morgan developed the philosophical concept of emergence.

Veblen did not use the concept of emergence explicitly. Nevertheless, it is striking that after 1896 the concept of the "natural selection of institutions" began to appear explicitly in his work. Furthermore, some passages in his writings read almost as rephrased versions of Morgan's texts (Hodgson, 1998).

Following Veblen, the concept of emergence assumed a marginal existence in Anglophone social science. There are a few rare examples from economics. The institutional economist Morris Copeland (1927) mentioned Morgan's concept of emergent properties. While at Columbia University, the New Zealand economist Ralph Souter (1933: 111) discussed the importance of emergent properties, acknowledging its precedent in Morgan and Alexander. He criticized the Austrian school of economics for its omission of emergent properties: Austrian economists proclaim an individualistic ontology and methodology, and insufficiently acknowledge the existence of emergent properties at the systemic level. The English institutional economist John A. Hobson (1936: 216) wrote in his book on Veblen:

"*Emergent evolution brings unpredictable novelties into the processes of history, and disorder, hazard, chance, are brought into the play of energetic action.*"

A more significant aftermath of Morgan's influence on Veblen was inexplicit, yet massive in its reverberations. The institutional economist Wesley Mitchell, formerly a student of Veblen, established himself as the leading American economist in the interwar period. At this time, macroeconomics was not yet established as a subdiscipline; mainstream economics was microeconomic in character.

Mitchell tried to break from this individualist foundation. In his 1924 Presidential Address to the American Economic Association, Mitchell (1937: 26) argued that economists need not begin with a theory of individual behavior but with the statistical observation of "mass phenomena." Mitchell (1937: 30) explained that this was possible "because institutions standardize behavior, and thereby facilitate statistical procedure." Mitchell thus hinted at a process of social conformism that stabilized behavior in an institutional context. To rephrase this in the language of emergence and complexity: Mitchell and others saw complex systems involving positive feedback effects that led to relatively stable emergent phenomena at the macroeconomic level.

Mitchell and his colleagues in the US National Bureau for Economic Research in the 1920s and 1930s played a vital role in the development of national income accounting. They suggested that aggregate, macroeconomic phenomena have an ontological and empirical legitimacy. Arguably, these developments prepared some of the groundwork for the Keynesian revolution in economics. Through the development of national income accounting, the work of Mitchell and his colleagues helped to establish modern macroeconomics, and in particular influenced and inspired the macroeconomics of Keynes (Mirowski, 1989: 307; Colander & Landreth, 1996: 141). Although the concept of emergence was implicit rather than explicit in these intellectual developments, we can trace the origins of this line of thinking in Veblen's break from reductionism and his establishment of institutions as units of analysis.

Elsewhere, the concept of emergence found a small refuge in sociology. Talcott Parsons came to Harvard University in 1927 and was influenced there by Whitehead. Parsons took on board aspects of Whitehead's organicist ontology and saw the existence of emergent properties as "a measure of the organicism of the system" (1937: 749). However, although Parsons played a role in preserving the concept, his use of the term was idiosyncratic and unclear.

As noted above, the idea of emergence was largely submerged in the positivist and reductionist phase of Anglo-American science in the interwar period (Ross, 1991). Although his use of such terminology was at odds with Morgan's original concept, Parsons helped to keep the concept of emergence alive during this difficult period.

The Re-Emergence of Emergence

After the Second World War, Michael Polanyi (1967), Sir Karl Popper (1974), Ernst Mayr (1985), and several others rehabilitated the idea of emergent properties. The concept had never entirely disappeared, but it took the decline of positivism to provide an opportunity for its redevelopment. One of those involved in this process was the great polymath Michael Polanyi. Perhaps significantly, Polanyi had worked in both the natural and the social sciences. His classic book on tacit knowledge has a chapter titled "emergence," in which he wrote:

"*...you cannot derive a vocabulary from phonetics; you cannot derive the grammar of language from its vocabulary; a correct use of grammar does not account for good style; and a good style does not provide the content of a piece of prose. ... it is impossible to represent the organizing principles of a higher level by the laws governing its isolated particulars*" (Polanyi, 1967: 36).

Another person who played a crucial part in the rediscovery of the concept was the philosophically inclined biologist Ernst Mayr. He argued that the characteristics of a complex system:

"*...cannot (not even in theory) be deduced from the most complete knowledge of the components, taken separately or in other partial combinations. In other words, when such systems are assembled from their components, new characteristics of the new whole emerge that could not have been predicted from a knowledge of the components*" (Mayr, 1985: 58).

Like his predecessors, Mayr established that the existence of emergent properties at a particular level of reality means that explanations cannot be reduced entirely to components and phenomena at lower levels. He wrote:

"*Recognition of the importance of emergence demonstrates, of course, the invalidity of extreme reductionism. By the time we have dissected an organism down to atoms and elementary particles we have lost everything that is characteristic of a living system*" (Mayr, 1985: 58).

We are reminded of the words of William Wordsworth in his poem "The Tables Turned":

Sweet is the lore which Nature brings;
Our meddling intellect
Mishapes the beauteous form of things: –
We murder to dissect.

After these statements by Polanyi, Popper, and Mayr, the development of the story of emergence was to take another turn. By the 1980s, the development of computer technology had greatly facilitated the simulation of nonlinear dynamic systems. This led to a related development known as chaos theory.

Working on nonlinear mathematical systems, chaos theorists have shown that tiny changes in crucial parameters can lead to dramatic consequences, known as the "Butterfly Effect – the notion that a butterfly stirring the air today in Peking can transform storm systems next month in New York" (Gleick, 1988: 8). There are parallels here with the account of "bifurcation points" in the work of Prigogine and Stengers (1984). After behaving deterministically, a system may reach a bifurcation point where it is inherently impossible to determine which direction change may take; a small and imperceptible disturbance could lead the system in one direction rather than another.

Accordingly, chaos theory suggests that apparent novelty may arise from a deterministic nonlinear system. From an apparently deterministic starting point, we are led to novelty and quasi-randomness. Consequently, even if we knew the basic equations governing the system, we would not necessarily be able to predict the outcome reliably. The estimation of "initial conditions" can never be accurate enough. This does not simply undermine the possibility of prediction: in addition, the idea of a reductionist explanation of the whole in terms of the behavior of its component parts is challenged. As a result, the system can be seen to have emergent properties that are not reducible to those of its constituent parts.

Chaos theory was closely followed by another phase in the history of the concept of emergence. This work was centered at the Santa Fe Institute (Arthur, 1995; Arthur *et al.*, 1997; Anderson *et al.*, 1988; Kauffman, 1993, 1995; Holland, 1998; Waldrop, 1992). Instead of focusing largely on disorder and chaos, complexity theorists at Santa Fe and elsewhere also stressed the emergence of "order out of chaos" and the sustained behavior of complex systems "at the edge of chaos" (Cohen & Stewart, 1994). One clear outcome of the work of the Santa Fe Institute has been to bring respectability and prominence to the concept of emergence, even in disciplines where it had been neglected.

Part of the impact of this work has been achieved through the use of powerful, heuristic computer simulations. Many such simulations have created artificial social worlds, in which modeled agents interact in various ways, often to create surprising, systemic outcomes. Much of this work shows the emergence of order and other "higher-level" properties in complex systems. Reviewing the modeling of such "artificial worlds," David Lane (1993: 90) writes that a main thrust "is to discover whether (and under what conditions) histories exhibit interesting emergent properties." His extensive review of the literature in the area suggests that there are many examples of artificial worlds displaying such attributes.

The Critique of Reductionism

At this stage we can take stock. Although the concept of emergence is over 100 years old, its rise to prominence has not been steady. The first three decades of the twentieth century saw relatively sophisticated developments of the concept in the sphere of philosophy. Many of these earlier insights and debates have since been neglected and are worth revisiting. In the last two decades of the twentieth century, the concept of emergence was brought from the margins into the limelight of discussion of the evolution of complex systems. Computer simulations have powered much of this recent work.

However, these developments involve a powerful challenge to prevailing conceptions of how both the natural and the social sciences should work. The impact of this challenge is not yet widely appreciated. Furthermore, old habits die hard, and the old ways of thinking about science have strong adherents. The most important issue over which this debate between the old and the new science is articulated is the question of reductionism.

Reductionism sometimes involves the notion that wholes must be explained entirely in terms of their elemental, constituent parts. More generally, reductionism can be defined as the idea that all aspects of a complex phenomenon must be explained solely in terms of one level, or type of unit. According to this view, there are no autonomous levels of analysis other than this elemental foundation, and no emergent properties on which different levels of analysis can be based.

Consider biology. Although many biologists acknowledge the existence of emergent properties, there are many theorists and practitioners who still cling to the view that all biological phenomena can and should be explained in terms of "lower-level" components, such as genes.

Furthermore, reductionism is still conspicuous in social science today and typically appears as methodological individualism. This tends to be defined as "the doctrine that all social phenomena (their structure and their change) are in principle explicable only in terms of individuals – their properties, goals, and beliefs" (Elster, 1982: 453). It is thus alleged that explanations of socioeconomic phenomena must be reduced to properties of constituent individuals and relations between them. Allied to this is the sustained attempt since the 1960s to found macroeconomics on "sound microfoundations." This "microfoundations revolution" meant the rejection of much of Keynesian macroeconomics. Indeed, it is the antithesis of the approach discussed above, as developed by the institutionalist and Keynesian economists of the 1920s and 1930s.

Reductionism should be distinguished from *reduction*. Reduction involves the partial decomposition of elements at one level into parts at a different level. Measurement is an act of reduction. The general idea of a reduction to parts is not being overturned here. Some degree of reduction to elemental units is inevitable. Science cannot proceed without some dissection and some analysis of parts. However, although some partial reduction is inevitable and desirable, a complete analytical reduction is both impossible and a philosophically dogmatic diversion. What is important to stress is that the process of analysis cannot be extended to the most elementary subatomic particles presently known to science, or even to individuals in economics or genes in biology. A complete reduction would be hopeless and interminable. Reduction is necessary to some extent, but it can never be complete. As Popper (1974: 260) has declared:

"*...there is almost always an unresolved residue left by even the most successful attempts at reduction.*"

Essentially, if socioeconomic systems have emergent properties, then reductionism is confounded. Emergent properties by definition are not entirely explicable in terms of constituent elements at a more basic level. Accordingly, the idea of explaining the macro behavior of socioeconomic systems completely in terms of individuals and individual actions (methodological individualism) is misconceived. Similarly, explanations of macroeconomic phenomena completely in terms of microeconomic postulates (the microfoundations project) are confounded. There are strong arguments to suggest that neither methodological individualism (Udéhn, 1987) nor the microfoundations project (Rizvi, 1994) can ever be successful in reducing all features of the system to its micro components. Given this incipient failure, in explaining complex systems we may be obliged to rely on emergent properties at a higher (macro) level.

In general, reductionism is clearly countered by the notion that complex systems display emergent properties at different levels that cannot be completely reduced to or explained wholly in terms of another level. By contrast, antireductionism generally emphasizes emergent properties at higher levels of analysis that cannot be reduced to constituent elements.

Emergence Distinguished from Supervenience

The concept of supervenience was developed by Alexander Rosenberg (1976, 1985) in both economics and biology. In part, this alternative and weaker concept is a symptom of a reluctance by some to adopt the concept of emergence. Supervenience applies to the situation where the identity of two or more entities at the macro level does not assume identity at the constituent micro level, but identity at the micro level does guarantee identity at the macro level. In this case the macro level can be said to be supervenient. Accordingly, similar properties at the macro level cannot all be explained by a single set of micro-level components, but identical configurations of micro-level components all give rise to identical macro-level phenomena. The concept of supervenience is used to defend a qualified form of reductionism. Supervenience retains the ontological priority of the micro level over other, higher levels.

By contrast, modern concepts of emergence suggest that different outcomes are possible with near identical configurations and interactions of micro-level elements. As chaos theory suggests, tiny, seemingly insignificant differences can lead to quite different systemic outcomes. In chaotic systems, almost exact identity at the micro level does not guarantee identity at the macro level, and supervenience is eluded. Notably, the supervenience concept was developed before chaos theory posed a severe challenge to reductionism. The possibility of a high degree of context sensitivity undermines the application of the supervenience concept.

Accordingly, in biology, identical genes do not lead to identical organisms or behaviors. The biologist Conrad Waddington (1975: vi) showed that in evolution the genetic makeup of organisms (their genotype) does not generally give rise to similar characteristics (phenotypes). This is partly because "genotypes, which influence behavior, thus have an effect on the nature of the selective pressures on the phenotype to which they give rise." This introduces "an inescapable indeterminism" in evolutionary theory. In a similar vein, Cohen and Stewart (1994: 3) give a related set of examples and cases. They argue that "simplicity of form, function, or behavior emerge from complexities on lower levels because of the action of external constraints."

The fact that different outcomes are possible with near identical configurations and interactions of micro-level elements confounds the concept of supervenience. It is no longer tenable to infer from almost exact identity at the micro level some identity at the macro level.

Conclusions

One of the reasons that the concept of emergence is important for social science is that it provides a necessary means to focus on higher-level units and relations and to avoid the potentially intractable problem of analytical reduction to lower-level units.

However, while emergent properties provide indispensable hooks to bring analysis up to a higher level, we must never lose sight of the dependence of these higher-level properties on lower-level units. Indeed, if it were possible to explain a higher-level phenomenon entirely in terms of lower-level units, then we should do so (Bunge, 1980). Biology cannot ignore chemistry and macroeconomics cannot ignore the microeconomics of individuals and firms. Emergence does not give license to neglect the constituent elements of which an entity or structure is composed.

What is required is the development of a methodology that is sensitive to the "marks of emergence in their most telling form." As Kyriakos Kontopoulos (1993: 22–3) elaborates further, the marks of an emergent property include its novelty, its association with a new set of relations, the stability and boundedness of these relations, and the emergence of new laws or principles applicable to this entity. We find emergent properties in any complex, evolving system, throughout the natural and the social realm.

The theory of emergence offers a nonreductionist account of complex phenomena. Indeed, with emergent properties a reductionist account is impossible. We are just beginning to grasp the full implications of this, although, as we have seen, the concept of emergence has been around for more than 100 years.

At the beginning of the twenty-first century, we are presented with an exciting agenda of research in which philosophical concepts such as emergence can find their place and impact within the flowering sciences of complexity. At least as far as the social sciences go, we are just at the beginning of this process.

References

Alexander, S. (1920) *Space, Time and Deity* (2 Vols), London: Macmillan.
Anderson, P. W., Arrow, K. J., & Pines, D. (eds) (1988) *The Economy as an Evolving Complex System*, Reading, MA: Addison-Wesley.
Arthur, W. B. (1995) "Complexity in economic and financial markets," *Complexity*, 1(1): 20-25.
Arthur, W. B., Durlauf, S. N., & Lane, D. A. (eds) (1997) *The Economy as an Evolving Complex System II*, Redwood City, CA: Addison-Wesley.

Bunge, M. A. (1980) *The Mind-Body Problem: A Psychobiological Approach*, Oxford, UK: Pergamon.
Cohen, J. & Stewart, I. (1994) *The Collapse of Chaos: Discovering Simplicity in a Complex World*, London/New York: Viking.
Colander, D. C. & Landreth, H. (eds) (1996) *The Coming of Keynesianism to America: Conversations with the Founders of Keynesian Economics*, Aldershot, UK: Edward Elgar.
Comte, A. (1853) *The Positive Philosophy of Auguste Comte* (2 Vols), trans. H. Martineau from the French volumes of 1830–42, London: Chapman.
Copeland, M.A. (1927) "An instrumental view of the part-whole relation," *Journal of Philosophy*, 24(4, 17 Feb): 96-104.
Dorfman, J. (1934) *Thorstein Veblen and His America*, New York: Viking Press.
Elster, J. (1982) "Marxism, functionalism and game theory," *Theory and Society*, 11(4): 453–82. Reprinted in J. E. Roemer (ed.) (1986) *Analytical Marxism*, Cambridge, UK: Cambridge University Press.
Gleick, J. (1988) *Chaos: Making a New Science*, London: Heinemann.
Hobson, J. A. (1936) *Veblen*, London: Chapman and Hall.
Hodgson, G. M. (1998) "On the evolution of Thorstein Veblen's evolutionary economics," *Cambridge Journal of Economics*, 22(4, July): 415-31.
Holland, J. H. (1998) *Emergence: From Chaos to Order*, Oxford, UK: Oxford University Press.
Kauffman, S. A. (1993) *The Origins of Order: Self-Organization and Selection in Evolution*, Oxford, UK/New York: Oxford University Press.
Kauffman, S. A. (1995) *At Home in the Universe: The Search for Laws of Self-Organization and Complexity*, Oxford, UK/New York: Oxford University Press.
Kontopoulos, K. M. (1993) *The Logics of Social Structure*, Cambridge, UK: Cambridge University Press.
Lane, D. A. (1993) "Artificial worlds and economics," Parts I and II, *Journal of Evolutionary Economics*, 3(2, May): 89–107 and 3(3, Aug): 177-97.
Lewes, G. H. (1875) *Problems of Life and Mind*, Vol. 2, London: Trubner.
McDougall, W. (1929) *Modern Materialism and Emergent Evolution*, London: Methuen.
Mirowski, P. (1989) *More Heat Than Light: Economics as Social Physics, Physics as Nature's Economics*, Cambridge, UK: Cambridge University Press.
Mitchell, W. C. (1937) *The Backward Art of Spending Money and Other Essays*, New York: McGraw-Hill.
Morgan, C. L. (1896) *Habit and Instinct*, London/New York: Edward Arnold.
Morgan, C. L. (1927) *Emergent Evolution*, 2nd edn, London: Williams and Norgate.
Morgan, C. L. (1932) "C. Lloyd Morgan," in C. Murchison (ed.) *A History of Psychology in Autobiography*, Vol. 2, New York: Russell and Russell: 253-64.
Morgan, C. L. (1933) *The Emergence of Novelty*, London: Williams and Norgate.
Polanyi, M. (1967) *The Tacit Dimension*, London: Routledge and Kegan Paul.
Popper, K. R. (1974) "Scientific Reduction and the Essential Incompleteness of All Science," in F .J. Ayala & T. Dobzhansky (eds) *Studies in the Philosophy of Biology*, London/Berkeley, CA/Los Angeles, CA: Macmillan/University of California Press: 259-84.
Prigogine, I. & Stengers, I. (1984) *Order Out of Chaos: Man's New Dialogue With Nature*, London: Heinemann.
Rizvi, S. A. T. (1994) "The microfoundations project in general equilibrium theory," *Cambridge Journal of Economics*, 18(4, Aug): 357-77.
Rosenberg, A. (1976) "On the interanimation of micro and macroeconomics,"

Philosophy of the Social Sciences, 6(1): 35-53.
Rosenberg, A. (1985) *The Structure of Biological Science*, Cambridge, UK: Cambridge University Press.
Ross, D. (1991) *The Origins of American Social Science*, Cambridge, UK: Cambridge University Press.
Sellars, R. W. (1922) *Evolutionary Naturalism*, Chicago: Rand-McNally.
Souter, R. W. (1933) *Prolegomena to Relativity Economics: An Elementary Study in the Mechanics and Organics of an Expanding Economic Universe*, New York: Columbia University Press.
Sperry, R. W. (1991) "In defense of mentalism and emergent interaction," *Journal of Mind and Behavior*, 12(2): 221-46.
Tilman, R. (1996) *The Intellectual Legacy of Thorstein Veblen: Unresolved Issues*, Westport, CN: Greenwood Press.
Udéhn, L. (1987) *Methodological Individualism: A Critical Appraisal*, Uppsala, Sweden: Uppsala University Reprographics Centre.
Waddington, C. H. (1975) *The Evolution of an Evolutionist*, Edinburgh/Ithaca, NY: Edinburgh University Press and Cornell University Press.
Waldrop, M. M. (1992) *Complexity: The Emerging Science at the Edge of Order and Chaos*, New York: Simon and Schuster.
Whitehead, A. N. (1926) *Science and the Modern World*, Cambridge, UK: Cambridge University Press.

4. Archetypal Dynamical Systems and Semantic Frames in Vertical and Horizontal Emergence

William H. Sulis
Departments of Psychiatry and Behavioral Neuroscience, and Psychology, McMaster University, CAN

From Vertical to Horizontal Emergence

Archetypal dynamics is a formal framework for dealing with the study of meaning laden information flows within complex systems. The main goal of archetypal dynamics is to study emergence in complex systems, but it carries sufficient descriptive and explanatory depth to serve as a foundational language for describing complex systems generally. This paper deals with the underlying philosophical concepts, introducing the basic mathematical representation, the tapestry, and provides an introduction to emergence through the idea of H-complexity.

Archetypal dynamics began with the realization that our current concept of emergence had become quite narrow, being restricted primarily to emergence from lower to higher spatio-temporal scales. Traditional examples of emergence include such situations as the formation of matter out of fundamental particles, of organisms out of cells, of species out of individuals, of minds out of neurons. These are all examples of what can be termed *Vertical Emergence*, as they all involve the seeming appearance of one class of phenomena, occurring at one spatio-temporal scale, out of a different class of phenomena occurring at a higher spatio-temporal scale. The phenomenon which appears at the larger spatio-temporal scale, the *emergent*, is said to have emerged from the phenomenon occurring at the lower scale through a process of *emergence*. The concept of emergence has thus been restricted to situations involving upward causation, in which high level processes are in some sense determined by or caused by lower level processes.

This bias in focus reflects the disciplinary origins of recent attempts to distinguish the notion of a complex system from its close cousin, the complicated system. A complicated system is one composed of a large number of small sub-units that are dynamically coupled to one another and whose large scale behavior can, in principle at least, be understood as a direct consequence of these low level interactions, e.g., the well over one million parts of the Saturn V rocket used to place an Apollo crew into orbit. Although in principle, the equations of motion for such systems can provide a predictive model for their behavior, they are seldom solved in practice except to ensure solutions where small perturbations in the system's state give rise to only small deviations in behavior.

The discovery by Poincaré at the turn of the twentieth century of chaotic dynamics, and its rediscovery by mathematicians and physicists some 60-80 years later, demonstrated some fundamental limits on the practicality of behavioral predictions for complicated systems. In particular, there arose the awareness that large classes of systems exhibit sensitive dependence on their initial conditions, meaning that small deviations in state become amplified over time resulting in an inability to make long term predictions of behavior. In an effort to better understand the conditions under which chaotic dynamics might occur, researchers began to explore a wide range of simple models of chaotic systems, usually through computer simulations since the equations of motion for these systems are not exactly solvable. This led researchers to the discovery of systems whose large scale behavior appeared to be so far removed from that of its low level constituents as to constitute in their mind an entirely new phenomenon. These researchers arrived at a communal intuition that systems that were deemed to be complex exhibited behaviors and structures that appeared to transcend those at the lower levels of their constituents. Although not rigorously proven, this insight led these researchers to coin a new term, that of a *complex system* and of its attribute, *complexity* (Forest, 1990).

The early complex system models were themselves complicated systems, generally being composed of large numbers of smaller sub-units interacting in complicated ways. Thus there was an explicit recognition of at least two distinct spatio-temporal scales governing the descriptions and dynamics of these systems. Since the reductionist paradigm that has been so influential in the sciences, especially the hard sciences, asserts that a complete understanding of the behavior of a system at the lower spatio-temporal scale would be sufficient to enable one to predict and to describe the behavior of the system at the larger spatio-temporal scale, the discovery of dynamical chaos put a damper on the general applicability of reductionist principles, though it did not cause the majority of scientists to abandon it entirely. Although the discovery of the concept of the complex system seemed to deal a death blow to the reductionist ideal, in fact that did not happen because these researchers asked themselves whether there might exist some general principles that would determine when a system would be complex. Some headway in this regard had already happened in the study of chaotic systems, and since reductionism had been a fundamental principle for many of the researchers moving into this field, it was natural that a concerted effort would be undertaken to preserve it. With that goal in mind the Santa Fe Institute was born. In its early years a great many notions of complexity began to appear in the literature. Almost always, these notions and models involved coupling phenomena that appeared at distinct spatio-temporal scales. Most definitions focused on the issue of the loss of predictability as one moved up the spatio-temporal scales. Thus the main focus of these researchers has been on vertical emergence and the general notions of emergence that have come out of this literature are mostly limited to situations of vertical emergence.

The concept of emergence, as pointed out by Goldstein (1999), is actually quite old, dating back to the nineteenth century, and much broader in scope. An expansion of the concept of emergence beyond that of vertical emergence yields new insights into its essential character. For example, the study of vertical emergence has focussed upon the loss of determinism imposed upon upward causation in vertically emergent complex systems. But why should one be limited solely to upward causation as a determinate principle? There are many examples in the literature of emergent downward causation. A striking example appears in neurobiology (Anderson, *et al.*, 1995; Sulis, 2000). There is abundant evidence that certain neurons modify the dynamical response characteristics of some of their receptors in response to patterned stimulation from upstream neurons. These receptors are modified if the upstream neurons fire with a particular timing, or fire in conjunction with a background of Theta rhythm activity. Thus large scale patterns in the perceptual environment impose spatio-temporal patterns on the firing activity of sensory neurons, and these patterns in turn trigger off functional changes in other neuronal populations. Thus a pattern at a large spatio-temporal scale induces changes in neurons at a much smaller spatio-temporal scale. A more arcane example derives from Jung's ideas about the influence of the collective unconscious on individual behavior (DeLaszlo, 1959). In Jung's psychology, the behavior of an individual is determined not solely by influences, conscious or unconscious, that can be considered as arising within an individual, but in addition by influences that arise from large scale patterns present within particular cultures as well as from historical and evolutionary factors. The presence of downward causal influences confounds the question of which phenomenon came first, the lower or the higher, and which phenomenon actually emerged. Thus even in the case of vertical emergence, causal influences and determinate processes are none too clear.

Another rather subtle example of emergence is that of Goldstein's (2003) "self-transcending constructions." Here the emergent appears as a result of the system acting upon itself in a reflexive process. An example offered by Goldstein is Cantor's Diagonalization method. Another example is the transcendental construction giving rise to Russell's Paradox. (See Appendix I for a more complete technical discussion of both examples). Both of these examples are rather abstract, but the point is that applying an operation reflexively to some structure may give rise to a new structure that transcends the old in some fashion. Goldstein would say that the new structure has emerged out of the old by means of this self-transcending construction.

The idea of emerging 'out of' suggests that all of the factors required for the appearance of the emergent through the process of emergence reside within the original phenomenon. This excessive focus on processes of so-called self-organization ignores the presence of contextual factors, which, as noted by Cohen & Stewart (1994), often play a fundamental role in many emergent situations. One of their more striking examples is that of frog gastrulation. When we think of developmental processes, we think of DNA providing a blueprint that describes how all of the varied cellular components are to be differentiated and organized into a complete living organism. In the case of frog gastrulation, however, it is the presence of specific features within the environment that determine whether or not this process is successful. In order for a developing frog embryo to properly gastrulate, that is, to properly fold itself so as to create the gastrointestinal tract, it must exist within a fluid environment having just the right current flow in order to curl the embryo onto itself, thereby forming the tube. If the current is not present, the tube does not form and gastrulation does not occur. In effect, the DNA anticipates the presence of certain external factors in order to carry out its program. Thus in a great many cases emergence is a process that must occur within a particular context, and a knowledge of that context is necessary to understand emergence.

In fact, there is no single process by means of which emergence occurs. One cannot talk about emergence in isolation from contextual factors, whether at higher, lower or similar spatio-temporal scales. For that reason archetypal dynamics emphasizes the notion of an emergent situation, which includes emergents and their context, as opposed to emergence *per se*, which suggests an emphasis upon intrinsic factors alone, the latter being the typical connotations of the notion of self-organization.

Horizontal Emergence

The preceding examples have all involved ideas of vertical emergence, that is, emergence across spatio-temporal scales, although it should be apparent by now that the processes that give rise to such emergence are quite varied, and in general do not reside wholly within the precursor system. There is another very large class of emergent phenomena that is little recognized and that involves *horizontal emergence.* Horizontal emergence is most typical within the social sciences. In horizontal emergence, the emergent phenomena arise out of and reside at the same ontological level as the precursor phenomena. A familiar example of horizontal emergence is the emergence of culture. The media are awash with images drawn from youth subcultures, which tend to change with each generation. A few individuals within the current cultural milieu develop attitudes and behaviors which begin to distinguish them from others within the larger culture. Over time, others become aware of these new attitudes and behaviors, and if they resonate sufficiently with their psyches they may begin to adopt some of these for their own. In general as they begin to further delineate these differences, the larger culture reacts to these through sanctions prohibitions or sometimes withdrawal or even rejection. These responses serve to further distinguish the new from the old. If enough people adopt the trends of this emerging subculture it may become self-sustaining. Frequently though it will fade away to be replaced by another candidate. If the new culture does persist, it may give rise to a new social unit. It is important, however, to note that both new and old culture exist on the same spatio-temporal scales and are supported by the same kinds of people. It is also important to note that these cultures may become so distinguished as to be completely foreign to one another. Without shared frames of meaning through which to interpret attitudes and actions, misunderstanding and conflict are quite likely.

Another example in which meaning plays a profound role is the diversification of languages. Each language provides a means of communicating meaning between people. In order to use a language effectively, it is usually insufficient to merely learn the rules of grammar and some vocabulary. Anyone who has used those handy travel guides has learned this lesson first hand. In order to communicate effectively one must adopt or at least share in the frame of meaning that is ascribed to words and phrases and sentence constructions within the language. Languages organize thought into very different patterns, with very different emphasis placed on linguistic components. That is why the learning of language within naturalistic settings proceeds more effectively than rote learning from a text. Different languages constitute distinct semantic universes. Nevertheless, new languages do not arise *de novo* but rather evolve over time from earlier languages. As each language comes to distinguish itself as a unique entity it emerges from the preexisting linguistic context. Again all of this takes place at similar spatio-temporal scales.

Other examples of horizontal emergence include speciation and the evolution of social roles, both of which clearly depend upon contextual factors. The phenomenon of horizontal emergence is not limited to the social sciences. For example, quarks and gluons occupy more or less the same spatio-temporal scales as hadrons, though they differ markedly in character and physical properties. Nevertheless, it is out of the interactions among quarks and gluons that hadrons emerge.

The study of horizontal emergence is particularly salient for the study of emergence more generally because it illustrates clearly how the distinct phenomena of a horizontally emergent situation require distinct frames of meaning for their understanding. Just as distinct frames of meaning are required to navigate among different cultures or to utilize different languages, so too distinct frames of meaning are required to talk about neurons and brain and mental phenomena, or about DNA and species, or about cells and organisms, or about fundamental particles and observable matter. Indeed, in every case of emergence, the emergents are distinguished by frames of reference, which provide meaningful answers to the six basic questions: who, what, where, when, how and why. These frames of reference not only define the emergents, they determine how the environment is to interact with them. They determine both intrinsic (local), and extrinsic (contextual) interactions.

Archetypal dynamics posits the existence of such frames of reference as the fundamental defining characteristic of an emergent situation. Emergent phenomena are not merely distinguished, but are in fact defined by their distinct frames of reference. In a situation of emergence, archetypal dynamics considers both phenomena to be emergents, the situation in which they appear is considered to be an emergent situation, the meaningful or causal influences that one exerts on the other are termed emergence processes, which can therefore be upward, downward, horizontal or reflexive.

The Role of Meaning in Emergence

The applicability of the role of meaning to more general emergent situations can be made clear in the case of a particular example of vertical emergence - the *Game of Life*, the cellular automaton invented by J. H. Conway (2001). In order to play the Game of Life, we must understand the following: the basic entities (who) which we can manipulate are the squares on the grid (where), whose colors we can select to be either white or black (what). We are allowed only to determine the initial configuration of colors. The subsequent evolution of the squares occurs in single time steps (when) according to a fixed, deterministic rule (how). If we violate any one of these understandings, we will not have a Game of Life. We may have some other game, or no game at all, but we will not have Life. In addition, in order to determine whether or not we win, we also need to have some notion of the goal of the game (why). In the case of Life, this is arbitrary (though usually specified for any particular play), but it need not be in other emergent situations.

In the search for the presence of non-repeating patterns, Conway and colleagues discovered a wholly remarkable property of Life: the ensuing configurations could serve as 'OR' gates, 'NOR' gates, 'NOT' gates, 'AND' gates, and so on. It was then possible to show that much larger patterns of these 'gates' could be constructed which could emulate the logical structure of any computer program, and thus emulate the behavior of any computer. It was quite a surprise to realize that the simple Game of Life carried within it the computational ability of a computer.

This computer simulation consists of logic gates and paths (who), arrayed in a grid (where), whose states are given by certain patterns of square colorings (what), which evolve in discrete time steps (when), according to the rules of logic (how), and whose goal is to carry out particular logical computations (why). It is clear that these simulation phenomena arise at a level higher than that of Life, since we are dealing with patterns of squares instead of squares, and obey a different dynamic, logical instead of arithmetical. Moreover, if we are to utilize this simulation effectively, we must impose an initial configuration which encodes both the data and the computation of interest, and whose subsequent evolution must follow logical principles and be interpretable as implementing the computation intended. In addition, we must seek a final configuration which can be meaningfully understood as encoding the result of our computation, and we must have at hand some scheme for carrying out this encoding and decoding of patterns. Life, together with its computational simulations, constitutes an emergent situation, with Life and the simulation being emergents, and the selection process constituting the emergence process whereby we obtain the computational emergent from the Life emergent.

Semantic Frames and Information

With this example of emergence in the Game of Life in mind, we can go onto to describe the role of Archetypal Dynamics in concerning itself with the epistemological aspects of the emergent situation rather than the emergent processes that give rise to it. It asks, "What fundamentally distinguishes one emergent from another?" In particular, archetypal dynamics focuses upon the role of meaning laden information in distinguishing between emergents. The frame of reference as conceived in archetypal dynamics is termed a *semantic frame*. It is termed a 'frame' because it is an organizing principle, one that parses phenomena into distinct entities, modes of being, modes of behaving, modes of acting and interacting. It is termed 'semantic' because it ascribes meaning to all of these in a coherent and consistent manner. The semantic frame shapes and guides our interactions with these phenomena so as to maintain this thread of meaning. The semantic frame provides answers, in whole or in part, to the six basic epistemic questions:

1. Who are the relevant entities?
2. What are their states, actions, reactions and interactions?
3. Where does this all play out?
4. When do these events occur?
5. How do these events play out? What are the rules governing them?
6. Why do these events transpire? What are the goals of entities and their interactions?

The semantic frame is to be understood as a primitive construct, much like a set. A semantic frame both organizes and exerts its influence through the flow of meaning laden information. Information within the context of archetypal dynamics is not to be understood in the sense of Shannon, that is, concerned only with the quantity of information, but rather in its active sense, as 'the act of informing' or 'the state of being informed', which means 'to give a special quality or character, to imbue, to instill, to impart by gradual instruction' (*Webster's Dictionary*). Information is to be understood as an efficacious agency that is capable of affecting the characteristics, and ultimately the behavior, of some other. Information may influence the course of events but it does not necessarily cause them. Information possesses content and elicits meaning.

In the ontology of archetypal dynamics, the most basic unit of information is termed an *informon*. An informon simply refers to any aspect of reality that possesses the capacity to inform. An informon is an aspect of reality that exists prior to any semantic frame. An informon must be attached to, coupled to, or in resonance with, a semantic frame in order to be given form, meaning, and behavior. Consider an analogy. Suppose that one had in their possession two reels of film from two movies. The story and backstory of the movies to which these two reels belong, together with the culture of film, constitute the semantic frames for these reels. They give meaning, coherence and consistency to the reels, and tell us what to do with them. Now suppose that each reel is unrolled and cut up into individual cells, and that the cells are then scattered on the floor. One is now faced with seemingly unrelated bits of experience. Each cell represents a single informon. The semantic frames of the movies allow one to give meaning to the chaos of these single cells, suggesting relationships between them which make it possible to organize the cells into coherent, consistent and meaningful patterns.

Archetypal dynamics posits that information and informons manifest in two distinct ontological forms, *active* and *passive*, and two distinct epistemological forms, public and private. The ontological distinction is inspired by the data/program distinction in computer science, and concerns whether the information in question does or does not require the active participation of the informons being related. Passive information is information that informs about properties of informons, or about fixed relationships between informons. Passive information is merely descriptive, like data. Active information, on the other hand, informs about actions and derives from actions. Active information denotes an action of one informon upon another, or an action by or upon some observer or user. Active information instructs, or imposes some choice, or impels some action, like a program.

The epistemological distinction is inspired by the hidden/public distinction in computer science. It concerns whether or not the information in question can impart knowledge to the systems, users and agents coupled to a particular semantic frame. Public information is information that is available to all of the informons comprising a system or user/agent, just like public data, or public programs. Such information is available to any user to guide their subsequent actions. Private information, on the other hand, is information that is not explicitly available to the informons of a system or a user/agent. Nevertheless it is information which influences the behavior of a system or user/agent. This is akin to hidden data like systems files, or hidden programs like systems programs, that are necessary for the operation of the computer system but inaccessible to the user. The public-private distinction is not an artifice, nor does it reduce to a subjective-objective distinction. It has an ontological basis. In physics for example, quark-quark interactions are private, whereas hadron-hadron interactions are public.

Realization, Interpretation and Representation

Archetypal dynamics posits that real systems relate to semantic frames in one of three primary modes: *realization*, *interpretation*, and *representation*. It is through these modes that one can gain an understanding of the semantic frame in spite of the belief that semantic frames, being conceptual primitives, cannot be fully formalized or defined.

In addition, archetypal dynamics asserts that real systems may relate to semantic frames in either an active or a passive manner. The passive mode arises from the idea that a semantic frame, as an organizing principle for information, must be about something. It provides definition and description for that something. This does not require the active participation of that which is being defined or described, merely its presence or potentiality, real or imagined. That something for which the semantic frame is about is termed a 'realization'. A realization may be real, such as an organism, or a physical structure or machine, or it may be ideational, such as a theory or a novel. A realization need not be material but it must be efficacious, that is, it is must be able to make a meaningful difference. A realization of a semantic frame is termed a 'system'.

Realizations

A semantic frame may have many different realizations, just as a theory in mathematical logic may have many different models (i.e., domains of true interpretation of the logical symbols). Likewise, a system may realize several different semantic frames. If these frames are independent of one another in the sense that one cannot be meaningfully reduced to other, then that system is said to be complex. A semantic frame whose realization is some system is termed an *external frame* for that system. The external frame tells us how to cut the system into meaningful units, and then provides a meaningful description of how those units relate to one another. The relationship between a system and its external frame is weaker than that between theory and model in formal logic where one can form one-to-one connections between the structures and relations of the model and the formal elements of the theory. Instead, archetypal dynamics suggests that the relationship between system and external frame is more akin to that of a resonance or a coupling, in other words, the 'goodness of fit' between a system and its frame reflecting the degree of this resonance. A system may couple to an external frame locally or globally. The stronger and more global the resonance, the more that the semantic frame captures the essence of the system, and the more its dynamics is captured by or influenced by that frame.

Social systems provide a ready example of this. Consider health care systems. In times past, the emphasis in health care systems was placed upon the delivery of health care. One may remember when it was argued that human life possessed inestimable value, and that all possible resources should be devoted to efforts to preserve life. Every effort was directed towards the delivery of care, resulting in large numbers of doctors and nurses, the ready availability of hospital beds and of diagnostic equipment. Delays in the provision of medical care were short. Doctors even delivered care in the home. The semantic frame within which health care personnel and planners operated was one that was health centered. In the past two decades, however, there has been a major shift in health care's semantic frame. More and more, health has been marginalized as a concern and economics has become the dominant focus. The central concern issue has shifted to how to deliver the least expensive health care, with the result that the standards of health care

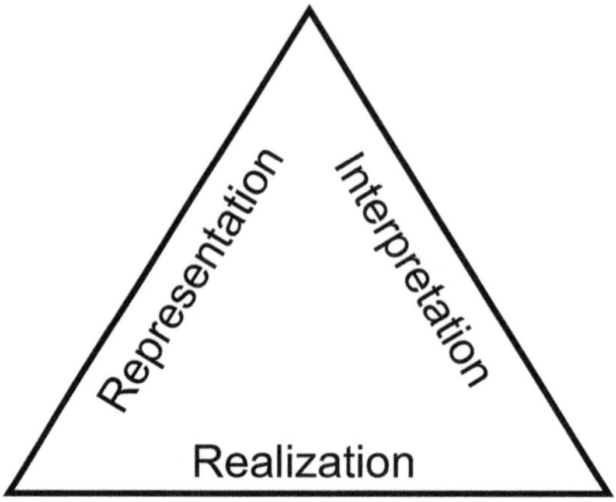

Figure 1 *Triadic Composition of Semantic Frames*

that are tolerated have progressively declined. There are too few doctors and nurses, too few hospital beds, too long delays in obtaining treatment, premature discharges from hospital, and even though the death rate has not changed, the amount of human suffering has.

Health care delivery is an example of a complex system, one that resonates to varying degrees with several independent semantic frames. The effectiveness of a frame in operating within such a system depends upon the degree of resonance that the system has with the frame. A poor fit results in a very poor interaction with the system. A strong fit results in a very effective interaction. Quite often it is necessary to consider many frames simultaneously when dealing with a system. Attempting to place the onus upon a single frame can lead to inefficiency or to disaster. Based upon the past twenty years experience, it would seem that the over emphasis upon the economic frame has not been beneficial to the delivery of health care to consumers, though it may have been beneficial to governments and insurers.

The Game of Life realizes two external semantic frames: the 'Life frame' and the logic gate or 'computational frame'. Each frame provides a complete description of its particular emergent, sufficient to enable one to interact with that emergent in such a way as to completely explore its dynamics and range of behaviors. These two frames appear to be independent of one another, suggesting that Life fits the criterion for being complex. Moreover the Game of Life couples globally to the Life frame, but only locally to the logic gate frame.

Interpretations

The active mode draws again on inspiration from computer science. A user sits before a computer with some particular problem in mind. That problem will be formulated in terms of data and a program that can be implemented by the computer to provide a solution to the problem.

The translation of the original problem into a form that the computer can manipulate requires a semantic frame governing the nature, structure and functionality of a computer, together with an understanding of how to translate real world problems into entities within that semantic frame. The user must interpret the problem through the semantic frame into something accessible to the computer. Based upon this idea, the active mode of a system relating to a semantic frame is termed 'interpretation'.

Interpretation is an active process, and involves an active relationship with the semantic frame that underlies it. Interpretation gives rise to action, and the actors that implement the action may reside within or outside some system. Those that reside within the system are termed 'agents', while those residing outside the system are termed 'users'. In both cases, the agent/user is the instrument of interpretation, and the results of its interpretation are then manifest through some action either within or upon the system, thereby realizing the interpretation. The semantic frame that a user/agent interprets is termed an internal frame for the user/agent. It is essential to keep in mind that the external frame for a user/agent need not be the same as its internal frame. Indeed, one might argue that it is the existence of separate internal and external frames that defines an entity as a potential user/agent and not merely a system.

Meaning, according to the archetypal dynamics viewpoint, is intimately related to action. Meaning, as a dynamical influence, holds little relevance in situations in which there is no choice, such as in strict deterministic systems or strict stochastic systems. It is only in situations in which a choice must be made that the meanings of individual events and choices become important. Since archetypal dynamics is concerned with meaning laden information, it acquires utility only in situations in which choice plays an important role in the dynamics of systems, which is why the underlying dynamics of archetypal dynamics is based upon the idea of a combinatorial game (see, Berlekamp, *et al.*, 1982; Nowakowski, 1996, 2002.)

Representations

Archetypal dynamics makes a clear distinction between realizations and interpretations of semantic frames, and their representations. A 'representation' is understood to be a particular situation, involving a representational frame which serves as a kind of universal object, whose realizations may be associated in some manner with the realizations and interpretations of some other semantic frame. The representation is not itself a realization or interpretation. Realizations and interpretations are ontological constructs, whereas representations are epistemological constructs like symbols and signs. A representation is a symbolic picture, a caricature, a cartoon, to help us to understand the reality. Representations are essential to the pursuit of any formal study, but they are not realizations or interpretations of the semantic frame under consideration, and they must be carefully interpreted in order that a useful understanding of the semantic frame is obtained and paradoxes avoided.

Tapestries and Coherence

Archetypal dynamics stipulates that semantic frames act through the organization of informons into meaningful patterns and interactions. A semantic frame thus establishes meaning laden relations among a collection of informons. Semantic frames provide the glue that allows informons to cohere to one another. Although semantic frames may not be directly formalizable, the existence of coherence and of coherent relations between informons provides an opportunity to formalize one aspect of semantic frames.

Coherence may be defined as: 1) holding or sticking together, 2) marked by logical consistency or as congruous, logically connected, or understandable. It is important to also notice that Archetypal dynamics broadens the notion of coherence further to mean 'meaningfully related'. Two informons are said to be coherent provided that there is a semantic frame that provides a meaningful relationship between them. In the absence of such a meaningful relationship, two informons are said to be incoherent or to be randomly or noisily related. Noise consists of informons that are incoherent or noisily related to one another.

Consistency is often associated with coherence, but archetypal dynamics treats the term consistent in a strictly logical sense in order to indicate that the coherent relations also preserve some kind of logical truth or valuation. Coherence is a local notion, whereas consistency is a global notion.

Coherence, in as much as it provides a link between informons, permits a natural and simple formal representation through the means of the mathematical construct of a graph. The natural representational structure for archetypal dynamics is a particular form of a graph termed a *tapestry*, (defined in formal notation in Appendices II and III below). The tapestry structure assigns a specific meaning to specific relationships, expressing all of those relationships that are understandable, and therefore all of those relationships that are coherent. The focus upon coherence enables us to construct formal representations of the meaningful linkages imposed upon informons by semantic frames without explicitly defining notions of meaning and of semantics. This avoids many of the pitfalls that have plagued other attempts to express semantic relationships, such as Montague semantics (Thomason, 1970) , Situation Theory (Barwise & Perry, 1983), modal logic (Blackburn, *et al.*, 2001) and circular logic (Barwise & Moss, 1996). These approaches in general posit a universe of meaning that is complete and fixed. Archetypal dynamics, on the other hand, views meaning as ongoing process arising from interpretation. Because meaning in archetypal dynamics involves context, it is always in a state of flux, incomplete and changeable. Meaning is a local construct, requiring an interplay between semantic frame, agent/user, system and context.

Loci and Nexi

An informon is a minimal instance that is capable of serving as a unit of information, whether it be in a realization or an interpretation. Informons acquire meaning through interpretation via some semantic frame, and so they form the substrate upon which a semantic frame acts. Informons are organized through coherent linkages, which themselves acquire meaning via the semantic frame. These linkages, being meaningful in their own right, are also conveyors of information, and thus are themselves informons. Informons therefore divide into two fundamental classes depending upon the role that they play. When an informon serves as a dependent element of some coherent relationship, it is termed a 'locus'. When an informon serves as a defining instance of some meaningful relationship it is termed a 'nexus'. It is important to note that such distinctions are context dependent, and that a particular informon may serve as both locus and nexus in different relationships.

Traditionally, relations among elements are treated as if imposed from some divine vantage point, being absolute and eternal. Archetypal dynamics, however, views relations as simply a particular class of informon, and therefore relations have a local, transitory quality, arising from within a realization or interpretation rather than being imposed upon it from the outside. These relations may arise out of dynamical interactions that exist intrinsically, independent of any particular semantic frame, or they may arise through the influence of the semantic frame for which they serve as either realization or interpretation. Relations are thus given a dynamic quality in archetypal dynamics, and this inspires the somewhat unusual approach to representing relations by means of tapestries. This dual nature has a counterpart in computer science as previously mentioned. A computer can act upon a file in one of two distinct ways. It may treat it as data, in which case it can act to alter the structure of the file and thereby the information contained within it. It may treat it as a program, in which case it will interpret the structure of the file as being a set of instructions to be carried out. A compiler serves a dual role, treating as data files those files that actually contain programs to be carried out by another computer. Thus files can play either role, or sometimes both, depending upon the specific context. This duality of role is extended to the notion of informon in an analogous manner.

If this approach to relations still seems bizarre and without merit, consider that quantum mechanics suggests that at the smallest scales such fundamental relations as space and time no longer have the same meaning they have classically. For example, the Heisenberg Uncertainty relation suggest that the measurement of one relation may render another relation immeasurable, and so, to that extent at least, it no longer exists. Furthermore, social roles come and go, especially those that depend upon the vagaries of technology. The context defines what relations can arise and whether or not they in fact exist. There is a pernicious tendency to reify and immortalize relations which obscures their ephemeral and parochial nature. However, archetypal dynamics accepts the processual and open-ended nature of relations as a fundamental attribute of all informons, whether locus or nexus.

Loci and nexi, being informons, come in four flavours: active/passive versus public/private. Epistemological constraints force the following restrictions on locus-nexus relationships. A public nexus relates public loci. A private nexus relates loci, at least one of which must be private. A nexus, whether active or passive, will form connections between either all active loci, or all passive loci, but will not mix types.

As a final note on informons, they are to be understood as epistemic units that make no reference to whatever it is that underlies them or gives them existence. Entities, on the other hand, refer to those ontological constructs that serve as generators of informons. This distinction becomes relevant when discussing the more detailed dynamics of tapestries, a subject that will be left for later papers.

General Tapestries

The preceding sections have introduced the main philosophical ideas underlying archetypal dynamics and so the focus now turns toward the main formal representational structure pertaining to emergence: the tapestry. The general tapestry serves as a representation for realizations of semantic frames. A generalization of the tapestry concept, the formal tapestry, serves as a representation of interpretations of a semantic frame. A variant of a tapestry, the constructor, serves as the basic element of tapestry dynamics, enabling the definition of the weaving process, which is carried out in the manner of a combinatorial game (these latter aspects are described in Sulis, in preparation 1 and 2). Here the focus will be on general tapestries and the notion of H-complexity. As formal representations of informons and their coherent linkages, the utility of tapestries arises from both their descriptive power and their links to semantics, modal logic, graph theory, group theory, category theory, knot theory, and combinatorial game theory, thus providing access to a large body of mathematical ideas and results.

The formal definition of a tapestry is given in Appendix II. Essentially a tapestry consists of a pair of graphs. One graph consists of struts that convey membership in a particular instance of a coherence relation. The other graph consists of relators that convey the particulars of dependencies which express a coherence relation. Informons that serve as vertices of relators are called 'loci'. Informons that serve as initial vertices of struts are termed 'nexi'. A nexus denotes an informon that determines a particular relation among a collection of loci. This is consistent with the view that relations, being conveyers of information, must themselves be informons, but recognizes the distinction between being a conveyer of a relation and being an element that is related within a relation.

A given informon may therefore be either a locus, a nexus, or both, depending upon its connectivity. The labels tie together struts and their associated relators so as to ensure semantic consistency, and can serve as linkages between meaning laden elements in on system and those in another. Although this will not be explored further here, it is possible to assign labels that are derived from some other mathematical object, a group for example, with the relation thereby carrying over some of the group theoretic information. For instance, such might be true if the relation conveyed a distance relation. Note that multiple nexi may define relations on a single label α, and a given nexus may define relations on multiple labels.

In most formal models of relations there is a distinction between the relation itself and the objects that are related. In logic, this distinction gives rise to various orders of logic, with each order being able to express concepts related to all orders below it but none above it. These order distinctions enable the expression of many subtleties but also induce many limitations. The tapestry approach enables relations of multiple orders to be expressed within a common framework. This gives the tapestry structure considerable expressive power.

In general, the collections of informons, struts, relators, and labels will each be divided into four distinguished subsets corresponding to the active/passive, public/private distinctions and which must adhere to certain coherence conventions which will not be discussed here. The interested reader can examine these elsewhere (Sulis, 2002; Sulis, submitted; Sulis, in preparation 1 and 2).

H-Complex Tapestries

We need to recall that the stated, ultimate goal of archetypal dynamics is to shed light on the phenomena of emergence. Although, we've been describing different forms of emergence, particularly the distinction between vertical, horizontal, and reflexive emergence, in all of the cases emergents are distinguished through distinct, independent semantic frames which provide answers to the fundamental epistemological questions of who, what, when, where, how, and why that define each emergent. The presence of distinct emergents thus implies that a system admits description through distinct, independent semantic frames, which in turn suggests our definition of a complex system which is capable of supporting emergent behavior, as a system able to realize multiple independent semantic frames.

Horizontal emergence offers itself perhaps as the easiest to grasp through this formalism of semantic frames. One reason is that the phenomena of horizontal emergence appear at the same spatio-temporal scales, and generally involve entities of similar kind. This enables us to avoid an infinite regress when engaging in a discussion of the concept of an entity. Accordingly, let's apply the notion of the realization of multiple independent semantic frames to horizontal emergence: if a system realizes multiple independent semantic frames all at the same spatio-temporal scales, indeed, at the same scales whatever, this suggests that the basic informons of the system admit distinct interpretations through multiple independent semantic frames. This further means that these informons can be represented by distinct tapestries, one for each of these independent semantic frames.

Although not absolutely necessary, since this system could represent a special case, it nevertheless seems a reasonable assumption that the tapestries expressing emergent systems are non-isomorphic. The reason is that it is commonly understood within the mathematical community that for two mathematical entities to be isomorphic means that they differ, essentially, only in the names given to the entities involved, and since names can be chosen arbitrarily, this means they represent the same thing. If it was the case that two semantic frames were never to give rise to non-isomorphic tapestries then, at least within the archetypal dynamics framework, we would have no means of distinguishing between them, and so for all intents and purposes they would appear as if they were the same. It appears sensible consequently to assume that we are dealing, in the case of horizontal emergence, with a situation in which these frames are distinguishable and thus representable by non-isomorphic tapestries.

Since we are considering horizontal emergence and the same basic collection of informons and linking relations, this suggests that we have one underlying tapestry which represents the informons and their linking relations but somehow we have this single tapestry representing distinct semantic frames. That would imply one tapestry appearing in non-isomorphic forms. That would, however, be nonsensical. But recall that the role of a semantic frame is to give meaning to entities and relations, and that in the tapestry construction specific meaning is conveyed through the use of the meaning labels. The struts and relators provide the structural information but the meaning information lies in the labels. So this opens up one possibility, namely, whether it is possible to start with the same basic tapestry structure, including the same basic collection of struts and relators, and on this structure impose distinct labelings which are consistent with this structure and yet result in the creation of distinct non-isomorphic tapestries.

From a mathematical standpoint this may be formulated as follows. Given some tapestry T, we may strip away all of the labels leaving a bare tapestry, $B(T)$. We now ask whether it is possible to find a new tapestry T', which is non-isomorphic to T yet such that $B(T) = B(T')$. A tapestry with this property is said to be H-complex. A tapestry for which no such non-isomorphic relabeling exists is termed H-simple.

This turns out to be a good definition and there are already a number of basic results concerning the existence and nonexistence of H-complex and H-simple tapestries (Sulis, submitted). Although there are an infinite number of H-simple tapestries, the conditions for their existence are quite strict and limiting. As a consequence, it is far more likely that any naturally occurring tapestry will be H-complex, and thus the conditions for the possible occurrence of horizontal emergence should be ubiquitous.

As an aside, one should note that the definition of H-complexity as stated above is actually too general, resulting in a rather narrow collection of H-simple tapestries. To correct this, it is necessary to introduce the idea of relative H-complexity. The basic idea is to associate the general tapestry with a formal tapestry that serves as a system of constraints, and to require that any relabeling of the general tapestry be consistent with this formal tapestry. This is akin to requiring that any relabeling of the tapestry satisfy a collection of symmetries. This issue is discussed further in Sulis (submitted).

Conclusion: A Scenario for Horizontal Emergence

The mere statement of a mathematical condition for H-complexity does not at all prove that it has anything to do with horizontal emergence as it appears in real systems. So let us consider one possible scenario within which horizontal emergence might appear. Suppose that we have a large population of agents acting within a system. Each agent will interpret the events surrounding it according to some internal semantic frame and so each agent could, in principle, construct a tapestry which represents these events with meanings associated to its semantic frame. In a homogeneous world, all of the agents would be acting under the influence of the same semantic frame, and would be witness to more or less the same set of events, at least in principle. Even if these agents did not have access to all of the information available, they would have access to local information and one would at the very least expect that those events that were perceived in common by groups of agents would give rise to one and the same tapestry.

Although it is not a prerequisite of the theory, let us assume that the semantic frame of the agents has its origin within the physical-dynamical properties of each agent. One mechanism through which such a physical-semantic coupling might come about is through TIGoRS (Transient Induced Global Response Stabilization described in (Sulis, 1995, 2000, 2002). In TIGoRS, a non-autonomous dynamical system exhibits a stable dynamical response to specific patterned inputs. Depending upon the particular dynamical system, these inputs may have patterns that are structured or unstructured (random). Thus each such dynamical system parses its experience uniquely into patterns that stabilize and patterns that do not. It has been shown that such a parsing enables the dynamical system to function as a grammatical rewriting system. In other words, the system treats each pattern as if it were a unit of information and generates a new unit of information expressed behaviorally and linked specifically to that unit which it had received.

Even in the absence of any intrinsic property of adaptation one can still assume that agents are subject to fluctuations in structure and dynamics. It is plausible that at some time, agents are created that possess an intrinsic semantic frame which differs in essential ways from that of the larger population. In many cases these will lead to only transient aberrations since these frames will not couple in any substantive way with the larger system. But if the system is such that its tapestry is H-complex, then it possesses an inherent ambiguity. This means that over time it is possible that agents could be created whose semantic frames generate tapestries that resonate with the system, meaning that they are compatible with the bare tapestry of the system and yet are no longer isomorphic with that generated by the population as a whole. One expects a stable resonance to be created in such a circumstance.

Now suppose over time that more such agents are created, whether through reproduction mechanisms or adaptation or some other selection mechanism. It is natural that such agents, sharing a common semantic frame, and perceiving the system through a common tapestry, come over time to interact more with one another than with the other agents. It is possible that the other agents, no longer sharing a common frame of reference, might begin to distance themselves from these agents. Either way it is possible that there come to exist two distinct sub-collections of agents with distinct semantic frames coexisting within the same system. It is possible that over further time, structural and dynamical changes may cause alterations in patterns of behavior that generate informons that are no longer compatible with a shared underlying bare tapestry, but generate wholly distinct tapestries. In this event, two entirely distinct systems are created, that is, emergence.

The scenario presented here is purely a thought experiment and meant solely to suggest that the H-complexity notion shows how horizontal emergence may take place. It is not meant to suggest that this is how it happens in general, or even at all. Hopefully, the reader will think it is sufficiently plausible to warrant further investigation.

References

Anderson, P. B., Emmeche, C., Finnemann, N. O. and Christiansen, P. V. (1995). *Downward causation: Minds, bodies and matter*, Aarhus, Denmark: Aarhus University Press.

Barwise, J. and Perry, J. (1983). *Situations and attitudes*, Cambridge, MA: MIT Press.

Barwise, J. and Moss, L. (1996). *Vicious circles*, Stanford, CA: CSLI Publications.

Berlekamp, E., Conway, J. H. and Guy, R. (1982). *Winning ways for your mathematical plays*, NY: Academic Press.

Blackburn, P., de Rijke, M. and Venema, P. (2001). *Modal logic*, New York, NY: Cambridge University Press.

Cohen, J. and Stewart, I. (1994). *The collapse of chaos*, New York, NY: Viking Press.

Conway, J. (2001). *On numbers and games*, 2nd edition, A. K. Peters (ed.), Natick, MA: A. K. Peters.

DeLaszlo, V. (1959). *The basic writings of C. G. Jung*, New York, NY: Modern Library.

Forrest, S. (1990). "Emergent computation," *Physica D*, 42: 1-11.

Goldstein, J. (1999). "Emergence as a construct: History and issues," *Emergence: Complexity Issues in Organization and Management*, 1(1): 49-62.

Goldstein, J. (2003). "The construction of emergent order, or, how to resist the temptation of hylozoism," *Nonlinear Dynamics, Psychology, and Life Sciences*, 7(4): 295-314.

Nowakowski, R. (1996). *Games of no chance*, New York, NY: Cambridge University Press.

Nowakowski, R. (2002). *More games of no chance*, New York, NY: Cambridge University Press.

Sulis, W. (1995). "Driven cellular automata" in D. Stein and L. Nadel (eds.), *1993 Lectures in complex systems*, Reading, MA: Addison-Wesley.

Sulis, W. (2000). "The neural basis of mental representations: Nonlinear perspectives" in I. Trofimova and W. Sulis (eds.), *Nonlinear Dynamics in the Life and Social Sciences*, Brussells: IOS Press, pp. 98-132.

Sulis, W. (2002). "Archetypal dynamics: An approach to the study of emergence" in J. Nation, I. Trofimova, J. Rand and W. Sulis (eds.), *Formal descriptions of developing systems*, NATO science Series II/121, Kluwer, pp. 185-228.

Sulis, W.(Submitted). "Archetypal dynamics 1: Horizontal emergence," submitted to *Nonlinear Dynamics, Psychology, and Life Sciences.*

Sulis, W. (In preparation 1). "Archetypal dynamics 2: The reality game."

Sulis, W. (In preparation 2). "Archetypal dynamics 3: Vertical emergence."

Thomason, R. (1970). *Formal philosophy: Selected papers of Richard Montague*, New Haven: Yale University Press.

Appendix I: Diagonalization and Russell's Paradox

To illustrate Goldstein's Diagonalization, one of the scenarios underlying his idea of a self-transcending construction, assume that one has constructed the collection of all maps from the natural numbers to itself. Suppose further that one can assign a natural number to each map. Denote the map corresponding to n by f_n. Now create an array a_{ij}, indexed by the natural numbers, by setting $a_{ij} = f_i(j)$. Now define a new map from the natural numbers to itself by defining $g(k) = f_k(k)-1$ if $f_k(k) \neq 0$, and 1 otherwise. Clearly g cannot be one of the original functions because if it could be assigned a number n then we would have either $g(n) = f_n(n) = f_n(n)-1$ or $1 = g(n) = f_n(n) = 0$. Thus, from the natural numbers one creates a new collection that transcends the original in size, or in other words, has transcended the set from which it emerged.

Another transcendental construction gives rise to Russell's Paradox. A set is just a collection of elements. From any set one can form a subset by simply taking some elements from the set and creating out of them a new set. It is easy to see that one can create sets that do not contain themselves as members. For example the set {1, 2} does not contain {1, 2} as a member as it contains only 1 and 2. So let us now construct a new object X which consists of all sets that do not contain themselves as members. We have just demonstrated that such an object must be non-empty, so it appears to be a good object. Suppose that it is a set. If $X \notin X$, then X is a set which does not contain itself as a member, so by the definition, $X \in X$. If $X \in X$, then again by the definition, X must be a set which does not contain itself as a member, so $X \notin X$. In both cases we achieve a contradiction, which forces [X] to be a new object which transcends the concept of set. Russell called such objects classes.

Appendix II: Tapestries

Here I present some of the technical details concerning tapestries and the concept of H-complexity.

Definition: A graph is a pair (V, E) where V is a set of vertices and E is a collection of two element subsets of V, that is, E consists of subsets of the form $\{a,b\}$ where a, $b \in V$. A graph is depicted as a diagram in which each vertex is represented as a point and each edge is represented as a line joining the points corresponding to its vertices.

Definition: A directed graph (digraph) is a pair (V, E) where V is a set of vertices and $E \subset V \times V$ is a set of edges. Each edge is represented as an ordered pair (a, b). There are two associated maps $i: E \rightarrow V$ defined by $i(e) = i(a, b) = a$, which assigns the initial vertex of e and $t: E \rightarrow V$ defined by $t(e) = t(a, b) = b$, which assigns the terminal vertex of e. A directed graph also has a diagram in which the line representing an edge is replaced by an arrow pointing from initial to terminal vertex.

Definition: A labeled graph (digraph) is a triple (V, E, L) where (V, E) is a graph (digraph), L is a set of labels, and there is a mapping $l: E \rightarrow L$ which provides a labeling of each edge.

Definition: A multigraph (directed multigraph) is a graph (digraph) in which more than one edge may link any two vertices. Usually the edges in a multigraph are labeled so as to distinguish them but this is not necessary in general.

Definition: A tapestry is a 4-tuple (I, S, R, Ω) where:

1. I is a collection of informons;
2. S is a collection of struts;
3. R is a collection of relators;
4. Ω is a collection of meaning labels;
5. TS = (I, S) is an Ω-labeled, directed multigraph;
6. TR = (I, R) is an Ω-labeled, directed multigraph;
7. Given any a, b in I and label α, there exist at most two α-labeled relators, r, r´ in R with $r: a \rightarrow \alpha b$ and/or $r': b \rightarrow \alpha a$;
8. For any a, b in I and any α-labeled relator r in R with $r: a \rightarrow \alpha b$, there exists exactly one n in I and α struts s, s´ in S such that $s: n \rightarrow \alpha a$ and $s': n \rightarrow \alpha b$;
9. Every n in I is either a vertex of some r in R or an initial vertex of some s in S, or both. That is, there are no isolated vertices.

A vertex of a relator is termed a locus. An initial vertex of a strut is termed a nexus. The terminal vertex of a strut may be a locus or a nexus. The set of loci is denoted L. The set of nexi is denoted N. Thus $I = L \cup N$. A vertex is locus or nexus depending upon context.

A tapestry is graphical if (I, R, Ω) is a directed graph, and proper if each nexus is associated with exactly one label.

For simplicity we may denote an edge $e: x \rightarrow \alpha y$ by $e_\alpha(x,y)$.

A given informon may therefore be either a locus, a neuxs, or both, depending upon its connectivity and context. A locus represents an instance of a relation while a nexus defines a relation. The struts associated with a nexus identify those loci that participate in the relation. The labels tie together struts and their associated relators so as to ensure semantic consistency. Note that multiple nexi may define relations on a single label α, and a given nexus may define relations on multiple labels.

In general, the sets I, S, R, Ω will each be divided into four distinguished subsets corresponding to the active/passive, public/private distinctions and the coherence conventions must be adhered to (see Sulis, 2002).

Appendix III: H-Complex Tapestries

Definition 1: Let T = (L, S, R, Ω) be a tapestry. The bare tapestry corresponding to T is the triple B(T) = (L, S, R) formed by stripping off all of the labels. The tapestry skeleton of T is the pair U(T) = (L, S $\cup$ R) which is the bare tapestry with the distinctions between struts and relators eliminated.

Definition: Let G = (V, E) and G´ = (V´, E´) be multigraphs. A multigraph morphism f: G $\rightarrow$ G´ is a pair (a, b) where a: V $\rightarrow$ V´, b: E $\rightarrow$ E´ are maps such that if $e = \{x, y\}$ and $e' = \{x', y'\}$ and b(e) = e', then x' = a(x) and y = a(y). A similar definition holds for directed multigraphs. In the case of graphs and digraphs one can eliminate the map b since there is at most only one edge per vertex pair in the case of a graph, and the ordering for a digraph distinguishes edges.

Definition: Let T = (I, S, R, Ω) and T´ = (I´, S´, R´, Ω´) be tapestries with labelings l, l´ respectively. Then a tapestry morphism is a map f: T $\rightarrow$ T´ such that f = (a, b, c, d) where a: I $\rightarrow$ I´, b: S $\rightarrow$ S´, c: R $\rightarrow$ R´, d: $\Omega \rightarrow \Omega$´ and:

1. (a, b): (I, S) $\rightarrow$ (I´, S´) is a directed multigraph morphism,
2. (a, c): (I, R) $\rightarrow$ (I´, R´) is a directed multigraph morphism,
3. d is a set morphism such that dl = l´(a, b) and dl = l´(a, c).

A tapestry morphism is an isomorphism if all of the associated maps are isomorphisms.

Definition: Let T = (I, S, R, Ω) be a tapestry with labeling l. Then a tapestry T´ = (I, S, R, Ω´) with labeling l´ is termed a relabeling of T if l $\neq$ l´.

Definition: A tapestry T is said to be H-complex if there exists a tapestry T´ which is non isomorphic with T such that B(T) = B(T´). A tapestry T is said to be H-simple if all tapestries having the same bare tapestry as T are isomorphic to T. Mathematically, a tapestry is H-complex provided that its bare tapestry admits multiple non-isomorphic labelings, and is simple otherwise.

5. Hopeful Monsters: Literary Teleology and Emergence

Victoria N. Alexander
Dactyl Foundation for the Arts and Humanities, USA

Introduction

Artists, as such, are *intentional* beings, self-directed and creative. Literary theory was sent into a state of crisis when teleological selfhood and authorial agency were called into doubt by Derrida (1966) and Barthes (1968). Tempered by deconstruction and post-structuralism, a few theorists (Freeman & Hosek, 2001; Alexander, 2002; Fletcher, 2004) are now trying to reassess the notion of intentionality using concepts from the complexity sciences. The idea of emergence in particular enables one to redefine teleological behavior and intentionality. Appropriate to the area of literary studies, this paper takes a poetic-semiotic approach to understanding emergence that builds off of as well as supplements a scientific approach.

I'd like to stress, first of all, that an artist's claims about creative intentionality are modest. I do not argue that it entails a break with causality or special divine dispensation. Artists simply argue, as complexity scientists do, that creativity is possible because laws are generated by systems, not the other way round. This idea is quintessentially Peircean. According to C. S. Peirce, all lawful regularities are really just "nature's habits." If this is so, we cannot know natural laws prior to their appearance in systems, so the link between determinism and prediction, figured so memorably by Laplace's all-knowing being, would be a false assumption. At any particular moment, an author knows all forces that set a story in motion, and all positions of all items of which the story is composed. An author is very like Laplace's being, and so she is in a good position to interrogate his conclusions. Despite her full knowledge of the present, she does not know what will happen next before she writes it. Every invented detail can resonate unpredictably with existing details, creating new effective contexts. The new perspective changes the meaning of the events so far depicted, and sends the story in a new direction, culminating in an ending that had been inconceivable on page one. In the end, the author often recognizes "a figure in the carpet" as James (1896) calls artistic intention, which retrospectively appears to have chosen "every word," dotted "every i," placed "every comma" (p. 365).

The conditions of Laplace's fantasy can never be met or tested, such is the usual complaint, but the crucial glitch in the logic is more subtle, for, as Bergson (1910) notes - borrowing roundly from Scholastic arguments on the nature of an omniscient God who experiences past, present, and future simultaneously - any being who has complete knowledge of conditions *and* the laws that guide them would, in effect, have already experienced the fullness of time and would *know* the outcome but would not be predicting it. That is, the story would already be writ. Although Bergson's argument is specifically designed to question the supposed theoretical predictability of conscious states, he comes near the conclusion reached by nonlinear dynamics researchers who contend that if every state change in a complex system can have a different rule attached (Crutchfield, 2002), then Laplace's being, who, like an author during the process of writing, is confined to a particular place in time, could not be in possession of all the rules that would enable him to predict events at any future state.

Due to their experience with creative processes, artists are naturally receptive to the idea of emergence, and science's long resistance to it has been part of the cause for the separation of the 'two cultures'. Inviting reconciliation, Crutchfield has outlined what he considers the "theoretical frontiers" of the complexity sciences. He asks questions usually only of interest to artists, "What is meaning? What is function? ... And how do these arise when they are not originally present?" (Crutchfield, 2003). I will try to describe the poetic-mechanics behind the origins of new 'meaning' and 'function' and show how this process generates 'new rules'.

Constraints in Art and Freedom

This paper does not attempt to overturn the argument, determinism = predictability, by using its mirror logic, indeterminism = unpredictability. Instead, I argue determinism is not at odds with unpredictability and that emergence is not equal to indeterminism. Creative acts are defined by their lawfulness as much as their novelty. As is commonly and wisely thought, the arbitrary constraints of form (e.g., genre, meter, rhyme scheme, and grammar) often force poets to invent ingenious lines, and these inventions, in turn, tend to affect the form. Following Wittgenstein, we observe that although every move in the language game must be played by the rules, some moves change the rules of the game.

To argue for artistic intentionality is not to argue for absolute 'free will', i.e., contra-causal or supernatural behavior. Having an arbitrarily free will would seem to preclude self-*directed* behavior anyway. The definition of intentionality requires a notion of constraint as well, or else freedom has no meaning. Emergents must be similarly disposed. They must proceed deterministically from their histories, and they must redefine their histories by being responsive to patterns that exist only in large interacting systems and are only available from long perspectives.

Defining Intention

The long and very rich history of teleology, the study of final cause or purpose, is a boon for literary/aesthetic theorists since it formalizes the appearance of intentionality by which we can describe an object as a work of art. Teleologists, like artists, claim the whole is greater than the sum of the parts because *telos* adds 'something more'. Historically, teleologists have represented their explanations with a reflexive series: parts $\Leftrightarrow$ whole $\Leftrightarrow$ parts or a cyclical series: $A \Rightarrow B \Rightarrow C \Rightarrow A$. Most teleologists have thought of *telos* as a product of feedback, not as a cause separate from or in the future of the process it guides. Although many may associate teleology with Christianity and claims that nature is progressive and "everything works out for the best," during the nineteenth century, there was quite a bit of fruitful scientific work in secular teleology (by, for example, Karl Ernst von Baer, Johannes Müller, Carl Bergmann and Lotze Leuchart), focusing on embryology, rational and functional morphology, and cell theory (Lenoir, 1989).

Some familiar accounts of teleology have considered intentionality in terms of conscious, *linear* 'goal direction'. This unfortunate metaphor conjures up an actually existing material object, such as a target, that is spatially and temporally removed from the agent. This image encourages the fallacy of backward causation, and teleology becomes misstated as the investigation of who does what *for the sake of* what. I believe agency hinges on different questions: Can a fortunate end-state emerge that *cannot* be predicted from the states that precede it? Does this emergent property, therefore, require that some *internal constraints* be posited to guide the interactions of all that contribute to whole? Combining the insights of teleologists with complexity scientists, I define *telic* systems as those systems showing order that is not pre-specified and that confers an additional functionality on the system. Such a view of intentionality is, as Freeman (2001) has argued, distinct from notions of strictly *conscious* and/or *predetermined* purpose and avoids the problems of Cartesian dualism.

Teleology can be a useful and extremely subtle tool for exploring what it means to create something truly new and yet meaningful. In particular, teleology has shown that intentionality has two distinct aspects: *directionality* and *originality*. *Directionality* is evident in self-organizing 'archetypes', from ubiquitous spiral shapes and target patterns to 'universal' grammars and 'strange attractors'. These patterns are neutral because they *result from* a dynamically stable whole; the patterns themselves do not function to maintain the whole. Upon first appearance, directional forms might be considered merely 'decorative art' in aesthetics or benign mutants in biology. In science, directional emergents have been dismissed as mere epiphenomenal side effects. In the 1950s, Belousov had difficultly publishing his results pertaining to spontaneous oscillating concentric patterns in chemical mixtures (Winfree, 1987). Had these merely pretty patterns been shown to cause unpredictable effects, they might have received more timely attention.

As novelist and lepidopterist Vladimir Nabokov noted, "eyespots," directional concentric patterns on a butterfly's wings (comparable to Belousov's patterns), may scare off a predator that mistakes them for owl's eyes, but their usefulness is not the cause of their existence since they can arise without the gradual nudge of natural selection. Elsewhere (Alexander, 2003) I've argued, using Nijhout's (1991) research in support of Nabokov, that concentric patterns are likely to appear on a butterfly wings given the way pigment diffuses from a point source. Wing patterns are self-organizing attractors that can jump from one form to another with single or few mutations. Natural selection is not required to gradually shape 'eyes' into existence. They can appear suddenly, fully eye-like, and can immediately serve a reproductive function. When the 'eye-spot' does come to have an accidental functionality, then it has gained the aspect of *originality*; it has acquired a meaningful interpretation within a specific context.

Nabokov's non-Darwinian evolutionary theory bears resemblance to Goldschimdt's (1940) hopeful monster theory, which argues that one or few mutations can lead to great changes in form and/or function, some fortunate, some not. His idea was derided for it seems to depend too much on improbable luck. Luck often invites speculation about supernatural causes, but if one is of a more scientific mind, one can consider telic causes instead. Efficacious coincidence, I contend, is teleology's proper object of study. The image of a hopeful monster concretely illustrates the key aspects of directionality and originality in telic phenomena, exemplifying as it does the fortuitous discovery of a meaningful function for previously neutral archetypical structure.

The distinction between directionality and originality is derived from Kant's conceptions of the aesthetic and the teleological aspects, which informed nineteenth-century arguments in biology between formalist and functionalists (e.g., Reichert, 1838; Owen, 1843). Formal causality involves a governing *eidos* as rational principle (e.g., harmonious proportions), and final causality involves a governing *eidos* as purpose (e.g., moral intentions). While directionality occurs when the individual elements of a *single* system each becomes synchronized with its immediate neighbors, collectively producing organized behavior, originality must involve *two* systems, one system being used by another.

If something is merely directional, like a spiral galaxy or any number of 'archetypical' patterns in nature, it may be beautiful, but it doesn't have the full intentional aspect that is so important in defining a work of art as art - or in defining an organism as teleological. Many abstract paintings seem merely directional to me, and (supposedly naïve) viewers may ask, What's the point? (which indeed may be the 'point'). In my opinion, art needs to satisfy this 'naïve' viewer's question.

If, however, a work exhibited much more originality than directionality, mistaking nonsense for sense, as in a dream, it would not be able to communicate a point. In James Joyce's *Finnegan's Wake,* "that 600-page crossword clue," as Amis (1999) puts it, every detail is poetically linked to something else, something idiosyncratic to the author. Amis observes that Joyce has little regard for the reader. In this particular work, Joyce's laws follow a singular logic, which the reader cannot access through the text itself. In my view, the audience must be able to sense both the directional and original aspects if it is to be interpreted as art, that is, without the intervention of artful critics. As I hope to show, noting both directionality and originality is as important in emergence studies as it is in teleology and aesthetics.

Directionality: Literary Genres

Teleologists focusing on the directional aspect argued that there exist archetypes from which animal forms are derived resulting in 'variations on a theme'. Many genetically unrelated species appear to have been formed according to the same general *ground plan*: that is, many species' major organs have relatively similar distribution schemes or developmental homologies. Contra Darwinian evolutionary theory, these similarities do not necessarily imply a common ancestor or similar functional adaptations, but common rules guiding development. The burgeoning fields of 'neutral' and 'structural' evolutionary theory now contend that there is a relatively small number of structural archetypes considering the multi-dimensional space in which they evolve (Fontana & Buss, 1994; Goodwin, 1994; Crutchfield & Schuster, 2002).

Hence, there are marsupials that are strikingly similar to wolves, cats, mice, and squirrels (Goodwin, 1994). Likewise, similar grammars have arisen in completely separate language families (see Pinker, 1994, for a counter argument), and the epic genre has arisen independently several times in several different cultures. To regard these 'universal' characteristics or laws as teleological, as I see it, is to think of them as arising from the systems, *not* pre-existing them.

Literary theory benefits from the study of teleology for we find that many genres are directional phenomena, and a genre's universality may have less to do with fitness selection than with other constraints. Genres often emerge out of the interactions of various writers, all of whom follow their own local rules, but do not have a standardized set of rules to guide them. Certain behaviors are selected, not because they are more functionally 'fit': they don't necessarily work better. They are merely more likely because some sets of behaviors can be interpreted as the same.

In *Design and Chance*, Peirce (1883-84), in an effort to generalize natural selection beyond the biological realm, eliminates the need for function or reproductive advantage, and makes his theory apply to all complex systems, not just those capable of increasing their numbers through self-replication. Using the analogy of a dice game, he demonstrates how order can arise spontaneously because there are more ways to roll seven than two. The degrees of freedom are reduced by merging equivalent outcomes five and two, four and three, and one and six, which are each *iconic* signs of sevenness. A similar kind of situation seems to occur in the directional process of new genre formation, where narrative elements that are coincidentally similar might be used in a modular fashion to construct stories with different details but similar structures. While some genres are born from a single progenitor or template, as is said to be the case of Walpole's *The Castle of Otrano* (1764) and the gothic tale, it might as easily go the other way, the 'template' emerging from various stories. A folktale often begins when a king, parent, boss, or friend leaves home for work, war, trade, or entertainment, and an interdiction is addressed to the young hero left behind: Don't go into the closet, woods, courtyard, or *etc.* As a folktale morphologist, Propp (1928) classified tales according to the way they relate to an archetype, rather than an ancestral form. Any number of generically similar stories might arise independently and spontaneously through the directional process I've described. This is to suggest less mysterious or more politically neutral causes for 'universals' and 'archetypes', than have been proposed (Campbell, 1948; Frye, 1957; Boyd, 1998; Carroll, 2002).

With a few exceptions, genre fiction is notoriously unoriginal, that is, it is not particularly artistic. Although self-organization is often cited as an example of emergence of radical novelty, there doesn't seem to be true intentionality in mere directional forms of emergence (see Goldstein, 2001, 2002 for a related critique of the practice of equating emergence with self-organizing phenomena).

Originality: Phenomenal Patterns

When an artist notes a self-organized pattern in his work (e.g., variations on a butterfly theme or a pastoral genre plot structure) *and then notices that it also coincidentally serves a function*, that is, he notices it has a new meaning within the larger context of the work, he is thrilled. It is for these moments that poets thank their Muses. A Muse is the artist's own intentionality becoming apparent. This section will focus on the original aspect - the function or meaning - of telic phenomena; in some case directionality may *not* be the cause of the form, i.e., the form may be purely accidental. I refer to such patterns as *phenomenal patterns.* Critics who interpret phenomenal patterns generally seek to identify the mechanisms behind the ephemeral *meaning(s)* of a literary work. Few can resist the temptation to argue that they are highly suggestive of an 'Author', an intentional being behind the text. Phenomenal patterns defy Barthes's (1968) argument against originality in *The Death of the Author* because one cannot attribute the function of phenomenal patterns to the forces of culture, language, or genre because they do not follow any known rules or regularities. Forster (1927) claims such patterns in *Moby Dick* give the novel its "prophetic" power. A coincidence (having to do with the word 'delight' that appears twice in unrelated circumstances which also involve the word 'destruction') seems to predict the demise of Ahab's ship. Kermode's (1979) term for a phenomenal pattern is an "occult structure," and he argues that the Gospel writers consciously employed them to make Christ's story more poetic and thus somehow more mysterious. Auerbach (1953) discusses "vertically linked figures," common in sacred texts, which imply an author who exists beyond the laws of time and space and who can poetically link events that have no local connection, e.g., when Moses raised his arms in the sign of a cross he prefigured Christ's crucifixion. Martin (1986) suggests the generic term, "odd textual conjunctions," to replace the many different terms in use. Dimmock (1997) borrows the term "stochastic resonance" from science.

A simple kind of phenomenal pattern is a pun, insofar as it suggests another meaning that reveals a more profound truth. Consider this pun on 'lie':

Hamlet: "*Whose grave's this, sirrah?*"

Clown: "*Mine, sir.*"

...

Hamlet: "*I think it be thine indeed, for thou liest in't.*"

Clown: "*You lie out on't, sir, and therefore 'tis not yours. For my part, I do not lie in't, yet it is mine.*"

Hamlet: "*Thou dost lie in't, to be in't and say it is thine. 'Tis for the dead, not for the quick; therefore thou liest.*"

Clown: "*'Tis a quick lie, sir; 'twill away again from me to you.*"

(Shakespeare, *Hamlet*, 1601, V.i.383)

Making a pun amounts to making a paradigm shift, which produces a meaningful side effect. What is stable and meaningful in one context can have a different meaning in another. If we consider any individual thing out of its usual context, it becomes apparent that it has no inherent end or function. A virus can make a very different use of another system, a white blood cell, than the one for which it evolved. A viral appropriation of a pre-existing structure vividly describes a way in which originality can arise. The virus makes the white cells function for it by making their activities *mean* something different. I consider this a *creative* act insofar as it involves a distortion of the rules. If I write a 'to do' list, then I notice that it happens to have fourteen lines, as sonnets do, I can interpret it as if it were a sonnet. I can ignore the to do list's essential features and utilize a phenomenal pattern instead - the fact that it has fourteen lines. Finding new functions in this way involves what Peirce called *iconic interpretation,* an ignoring of differences and a noting of similarities. But this is a special kind of iconic interpretation. Making use of a side effect entails not seeing what is essential (pertinent to the creation/development of the object in question) and seeing what is not essential. In Darwinian biology, side effects are not specifically 'programmed in' the genes. More generally, side effects are patterns that are not pre-specified by the initial conditions. When a side effect is meaningful, it is a phenomenal pattern.

Another kind of phenomenal pattern is an echo. In *Hamlet*, Shakespeare invented 'Polonius' as a name for the king's counselor. Polonius sounds like Poland. Hamlet kills Polonious. Fortinbras, Hamlet's foil, attacks Poland. The echo seems intended to inform us about the way we are supposed to view Hamlet's actions vis-à-vis Fortinbras's. One can suppose that either Shakespeare contrived this pattern for some reason or it is an unintended coincidence, i.e., unintended side effect (Alexander, 1999). In literary criticism, one can never be certain whether one has discovered or imposed the interpretation of a phenomenal pattern.

Finding new meaning in this way is not dissimilar to the way that some suppose they can read divine intention in various coincidental 'signs' in nature, such as finding a dead bird on one's doorstep as one leaves to catch an airplane. As I have already noted above, coincidences are especially relevant to the appearance of the original aspect of *telos.* Directionality can offer explanations for self-organizing patterns (e.g., various spiral galaxies that resemble each other) but not patterns that are merely coincidental (e.g., the big and the little dippers). According to Peirce, a resemblance can be governed either by a "real" law, indicating the intelligible regularities (habits, if not laws) of nature or by an "accidental" law. When an accidental law is found, a "synthetising" subject "introduces an idea not contained in the data, which gives connections, which they would not otherwise have had" (Peirce, 1890: 261).

Peirce's distinction can be linked to a distinction in teleology between *nonmentalism* and *mentalism* (Bedau, 1990), which I identify here with directionality and originality, respectively. Nonmentalistic teleologies are founded upon evidence of intrinsic constraints that guide and limit processes. That is, the stochastically interacting parts produce a lawful whole. Mentalistic teleologies are founded upon evidence that appears to necessitate a mind to arbitrarily link two different things - through some sort of coincidental similarity - which has some meaningful effect. In the remainder of this paper, I will apply Peirce's concept of an "accidental" law to his semiotics in order to demonstrate how mentalism might arise in language or in language-like interactions. For the most part, I will be interpreting natural systems instead of poetic systems, but what I will be doing is very similar to what I do as a literary critic when I interpret phenomenal patterns. In nature as in literature, it is these kinds of interpretations that suggest the mind of an author.

Charles Sanders Peirce's Semiotics

If emergence can be compared to intentional behavior, it can because emergence is unpredictable and even inanimate emergent systems seem to have minds of their own. This is one of the reasons why we find emergence so compelling as an area of study and why others regard the interest as unscientific. What is needed, then, is a way to naturalize the appearance of 'mentalism' in emergent processes. In Peirce's semiotics, very useful for these purposes, all interactions, including interactions between inanimate things, are considered forms of interpretation. Peirce (1894) classifies different sign relations according to whether they involve iconic, indexical, or symbolic interpretation. These distinctions correspond to his notions of firstness, secondess, and thirdness, respectively. The key to understanding emergence is using a triadic system precisely because it cannot be reduced to dyadic stimulus-response relations. Only a triadic semiotic can give a role to contextually dependent meaning in nonsentient systems, allowing for flexibility, unpredictability and, consequently, intentional behavior (Sebeok & Umiker-Sebeok, 1992).

The marks ||| are iconic of threeness and may be recognized as such without any prior learning. Following Peirce in the context of distinguishing between types of laws, one may say a group of stars that look like a big dipper is an *accidental* icon. A photograph of a person is a *real* icon. (In order to avoid unwanted connotations of 'real', I will henceforth use 'non-accidental' in its stead). Above, I've used die faces as examples of icons, and I've mentioned how iconic interpretation must play a role in directionality: different signs may be considered essentially the same. Whichever kinds of signs happen to form a large class of equivalents, even if accidentally iconic, will have a greater tendency to be selected.

Iconicity involves not seeing differences between sign and object. It's a kind of interpretation that doesn't take much thought. Iconic interpretation can occur with a bird's inattention to the difference between a camouflage moth and the bark on which it rests. Iconicity also comes into play in puns, as discussed above. Puns can be unconscious: one can mistake one thing for something that is similar. Puns can be entirely thoughtless. Examples of puns in the natural world are antibodies and antigens because the receptor sites happen to have a similar shape as the binding sites. Although the resemblance is accidental, one would not, in this instance, suppose that 'synthesizing' subject is necessary to link the two. As a result of the antibody-antigen selection process, the aspect of originality - a functioning immune system - appears.

While an icon is *similar* to the object of which it is a sign; an index is *contiguous* with the object it represents. It compels and directs one's attention to something. In the natural world, indices mark a connection. A symptom is a sign of a disease. An axon is the index of the neuron. Such indices are nonaccidental.

In language, pronouns are indices that point to and establish the exact object. Words such as 'that' as in "*that* one" or 'here' as in "put it *here*" derive meaning by being linked to specific contexts and situations. Such indices are not physically connected to the objects they represent - not all indices are effects of their objects - some are only in some sort of resistance or reaction with their objects. Therefore, anything that is used to measure something else, or to get some sense of something else, as when a photon hits a particle, is an index of an object. I classify this latter type as an accidental index.

A symbol is a sign of an *arbitrary* relationship that has been forced on us through habitual exposure. First, we understand *words* as arbitrary signs of objects through the convention of language. We learn that V is the sign for |||||. Second, we learn *grammatical rules*, relationships between words. Once we understand the grammar of Roman numerals IV, we can apply the rules in novel situations. (We can even understand that 'five' is meant by the incorrect form of VX.) Words are in some sense accidental signs of the objects they represent, but the grammar (the intelligible regularity) that emerges from the relationships between words is a governing habit, if not law.

Arbitrary Contexts and Contextual Change

One can see that iconic and indexical interpretations occur instantly: they do not require learning, but the situation is quite different with the symbol. It's a much more mysterious sign gained through habitual experience with things arbitrarily linked. It emerges out of icons and indices. Curiously, the symbol seems to be able to take special advantage of what I have been calling 'accidental', icons and indices, that is, icons that only coincidentally seem similar to an object (because of a side effect) and indices that only seem contiguous with an object (because of an arbitrary association between an object and a context). If an interpretation of an accidental icon or index occurs frequently enough, it becomes 'nonaccidental'. This is the fundamental principle behind Peirce's idea of natural law as habit. To make a connection with the Janus-faced aspects of emergence that I've been developing, if the interpretation of single accidental sign bestows an additional functionality on the system, we have originality; if numerous interpretations of accidental signs occur, we have directionality. The processes loops on: if the directional (self-organized) form then comes to serve a function, it has acquired a new aspect of originality, and on and on, ever acquiring greater complexity.

An application of Peircean semiotics to the study of emergents is easily made, for emergents have already been discovered to rely upon arbitrary or accidental associations such as we find in symbol use. Goldstein (2001) has urged emergentists to consider the order found in the *containers* of complex systems. The constraints imposed by the context of interactions become *arbitrarily* associated with the interaction. Thus, the process appears to be a type of symbolic interaction. In an effort to describe symbol use (and hence mentalism) in inanimate systems, I will examine containers of the chemical process of butterfly wing pattern formation, further developing my Nabokov-butterfly theme. Butterfly wing patterns have been characterized as archetypical in nature, that is, in the 1920s, lepidopterists (Schwanwitsch, 1924; Süffert, 1927) discovered a 'template' or 'groundplan' from which, they claimed, all variations are derived. We now understand the pattern formation to be self-organizing, and the groundplan is an example of an attractor. According to Nijhout (1991), two-dimensional models of wing pattern development require two substances, an activator and an inhibitor with equal distribution. The activator's by-products (i.e., its syntheses) will tend to increase their own production, but the inhibitor will neutralize the activator's by-product, resulting in a steady state. However, the steady-state behavior will go out of balance along the edges at the container boundary where the activator is in contact with the edge not with the inhibitor. The activator produces more of itself such that its neighborhood becomes made up only of other activator molecules. Above I mentioned that directionality results partly because iconically equivalent signs can form a large class (various combinations of die faces) which has a higher probability of being selected, but directionality also sets in because, as Peirce (1883-84) metaphorically puts it, the dice "are worn down in such a way that every time a man wins, he has a slightly better chance of winning on subsequent trials" (p. 220). What's been done before tends to be done again. In this chemical reaction, we see that it is the arbitrary boundary that constrains the interactions of the activator molecules, allowing habituation or sameness in the environment to set it. Activation leads to more activation. Exemplifying Prigogine symmetry breaking, activation along the boundary of the butterfly wing spontaneously increases, giving rise to differentiation, wave-like dynamics, and overall patterns.

Chemical reactions and the container can become arbitrarily associated with one another and this may be compared to the way a word becomes arbitrarily associated with an object, allowing habituation, eventually leading to grammar-like 'intelligible regularities' or self-organization. This analysis of an 'inanimate symbolic system' is just a first step, but I believe this is the sort of thing we should be looking at if we want to understand emergence in terms of language-like interactions.

What's missing in this example is insight into the generation of originality. As noted above, mere self-organization doesn't seem to satisfy the requirements of emergence *per se*; the self-organizing pattern must bestow an additional functionality upon the system. To understand the generation of radical novelty, we also need to look at contexts and constraints, but we need to widen the perspective to include two systems interacting with each other or one system in an environment. The self-organized pattern must itself become a sign that can be interpreted or, more precisely, *mis*interpreted.

Building on Cantor's example of radically novel number generation, Goldstein (2005) has noted that emergence must also involve following and *negating*, which opens up a space for *radical* novelty-not mere ordinary change. Rules must be simultaneously, yet unparadoxically, used and transcended. Deacon (1997) has reached a similar conclusion in his application (and, I would argue, *extension*) of Peircean semiotics in his study of the emergence of grammar in hominids. I believe that both Goldstein's and Deacon's arguments can be related to my thesis linking originality (i.e., radical novelty) and accidental signs. Accidental signs acquire new meanings (without breaking rules) when the context shifts: accidental signs can be imported and used in other contexts *because they are accidental*, i.e., not physically tied to the context of origin.

Deacon argues that other species may be capable of indexical and iconic interpretation - they even can learn a number of words - but grammar seems much more difficult: it requires a negation of what I call 'accidental indices'. As we have learned above, a nonaccidental index is a sign of a causal relationship; smoke is an index of fire. Deacon claims symbolic thinking first makes use of indexical associations by utilizing a *chance* contiguity rather than an inevitable one. The word fire appears in the context of the object fire, thus sign and object become indexically associated through repetition. The pseudo-indexical associations between a word symbol and a particular object gained by experience then have to be 'unlearned' in order to develop grammar, i.e., to understand the relationships between words. 'Fire' must come to represent more that just *that* fire: it also must iconically signify other fires or fire-like things: importantly, it also must be recognized as a noun. Once habituation has set in with a pseudo-index, that is, a word has acquired conventional meaning, it may be imported to different contexts, loosely tied to its original context by memory, no longer contiguous with it. The change of context allows the possibility of finding a radically new function for the sign by following the symbolic meaning and negating the indexical association. This is precisely what happens to puns that change meaning as the context shifts.

The symbol is efficacious for abstract and imaginative thought because of its loose ties to the world, its arbitrary connections between words and objects. The decoupling of the word-symbol from its false indexical associations or the particular experience in which it was learned enables subjectivity, a decoupling of one's experience of things from the things themselves. Subjective meaning entails the symbolic grammar of one's experiences, i.e., things are understood only by the way they relate to other things.

To the extent that language is grounded, it must make use of non-accidental signs. But the power of language to self-organize and create, I argue, must arise from the utilization of chance similarities and arbitrary contiguities. The mysterious power of the symbol to generate mentalistic interactions appears to come from its use of *accidental* icons and indices, which, when they appear in the literary realm, I call phenomenal patterns.

Conclusion

Although teleologists were on the right track in pursuit of the mysterious causes of emergence, they allowed the label 'final cause' to substitute for a true understanding of the mechanics, involving efficacious coincidence of similarity or contiguity. In this paper, I set out to answer the question of how the emergence of new functions and meanings can generate 'new laws' and thus make artists capable of both directional and original behavior, i.e., intentionality. I have tried to show that new laws result from symbolic sign relations that make use of accidental icons and indices. Although my understanding of intentionality may be linked to luck, I do believe that artists can claim responsibility for meaning that arises in this way precisely because artists, insofar as they are worthy of the designation, are keenly aware of the way phenomenal patterns function in their work, and they intuitively understand the laws they generate.

References

Alexander, V. N. (1999). "Polonius and Poland, a coincidence?" *English Language Notes,* ISSN 0013-8282, 36: 8-13.

Alexander, V. N. (2002). "Narrative *telos*: The ordering tendencies of chance," Ph.D. dissertation, Graduate Center, City University of New York.

Alexander, V. N. (2003). "Nabokov, teleology, and insect mimicry," *Nabokov Studies,* ISSN 1080-1219, 7: 177-213.

Amis, M. (1999). "Nabokov and literary greatness," Nabokov's 100th Birthday Celebration, Town Hall, New York, 15 April.

Auerbach, E. (1953). *Mimesis: The representation of reality in Western literature*, W. R. Trask (trans.), Princeton, NJ: Princeton University Press, ISBN 0691060789.

Barthes, R. (1968). "The death of the author," in *Image, music, text*, New York, NY: Hill and Wang, ISBN 0006861350 (1993): 142-148.

Bedau, M. (1990). "Against mentalism in teleology," *American Philosophical Quarterly,* ISSN 0003-0481, 27: 61-70.

Bergson, H. (1910). *Time and freewill: An essay on the immediate data of consciousness*, F. L. Pogson (trans.), London, UK: G. Allen & Company, ISBN 1564595935 (1997).

Boyd, B. (1998). "Jane, meet Charles: Literature, evolution, and human nature," *Philosophy and Literature,* ISSN 0190-0013, 22(1): 1-30.

Campbell, J. (1948). *The hero with a thousand faces,* Princeton, NJ: Princeton University Press, ISBN 0691017840 (1972).

Carroll, J. (2002). "Human universals and literary meaning: A sociobiological critique of *Pride and Prejudice, Villette, O Pioneers!, Anna of the Five Towns*, and *Tess of the d'Urbervilles*," *Interdisciplinary Literary Studies,* ISSN 1524-8429, 2: 9-27.

Crutchfield, J. P. (April, 2002). Personal communication. "While linear systems can be said to act according to the same rule as they pass from one state to the next, as nonlinear systems pass from one state to the next, they also do so according to rule, but each new state can have attached a different rule. (Linear systems, in some sense, have the same rule attached to each state.)"

Crutchfield, J. P. and Schuster, P. (eds.) (2002). *Evolutionary dynamics: Exploring the interplay of selection, neutrality, accident, and function*, New York, NY: Oxford University Press, ISBN 0195142659.

Crutchfield, J. P. (2003). "Complex systems theory?" Santa Fe Institute Retreat, Bishop's Lodge, Santa Fe, 25 October.

Deacon, T. W. (1997). *The symbolic species: The co-evolution of language and the brain*, New York, NY: Norton, ISBN 0393038386.

Derrida, J. (1966). "Structure, sign, and play in the discourse of the human sciences," in R. Macksey and E. Donato (trans.), D. H. Richter (ed.), *The critical tradition: Classic texts and contemporary trends*, Boston, MA: Bedford Book, ISBN 0312101066 (1989), pp. 959-971.

Dimock, W. C. (1997). "A theory of resonance," *PMLA,* ISSN 0030-8129,10: 1046-59.

Fletcher, A. (2004). *A new theory for American poetry: Democracy, the environment, and the future of imagination*, Cambridge, MA: Harvard University Press, ISBN 0674012011.

Forster, E. M. (1927). "Prophecy," in *Aspects of the Novel*, reprint 1956, New York, NY: Harcourt Brace, ISBN 0156091801, pp. 125-147.

Freeman, W. J. (2001). *How brains make up their minds*, New York, NY: Columbia University Press, ISBN 0231120087.

Freeman, W. J. and J. R. Hosek. (2001). "Osmetic ontogenesis, or olfaction becomes you: The neurodynamic, intentional self and its affinities with the Foucaultian/Butlerian subject," *Configurations*, ISSN 1063-1801, 9: 509-541.

Frye, N. (1957). *Anatomy of criticism: Four essays*, Princeton, NJ: Princeton University Press, ISBN 0691060045.

Fontana, W. and L. Buss. (1994). "What would be conserved if the tape were played twice?" *Proceedings of the National Academy of Sciences,* ISSN 0027-8424, 91: 757-761.

Goldschmidt, R. B. (1940). *Material basis of evolution*, New Haven: Yale University Press, ISBN 030002822 (1982).

Goldstein, J. (2001). "Emergence, radical novelty, and the philosophy of mathematics," in W. Sulis and I. Trofimova (eds.), *Nonlinear dynamics in the life and social sciences*, Amsterdam, Holland: IOS Press, ISBN 4274904083, pp. 133-152.

Goldstein, J. (2002). "The singular nature of emergent levels: Suggestions for a theory of emergence," *Nonlinear Dynamics, Psychology, and Life Sciences*, ISSN 1090-0578, 6 (4): 293-309.

Goldstein, J. (2005). "Emergence, creative process, and self-transcending constructions," in K. Richardson (ed.), *Managing organizational complexity: Philosophy, theory, and practice,* Greenwich, CT: Information Age Press, ISBN 1930608217, pp. 63-78.

Goodwin, B. (1994). *How the leopard changed its spots: The evolution of complexity*, New York, NY: Scribner, ISBN 0025447106.

James, H. (1896). "The figure in the carpet," in F. Kermode (ed.), *The figure in the carpet and other stories*, New York, NY: Penguin, ISBN 0140432558 (1986), pp. 355-400.

Juarrero, A. (1999). *Dynamics in action: Intentional behavior as a complex system*, Cambridge, MA: MIT Press, ISBN 0262100819.

Kermode, F. (1979). *The genesis of secrecy: On the interpretation of narrative*, Cambridge, MA: Harvard University Press, ISBN 0674345258.

Lenoir, T. (1989). *The strategy of life: Teleology and mechanics in nineteenth century German biology*, Chicago, IL: University of Chicago Press, ISBN 9027713634.

Martin, W. (1986). *Recent theories of narrative*, Ithaca, NY: Cornell University Press, ISBN 0801417716.

Nijhout, H. F. (1991). *The development and evolution of butterfly wing patterns*, Washington, WA: Smithsonian Institution Press, ISBN 0874749212.

Owens, R. (1843). *Lectures on the comparative anatomy and physiology of the invertebrate animals*, London, UK: Longman, Brown, Green and Longmans, ASIN B0008CHFFI.

Peirce, C. S. (1883-84). "Design and chance," in N. Houser and C. Kloesel (eds.), *The essential Peirce: Selected philosophical writings*, Vol. 1, Indianapolis, IN: Indiana University Press, ISBN 0253207215 (1992), pp. 215-224.

Peirce, C. S. (C. 1890). "A guess at the riddle," in N. Houser and C. Kloesel (eds.), *The essential Peirce: Selected philosophical writings*, Vol. 1, Indianapolis, IN: Indiana University Press, ISBN 0253207215 (1992), pp. 245-279.

Peirce, C. S. (1894). "What is a Sign?" in N. Houser and C. Kloesel (eds.), *The essential Peirce: Selected philosophical writings*, Vol. 2, Indianapolis, IN: Indiana University Press, ISBN 0253211905 (1998), pp. 4-10.

Pinker, S. (1994). *The language instinct: How the mind creates language*, New York, NY: William Morrow & Co, ISBN 0688121411.

Propp, V. (1928) *Morphology of the folktale*, reprint 1968, Austin, TX: University of Texas Press, ISBN 0292783760.

Reichert, C. B. (1838). *Vergleichende Entwickelungsgeschichte des Kopfes der nackten Amphibien.* Konigsberg, quarto.

Sebeok, T. A. and Umiker-Sebeok, J. (eds.) (1992). *Biosemiotics: The semiotic web*, New York, NY: Mouton de Gruyter, ISBN 3110129477.

Schwanwitsch, B. N. (1924). "On the groundplan of the wing-pattern in nymphalids and certain other families of rhopalocerous Lepidopetra," *Proceedings of the Zoological Society of London*, ISSN 0370-2774, 34: 509-528.

Shakespeare, W. (1601). *The tragedy of Hamlet, Prince of Denmark,* H. Jenkins (ed.), New York, NY: Routledge, ISBN 0415026830 (1995).

Süffert, F. (1927). "Zur vergleichende Analyse der Schmetterlingszeichnung," *Biologisches Zentralblatt* 47: 385-413.

Winfree, A. T. (1987). *When time breaks down: The three-dimensional dynamics of electrochemical waves and cardiac arrhythmias*, Princeton, NJ: Princeton University Press, ISBN 0691084432.

6. The Challenge of Complexity: Unfolding the Ethics of Science In Memoriam Ilya Prigogine

Isabelle Stengers
Université Libre de Bruxelles, Brussels

The name of Prigogine, and of his collaborators, Paul Glansdorff, Grégoire Nicolis, René Lefever and so many others, is part of the very origin of the discovery by scientists of the huge, unexplored continent whose name is 'complexity'. When we wrote together the book which has been translated as *Order out of Chaos*, but whose French original title was *La Nouvelle Alliance*, this title was an answer to the molecular biologist Jacques Monod's announcement that "l'ancienne alliance est morte" : Men had to finally recognize they were alone in a world foreign to their music and values, because life, as understood by molecular biology, was only a matter of chance and blind natural selection. But *La Nouvelle Alliance* also had strong biblical resonances. This was not a matter of chance, since Jacques Monod's idea was that molecular biology was indeed striking a deadly blow against any faith, excepting only the scientific one. But it possible to go further, and find among the "new covenants" which are proposed in the Bible a strangely relevant one with regard to Prigogine's ethical stance. It is after the great flood, and God swears that whatever Mankind's sins, he will never flood away his world again. As a memento of this covenant, He establishes the rainbow.

Such a commitment, if accepted by scientists, who so often have explained away what they later discover to be very interesting and the source for many fruitful developments, is a challenge indeed. I will propose that this challenge can be associated with 'complexity'. It may seem obvious, since complexity gives the lie to the motto which was so often used in order to claim that everything is clear, at least in principle. "This is the same thing that we already understand, just more complicated." This was precisely Jacques Monod's claim: the study of bacteria had produced the secrets of life; the royal road, the only scientifically relevant one, had been opened. For the mouse or the elephant, or man, it would be the same questions, the same road.

Ethics, as I feel it is relevant for the scientific adventure, is first linked with keeping alive the sense of wonder. It is, already apparent in the eighteenth century, this sense of wonder that the philosopher Diderot proposed, in an imaginary conversation, to the great physicist D'Alembert. D'Alembert defended with calm certainty that matter is nothing more than inert masses and indifferent physical interactions, and that, as a result, "Time means nothing to nature." To quote Diderot passionately pleading for D'Alembert to resist the temptation to turn the open advance of knowledge into a closed vision, a scientific one maybe, but as dogmatic as the theological and religious ones Diderot was fighting against:

"You see this egg? That's what enables us to overturn all the schools of theology and all temples on the earth. What is this egg?... Instead of answering me, sit down, and let's follow these effects with our eyes from moment to moment. And first there's a point which oscillates, then a thread which grows and takes on color, flesh forms, a beak, the tips of wings, eyes, and feet appear, a yellowish material which unwinds and forms intestines. It's an animal. This animal moves, agitates itself, cries. I hear these cries through the egg shell... And then it breaks the shell. It comes out, it walks, it flies, it responds to a stimulus, it runs off, it comes closer, complains, suffers, loves, desires, rejoices. It has all your moods and goes through all your actions. Do you claim, with Descartes, that this is a purely imitative machine? But small children will make fun of you, and philosophers will reply that if that's a machine, then you are another machine. If you admit that between you and the animal there is merely a difference in organic structure, you'll be following good sense and reason, acting in good faith. But people will conclude from all this, in opposition to you, that from an inert material arranged in a certain manner, impregnated with another inert material, and subject to heat and movement, we get sensibility, life, memory, consciousness, passions, and thought."

At the end of the conversation, D'Alembert evades the argument with those final words: "I want to sleep. Good night." But it is Diderot who is holding the pen, and he chose not to give D'Alembert an easy night. What will follow is the famous Dream of D'Alembert, a wild dreaming speculation about swarming bees. I quote again, this time D'Alembert babbling in his sleep:

"Mademoiselle de L'Espinasse! Mademoiselle de L'Espinasse! Have you sometimes seen a swarm of bees going out of their hive?... The world, or the general mass of matter, is the large hive... Have you seen them move out to the end of a tree branch to form a long cluster of small winged animals, all hooked to one another by their feet?... This cluster is a being, an individual, an animal of some sort... Have you seen them?... If one of these bees decides somehow to pinch the bee to which it is hanging, what do you think will happen?... You don't know, but the philosopher knows... yes, he does. If you ever see him, he'll tell you that the second bee would pinch the one next to it, that in the entire cluster there would be as many sensations aroused as there are small animals, that everything will get aroused, shift itself, change position and shape, that a noise will arise, small cries, and that someone who had never seen a group like that arrange itself would be tempted to assume it was an animal with five or six hundred heads and a thousand or twelve hundred wings."

I will not claim that this philosopher already knew, more than two centuries before the beginning of the study of complex systems, how to define the behavior of a swarm acting like an emerging whole from the many interactions of its constitutive parts. There is a long way between an idea and the mathematical definition of a problem, between what Diderot describes and today's computer simulations of swarms and flocks. Science and philosophy are not rivals; Diderot's dream was that they would become partners, and that this would happen if, and only if, they shared, each with its own means, the same interest in the diversity and multiplicity of questions that nature and human societies may inspire in us.

Equilibrium

It was the great achievement of thermodynamics to describe equilibrium states as characterized by the extreme, minimum or maximum, value of an adequate potential function, and this achievement brought to light the importance of a new, quite abstract, physical function - entropy. I recall the event the definition of entropy constituted for nineteenth century physicists because this is already a great example of scientific creativity. For Diderot, the philosopher, it was obvious that D'Alembert was wrong, that time means something for nature, that time matters. However the first definition of the way time may matter, the spontaneous increase of entropy associated to natural processes, was an answer that only mathematical physics could provide. Only the precise definition of the demands that a potential function must satisfy could lead to the conclusion that the relevant characterization of thermodynamic equilibrium states cannot derive from energy, the central function for understanding dynamical motion, but requires this new function Clausius named entropy and which he associated with the famous second law of thermodynamics. As we all know, if all natural processes conserve energy, they cause entropy either to increase or to remain constant.

It is understandable that so many physicists considered the case as closed. There was on the one side the indefinite diversity of particular macroscopic configurations and, on the other, their common fate, an equilibrium state that the physicist is able to characterize. Furthermore, this characterization did include the reason why, whatever the particular evolution, it would spontaneously lead to equilibrium and would be unable to have the system leave equilibrium. Indeed it is the second principle of thermodynamics that directly defines the equilibrium state as the stable one. Any evolution which would cause the system to deviate from a state defined by the extreme value of its defining potential function would do so in opposition to the second principle.

At the time Prigogine was trying to extend thermodynamics to permanent out-of-equilibrium situations, where active exchanges with its environment make it impossible for the system to reach equilibrium, many were satisfied with a simple elegant answer to the question of the relation between the second principle and the kind of stable order characterizing living beings. The answer was that such an order corresponds to a production of *negentropy* that is 'negative entropy'. This is indeed a perfectly satisfying answer since it corresponds to the old vitalist idea that one way or another, life must be defined against physics, the rules of physics corresponding to death, the loss of the active coherence of life.

Dissipative Structures

If Prigogine could not accept negentropy as satisfying the problem of life, defining life as an exception or as some kind of antagonist power opposing the general power of the second principle, it is not because he wanted to explain life in terms of physics, but because for him it was not a solution, just another name for the problem. In order to characterize non-equilibrium situations, you can no longer rely on the second principle to define stable states. All situations in an open, far-from-equilibrium system, that is, a system whose exchanges with its environment keep the processes going on, do satisfy the second principle. The question has changed: it is no longer that of deducing from the definition of the system the stable behavior it will take on. It is asking about the possibility that a stable behavior becomes an unstable one if pushed further away from equilibrium, and looking for the new kind of stable behavior the system can then reach. And as we know, there is no general answer to this problem. It depends on what the general definition of equilibrium thermodynamics could safely ignore - that is, the coupling among processes. Only nonlinear coupling can lead to the possibility of what was called 'bifurcation points', when a system undergoes a global change, when we deal with the emergence of the kind of coherent collective self-organized activity Prigogine and his co-workers named 'dissipative structures'.

The very naming of dissipative structures was important. For the first time a precise meaning was associated with the old idea of emergence and novelty. Since the nineteenth century the interpretation of phase transitions - for instance what may appear as the 'emergence' of a liquid from a gas - was clear, at least in principle. Even if physicists still struggle with the precise characterization of the transition itself, liquid, gas and solids are all equilibrium states, exemplifying the power of definition of their adequate potential functions. But the emergence of out-of-equilibrium order can be associated with a kind of surprise no explanation can eliminate. I still remember Prigogine explaining with an always renewed emotion the marvel that billions and billions of molecules, devoid of any global coordination, just randomly interacting with their neighbors, would enter together a collective coherent behavior, producing together time or spatial orders with macroscopic dimensions. "Look at this liquid periodically changing its color!"

Diderot wanted D'Alembert to recognize that we have to accept matter as characterized by a general property of sensitivity. But the scientific answer is much stranger indeed than the philosopher's imagination. No need for a new property. Sensitivity, if I may use Diderot's term, is emerging; it cannot be attributed to the molecules, the individual behaviors of which remain impeccably random. Sensitivity is, as such, a far-from-equilibrium property. And as a result the definition of equilibrium states has changed. At equilibrium what physicists define as 'correlations', measuring the kind of togetherness among randomly interacting constituents, are by definition of zero range and intensity.

As a philosopher, an inheritor of Diderot, I cannot but rejoice. But I wish also to carry on with Diderot's plea for D'Alembert to accept just looking at the egg. I would like scientists to accept looking at one particular aspect of the tale I just told. There is an old Chinese proverb saying that when the wise man points to the moon, the fool looks at the finger of the wise man. Unfolding the ethics of science means that it is also worth looking at the scientific finger. Not in order to suspect it, but in order to produce a more complete celebration of the achievement. Those billions and billions of out-of-equilibrium correlated molecules are not like the moon, a beautiful presence in the sky just waiting for us to contemplate it. Or, if it is like the moon, then we must celebrate together with the presence of the moon in the sky the greatness of biological evolution that produced not only eyes to see but also the aesthetic sense that beauty matters, that the presence of the moon is worth looking and feeling. This is presupposed by the finger of the wise man and by the Chinese proverb. For, regarding a scientific achievement, the more complete celebration should include the description of the new practical questions that scientists learned were relevant in order to successfully characterize what they address. Unfolding the ethics of science means here not being fascinated by the advent of a new scientific answer, as if a new moon was coming within sight, for everybody to contemplate. It means knowing that there is no new scientific answer without the creation of new relevant questions, and a new practical definition of the kind of relevant relationship those questions entail between scientists and what they address.

Far-from-Equilibrium

I have already emphasized that, far-from-equilibrium, the power of potential functions must be renounced. But I have not sufficiently emphasized the change it produces in the very meaning of what is a scientific, objective definition. Indeed what was so satisfying with the equilibrium thermodynamic definition was that it extended, but at the price of describing equilibrium states only, the kind of power then associated with dynamics. Today, dynamics is also transformed, as instability, the so-called sensitivity to initial conditions and the study of dynamic chaotic systems, have entered the picture. But the classical ideal fulfilled by both classical dynamics and thermodynamics was that it is sufficient to identify the system, in terms of a well-defined set of variables, in order to acquire the power of deducing its possible behavior. This is why D'Alembert could indeed say that "Time means nothing to nature." What the physicist deduces may well be an evolution, but this evolution is just a transition among states physicists are able to deduce from their definition. And in equilibrium thermodynamics, it can take time for a system to reach equilibrium, but the thermodynamic definition allows physicists to know that it does not matter: sooner or later, the system will get to the state they have the power to deduce from their definition.

Far-from-equilibrium physicists must adopt a rather different position. They have to pay attention to the details of the processes, and their coupling, knowing that something, which does not matter at equilibrium, may come to matter far from equilibrium. A change in the characterization of what is relevant, a 'becoming relevant', means that in general scientists have to question their definition, not to trust its power. The "finger" and the "moon" both acquire an open, problematic dimension.

Indeed, in abstract models, constructed in order to exhibit and explore features of far-from-equilibrium order, physicists know how to question. They know which couplings, highly nonlinear ones, are crucial; they can even deal with new questions. I still remember my wonder when it was shown that gravitation, the effect of which is only an insignificant noise at equilibrium, may become quite relevant far from equilibrium, entering into the definition of structures that would not have been possible without it - that is attaining a new meaning, no longer a general meaning. It is the very role conferred to gravity by the coupled processes that produces this meaning. We cannot say any longer that the system is defined as submitted to gravity, as I feel myself to be (if I loose equilibrium I will just fall down), we must say that gravitation participates, takes part in, and is not a mere factor but an actor in the production of what becomes possible far from equilibrium, just as it plays a part in a dancer's moves - she does not fall down, she knows how to play with gravity.

Thus, even when they themselves have defined their model, physicists may be surprised by what they discover. They must learn what their model is able to produce, not deduce it. At equilibrium the system was defined by its variables, while far from equilibrium it is the very functioning of the system which will determine how processes and interactions will matter, and matter together, because of their far-from-equilibrium togetherness. And this difference becomes quite obvious when we turn, for instance, to models reproducing what we know of a biochemical pathway. Here it is indeed worth looking at the fingers, at the careful tact needed in order to try and disentangle what a cellular process does entangle. Indeed the one who creates the model is not just describing all the possible far-from-equilibrium behaviors, but meeting the contrast between what is a living cell and what would spell its death sentence. How does a cell escape chaos, the possibility of which is inscribed in the equations describing the pathway? How does a regular behavior succeed at being stabilized? This becomes a vital part of the problem of the cellular pathway.

As a result, the entangled nonlinear set of reactions, which are the rules in biochemistry, can no longer be defined simply as a mean to transform and utilize food. What scientists have to understand has indeed this function of transformation, but they must also deal with what I would call a production of existence, the existence of an ordered pattern on which life depends. They discover the crucial fact that the speed of this reaction is low in comparison with that other one, that this nonlinear coupling is much more intense than that one. They can no longer see life and death as opposed, but are led to wonder, recognizing their proximity, at the very small transformation that would be sufficient for the pattern to be destroyed, for the cell to enter into a chaotic and probably deadly regime. Just as we wonder at and admire the skill of the dancer, understanding how close her gracious moves take her to the risk of falling down, seeing in the dance how close together the double possibility of harmony and failure come.

Physics and chemistry are thus no longer defined only by the power of their definitions; they can also explore the cases where this power is lost, where scientists' practical position in relation with what they address must change. The problem is no longer one of deduction but of wondering what is relevant and how. Scientists no longer address a system as explained by what they know about it, even if they know it perfectly well, because it is a model. Their questions imply an open situation: "what will it be able to produce?" "What kind of behavior will emerge ?" And the question must be asked each time, with each new situation. Far-from-equilibrium scientists are no longer in the position of generally defining the difference between what is possible and what is not possible - they are exploring it. And so do all others exploring purely kinetic models, where constituents are defined as active and sensitive and we can forget about equilibrium. Just like Diderot's bees, just like ants, the collective behaviors of which, as studied in his department, became one of Prigogine's favorite examples.

Such a transformation was a deep source of joy for Prigogine, because he felt that at last, physics, the practice he chose and loved, the one which kept him thinking and dreaming, was able to address a wide diversity of natural situations. He knew perfectly well that he had not explained life in physical terms, but that he had contributed to an understanding of material processes in which that life would no longer be considered as a statistical miracle, against the laws of physics, but as belonging to the same strange and wonderful world. Not to explain but to go from opposition to coherence, this was the fulfillment of Prigogine's quest. And it was also the confirmation of the guess once produced by Joseph Needham, the famous embryologist turned into an historian of Chinese civilization. Indeed, Needham wrote that if the life sciences and physics ever came to a unity, it would be physics that would have changed, much more than biology.

Unfolding the Ethics of Science

Unfolding and not creating a new ethics, as associated with the new science of complexity. Indeed, let us turn towards the past, towards the so-called Newtonian revolution which heralded the triumph of a science able to submit everything to general laws: the science of Laplace imagined the first of the long series of demons and of Gods in relation to which physicists have claimed to discuss the ultimate truth of the world. Such a triumph was not as simple as it may appear. It meant the defeat of philosophers, like Diderot, and of scientists, like the chemist Venel, who associated Newton with quite another understanding of science. This defeated party celebrated Newton's courageous acceptance of what he understood was relevant. Even if we cannot understand how it is possible that distant bodies attract each other, we cannot limit nature to what we conceive as possible. The chemist Venel, fighting against a reduction of chemistry to Newtonian physics, considered himself as a true follower of Newton, who "knew that nature achieves most of her effects by unknown means; and that the only real absurdity would be to wish to limit her, by reducing her to a certain number of principles of action and ways of operating."

Venel was defeated, and it would take the quantum revolution to understand that the chemical bond was indeed something very special. Repeating Joseph Needham's remark, chemistry and physics did indeed come to a unity, but it was physics which had to be deeply transformed in order to accommodate atoms as able to enter into chemical reactions.

But telling the quantum story like that is a bit unusual. The part chemistry played is usually forgotten, and physicists, as true inheritors of Laplace, prefer to talk about the final reduction of poor blind chemistry to the general rules of quantum physics, and to speculate about quantum reality. Here resounds with full power the triumphal reductionist motto. Chemistry is the same thing as physics, it is just more complicated.

Unfolding the ethics of science, as a challenge for complexity, may then be to tell tales - to make them like many rainbows, those rainbows the Biblical God created in order to remember he had to resist the power of destroying the beauty and diversity of the world. Complexity, as it started with the discovery and study of surprising properties, usually related to the irreducible importance of nonlinear relations, would thus not mark a rupture inside the sciences, the creation of an opposition between the past and the future. It would produce the opportunity to entertain a different relation with the past, emphasizing openness, surprise, the demand of relevance, the creative aspect of the scientific adventure, and not reduction to simplicity. True scientific simplicity is never reductive; it is always a relevant simplicity that is a creative achievement, like Newton's force of attraction. And, as we know from the case of unstable dynamical systems, this creation is rarely a final point. The true grandeur of science is not power but the demanding quest for relevance.

Finally, I would stress that the science of complexity also needs rainbows, in order to escape the temptation of power, in order to keep alive the learning process and the capacity not to turn a surprise into a triumph. It is also something I learned, working with Prigogine: the many people who came to his department asking the so-called new physics to define for them the good questions in their own field, be it ecology, sociology or economy. As the house philosopher I was often obliged to receive them and talk with them, and I often had to disappoint them, telling them that physics is no universal key, that nothing can take the place of the process of creation of relevant questions in each field.

I would propose that the many complex models that are proposed be first appreciated as enlightening abstractions, precious new tools for thinking. By this I emphatically do not mean ready-made instruments. Tools modify the ones who use them. To learn how to use a tool is to enter in a new relation with reality, both an aesthetic and practical new relation. I would say that those new tools put the ones who use them in a very interesting new practical position with regards to what they address. Indeed the problem is no longer simply how to explain; it includes the problem of what is to be explained. What are the good questions? What is the relevant way to address a pattern? Is it to be directly explained, for instance, or should the relevant question be what stabilizes this pattern rather than others, or rather than chaos? And if this is the case, how are we to understand this stabilization? As the result of the past, as may be the case for biochemical patterns? As a result of the way the pattern participates in its environment? Or both? Or would it not be as a result of the patterning of the very behavior of the individual components, as it is probably the case with social situations, which heavily depend on social habits. Our situation here today, for example, heavily depends on the nice social habit we all have of listening and keeping silent, but there are many schools where the unstable, chaotic collective assembly of untamed young people produces distress and not happy surprise.

I would thus claim that the tools afforded by complexity may free us from the intellectual habit of relying on the obvious identification of what is to be explained, as obvious as the moon in the sky, and of looking for causes that would be responsible, that would have the power to directly explain what we observe. This makes a great practical difference indeed, as it creates the need for a new imagination, the cultivation of new insights, and the explicit recognition that there are no good answers if the question is not the relevant one. Tools are demanding - they do not confer the power of judging, they ask for the choice of the right tool for the right situation. In other words they oblige us to think and wonder.

Accepting Duality

As we know , the production of light also means the production of darkness. It is not that the enlightened zone makes the rest measurably darker, of course, it is just that it feels like it. Maybe the man looking for his keys would have been able, one way or another, to imagine how to look for them elsewhere. But light made him feel that elsewhere was really too dark. Feeling matters in science also - the entire history of the scientific adventure testifies to this. So very often we see the same coexistence between some who quite clearly define the scope of new results or models - what those models demand in order to be relevant - and others who will say: "we know that quite well, but it's so dark, by comparison, out here where we are; and who knows, maybe we will make some progress if we trust your questions to apply to our problems."

Since Galileo and Newton, there has been such an ongoing double identity of science, the one which relates to my definition of ethics, where the production of relevant questions is always an event, a selective and demanding creation, and the one which answers to my definition of power, where the selective, creative event becomes a new definition of a so-called scientific or rational general approach, which must be used to enlighten regions that are still so dark. Those two identities are strongly correlated. Light produces its darkness. Indeed it should never be forgotten that in fact there is knowledge about those seemingly dark regions. It is just that this knowledge - the relevant questions people have learned to ask, often since before the advent of modern science - does not look like the kind of science we desire. The feeling of darkness is produced by the contrast, and it produces the need for some light, any light.

To take only one example, among the people who dislike and criticize market mathematical economy. There are many historians of economy, who know a lot about precisely those aspects of economy which mathematical economy simplifies away. But the mathematical economists' answer is always: "yes, you historians know a lot, but it is only storytelling, not science." If we take into account everything you know, all your experience about aspects that matter, our models will be useless, but we will then be back in the darkness from which we escaped, and the chance for true scientific progress will be lost.

Unfolding the ethics of science would then mean resisting this contrast, resisting the claim that abstracting away what appears as an obstacle against a scientific approach - if we take that into account we are lost - is the condition for scientific progress, and also resisting the kind of contempt so easily directed toward scientists who learn and experience the demands of their field. Resisting this double temptation may be a challenge that may be the more specific duty and need for the complexity field to accept.

First, it may be considered as a duty, since what is so important with complex systems is that they are characterized by features which were traditionally explained away as non-scientific, the kind of things that had to be eliminated away in order for scientific progress to proceed. It is a matter of ethics that complex-systems people do not side with the kind of conception that eliminated what they now know how to address. This is why the example of Prigogine is so precious to me. His work was inspired by a strong feeling of duty. He felt obliged by the world physics claims to understand and which should not be destroyed by this claim, and he felt obliged by the experience and knowledge embryologists had produced about the mysterious, spontaneous enfolding of living order, even if they did not know how to relate this enfolding to the usual laws of physics and chemistry.

But it is also a need. Indeed the many models of complex systems, as can now be produced with computer simulation, lead to such a proliferation of diverse behaviors that there is a strong temptation to use them as a kind of universal key, able to serve whatever purpose we like. Just give me the data, well-defined data, and I can make a model! Or, inversely, I have got an interesting behavior, who has got data which would fit with it? Such use of abundant models is nice and playful. What makes the models and their use potentially dangerous is the strong claim that the business of science is to explain away what is only subjective opinion and illusion. On that claim rests the authority associated with modern science and that authority is anything but playful. How are we to avoid taking a simulation as scientific theory, explaining away what the model had no need to take into account?

On Consciousness

I will make such forebodings more explicit by taking, as a final example, what appears now to be the last scientific fashion, the claim that "consciousness is just about the last surviving mystery", to use the words of Daniel Dennett. It strikes me as the worse sign towards the direction I fear, transforming precious tools for thinking into a universal source of answers.

Indeed, if we follow Dennett, it seems on the one hand, that we now have the answer for everything but consciousness, which means for instance that we would also know how to describe biological evolution or the relation between what we call genes and the development and differentiation of the vast, proliferating population of cells that together constitute the body, including the brain. It would be just a matter of finding a complex algorithm.

But on the other hand, the so-called 'mystery of consciousness' is described in a quite abstract manner, exactly in the same way that Leibniz, some three centuries ago, declared an impasse with no hope for escape. Never would something like experience emerge from a reality described in terms of a matter devoid of experience. Leibniz was thinking of Cartesian matter, but he would certainly have repeated the same thing about the complex pattern of neuronal interactions.

And, finally, what is completely ignored, as consciousness is reduced to simple experiences like "I see red, I feel hot" is that those experiences are probably not what matters. Whatever the story of evolution, redness and hotness, or even consciously seeing that table or that face, are probably not the kind of achievement that gives evolutionary meaning to the production of the possibility of conscious awareness. Consciousness as such - abstract consciousness, the abstract fact that we are conscious - indeed appears as a mystery if you feel that the knowledge we have produced about nature truly defines nature. But what is then forgotten is the finger, the very conscious finger that was needed in order to create and verify those definitions. And what is ignored, despised, and left in the darkness is consciousness as the stake for so many human questions and techniques. How to produce new kinds of conscious awareness? How to learn? How to pay attention? How to acquire new habits of thinking? How to concentrate or explore other kinds of experiences? Those are questions that matter, that have always mattered, and that the brain cannot answer as such, because those questions imply that humans are social beings, learning how to learn, experimenting how to become, not because of their brains, but because whatever we call the brain makes them able to relate with others, with tools, with language, with writing.

What Matters

The menace here is that the complexity field may play exactly the same role as the traditional sciences of simple, linear systems: "Let us begin with the complex, the rest is the same, just still more complex."

I feel a bit like Diderot in front of D'Alembert, so many 'D'Alembert's who will tell me that I am only a philosopher and that only philosophers may speculate outside the bonds of serious science. I could obviously answer like Diderot that small children may well make fun of the idea of conscious beings consciously trying to explain consciousness, and that philosophers are entitled to reply that if consciousness is the product of blind, algorithmic complex processes, then you, and also the entire scientific adventure - all your hopes to understand and explain - are also such a product. Why would I, as a product of complex algorithms, care to listen to other algorithmic productions? But I learned to love science with Prigogine, and to honor it as a demanding adventure, not something to deride. This is why I trust that the field of complexity may answer the challenge of resisting power and accept this new version of Joseph Needham's remark: if there is ever to be any coherence between the brain as a very, very complex system and the production of experience, it will be complex-systems theory that will have changed.

7. Ethical Implications of the Laws of Pattern Abundance Distribution

Stephan R. P. Halloy[1] & Jeffrey A. Lockwood[2]
1 New Zealand Institute for Crop & Food Research, Ltd., NZ
2 University of Wyoming, US

Introduction: The POLO Distribution

Pareto (1897) was perhaps the first to recognize that a complex, social parameter - individual wealth - followed the regular pattern that we now know as a power or lognormal function[1]. Given the former is

1. We should hasten to state that power and lognormal distributions are not mathematically the same. Halloy's (1998 and subsequent) use of the POLO concept is necessary to convey the notion that these two mathematical constructs are very close (Mitzenmacher, 2004; Arita, 2005), and that as yet the debate about which natural system fits power or lognormal distributions, and why, is not resolved (a question which may not be useful in the first place). Essentially, many natural distributions "convincingly mimic" power laws but are more lognormal-like (Perline, 2005). The POLO concept thus becomes essential to show that natural phenomena never exactly fit power or lognormal distributions, but do approximate both. Trying to resolve whether a natural distribution is one as opposed to the other (and then explaining them through theoretical mechanisms) is not as useful, at least in regard to the practical, philosophical and management implications of such distributions.

The power vs. lognormal dispute revolves around the perception that we can capture a snapshot of a system at one time and characterize it with a curve. This approach led to claims of universality of power distributions (Bak & Chen, 1991; Brown, *et al.*, 1993) and equally generalized claims for lognormals (Limpert, *et al.*, 2001). This relatively recent debate often disregards what can be learned from the richness of a long history of research showing the widespread prevalence of lognormal and power distributions (Weber, 1834; Fechner, 1860, 1897; Galton, 1879; McAlister, 1879; Kapteyn, 1903, 1916; Groth, 1914; Raunkiaer, 1918; Thompson, 1942; Hutchinson, 1953; Nobuhara & Numata, 1954; Limpert, *et al.*, 2001; Mitzenmacher, 2004). More important for our understanding of processes are two aspects of natural systems which are poorly considered in the power vs. lognormal debate: 1) systems are dynamic; their distributions are constantly changing and 2) systems have diffuse boundaries.

The key consequence of point 1 is that instead of a snapshot, we need a moving film to show in what direction a system is heading. In cases where this is done with unconstrained natural systems (and as we show in this paper with sizes of nations and populations), these tend to increasingly approximate a lognormal pattern when released any distance away from it (Halloy, 1999). The key consequence of point 2 is that power functions may transform into lognormals. The straight lines of power distributions peter off toward the edges, the tail of the distribution falls, again approximating a lognormal distribution (Harte, 2004). Some authors writing about power distributions, in their enthusiasm for demonstrating universality, have either ignored the 'lognormal' tails or struggled to explain

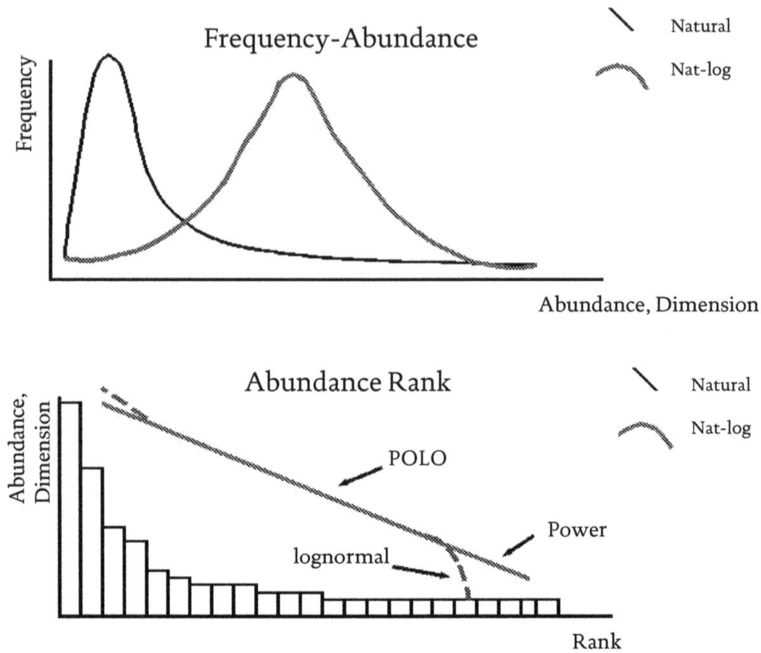

Figure 1 *The POLO (power, lognormal) pattern as manifest in frequency-abundance and abundance-rank distributions*

simply a particular case of the latter, Halloy (1998) provided some clarity by referring to power or lognormal distributions under the unified term of 'POLO' (see Figure 1). POLO distributions have been subsequently found to describe a wide range of social, political and economic values (Galton, 1879; Limpert, *et al.*, 2001; Hamada, 2004), including such unexpected measures as the sizes of cities (Zipf, 1949; Eeckhout, 2004). These patterns were also found to be widespread in biological, physical, and astronomical systems (Preston, 1948; May, 1975, 1981; Sugihara, 1980), leading to a plethora of synonyms and related terms (e.g., scale-free systems, edge of chaos, criticality) and debates about their meaning (Brown, *et al.*, 2000; Li, *et al.*, 2004; Halloy & Whigham, 2004; Arita, 2005).

In recent years, the POLO distribution has been elevated from a purely descriptive observation of a mathematical regularity into the realm

how these tails fit into the mechanistic explanation of power distributions. Other authors have stressed that the power fit for natural distributions is typical of parts of their range, thus implicitly showing the shifting nature of the tails of the distributions. In many cases, the power exponent has been calculated explicitly eliminating the curved parts of the distribution (Newman, 2005). Shiode and Batty (2005) have recently repeated the limitations of a 'universal' conception of power applications to natural curves, particularly in view of the approximation to various other models (i.e., the Yule and lognormal distributions), and recognize the more pragmatic approach of using the power function "largely because it represents a first attack on the problem of measuring Y and there are good stochastic models that are consistent with the kinds of distributions that we observe."

	Parameter	1960	1990
World, all nations	*n* units	146	191
	Land Area	4.06*	0.88*
	Population	0.49*	0.30*
USA, all territories	*n* units	68	57
	Land Area	1.55*	1.19*
	Population	0.77*	1.21*
USA, 50 states	*n* units	50	50
	Land Area	0.96*	0.96*
	Population	0.27	0.13

Table 1 *Distance from the lognormal in the abundance distribution of land area and population controlled by nation-states. NB Distance expressed as chi^2/n for the distribution including 0s at both ends. Asterisks indicate significant difference from lognormal at $P<0.05$.*

of a natural law, although diverse causative mechanisms are still being debated (Zipf, 1949; MacArthur, 1960; Bak, 1997; Halloy, 1998; Arita. 2005; Wright, 2005). In brief, the POLO abundance distribution appears to act as a 'pattern attractor', an archetype to which complex systems tend in the absence of countervailing forces or constraints. The attractive force which aggregates resources in the POLO distribution has been formalized through a gravity equation. Resources are attracted to agents in proportion to their existing 'mass' (expressed in terms of resources already possessed) and in inverse proportion to the 'distance' (expressed in terms of difficulty, which may be a function of spatial separation but need not be) to the resource to be obtained (Barabási & Albert, 1999; Halloy & Whigham, 2004; Soares, *et al.*, 2005). To illustrate the nature of POLO distributions and the underlying causative mechanisms, we consider the distributions of land, people, and wealth[2].

2. Geographic information for our analyses was obtained from *Hammond's World Atlas* (Hammond, 1964) for 1960 and the *Times Atlas of the World* (1993) for 1990. All colonies, protectorates, territories and other dependencies were lumped with the nation-state controlling them at the time. In keeping with the theoretical notion of an attractor (Halloy, 1998, 1999), we are not concerned with whether a distribution is or is not POLO in the classic statistical sense (i.e., with a given significance level), but rather we measure its distance from the POLO pattern attractor. Following Halloy and Barratt (1999), the distance to the POLO (ΔL) is expressed as the sum of squares of the differences between the actual frequency distribution and the fitted lognormal, divided by the number of agents ($\Delta L = chi^2/n$). Chi^2/n values for centered lognormals (most values in this paper) are calculated with 0s at both ends. For truncated lognormals (e.g., the business size curve), chi^2/n is calculated with 0s at the top end only.

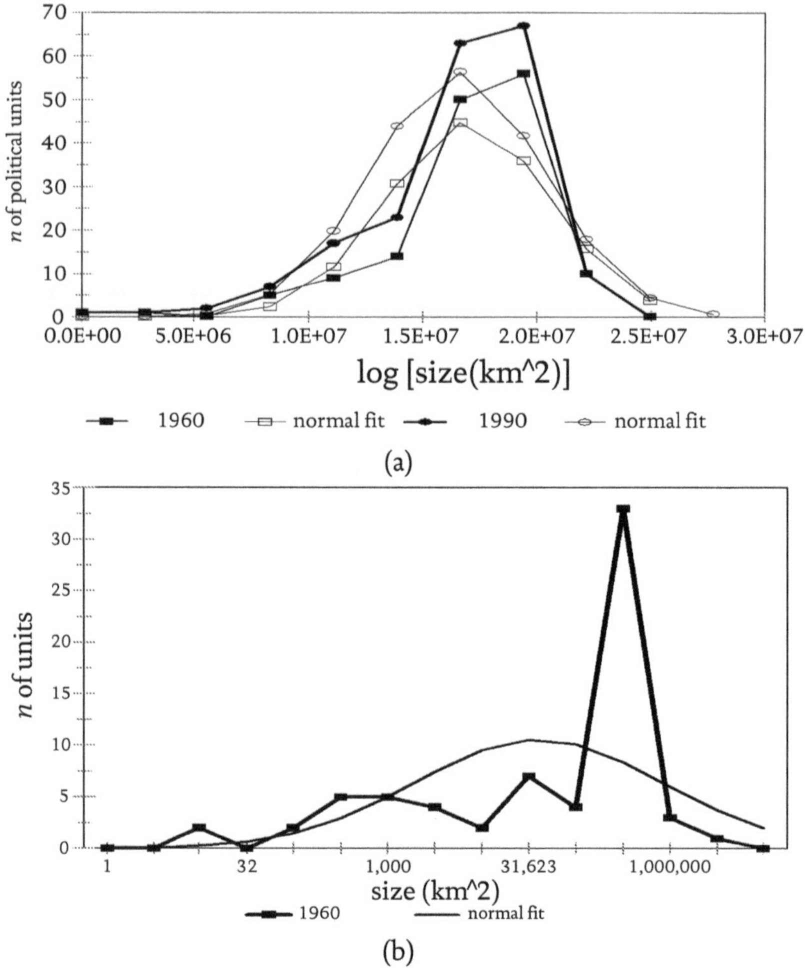

Figure 2 *Size distribution of (a) nations in the world and (b) states and territories in the USA. Chi²/n values for these are given in Table 1.*

Exemplars of the POLO Distribution: Land and Population Distributions

The attraction principle of complex systems suggests that the distribution of resources among competing peoples of the world should approximate a POLO pattern, if unconstrained. The amount of land controlled by nation states (agents) is a reasonable expression of resources. Countries have been historically gaining and losing large portions of land in a highly dynamic (mostly violent) game of competition. As predicted by theory, the areas of nations approximates a POLO distribution (see Figure 2a), maintaining a similar pattern over time (1960-1990) despite very large changes in the individual countries possessing land. Notably, the distance from the POLO decreased substantially from 1960 to 1990 (see Table 1), a fact which may be related to a release from the political constraints of colonization and the emergence of international competition and exchange. Both in 1960 and 1990 the distribution is biased toward an excess of large countries. Thus, unless outside force is applied (e.g., a strong United Nations or a shift of power from countries to multinational corporations), the next decades will likely see the continued breakup of large countries.

Unlike land, world population is a more fluid resource, particularly with the present population explosion, and our analysis revealed that population distribution approximates the POLO pattern better than land area. This is expected from measuring a 'resource' which is exchanged much more freely among nations. It also conforms to the prediction that small departures from the POLO (due to historical accidents or acts of human free will) tend to decrease with time unless other forces are at work[3].

Economic Resource Distributions Among Companies

The pattern shown by nations may become increasingly irrelevant to world affairs as the mechanism for power and resource distribution shifts to multinational corporations. Based on thousands of manufacturing companies, Stanley, *et al.* (1995) found that sales, as well as other economic parameters, almost perfectly fit a lognormal. The size of these organizations should also reflect a POLO distribution, as long as their development is not significantly constrained by social rules or ethical considerations. We found that the pattern of resources (wealth represented by market value) shown by the 1,000 largest corporations approximates a truncated lognormal distribution (see Figure 3).

---

3. Theory predicts that a POLO pattern will only be achieved if the resources can be exchanged (i.e., lost or gained) between the different agents. A constrained system without exchange would be frozen in time, which would likely give a historically deterministic or random pattern. In contrast to largely unconstrained land areas at the global level, political constraints may be severe within nations. For example, in the United States the boundaries of states were determined once in history and have been frozen in that condition. This 'freezing' occurred only after a period of shuffling in which the initial states modified their boundaries substantially to cater to the growing western frontier (see Figure 2b, Table 1).

As may be expected, chi^2/n is frozen at 0.96 for the 50 states rising to 1.55 in 1960 and 1.19 in 1990 considering all US territories (see Figure 2b). However, on a population base, the trend is quite different. Indeed, the freedom and ability of population movement in the US should represent a near maximum compared to international migration. The result is that the US population for the 50 states is not significantly different from the lognormal and has decreased its chi^2/n distance from 0.27 in 1960 to 0.13 in 1990.

The effort to maintain fixed boundaries is manifest in legislation and political structures which ensure that only the federal government can reallocate land among states. The balance of power among the states, the federal government and the people was clearly a matter of great concern to the framers of the Constitution. It was achieved through a congress including representatives based on a proportion of the population and a fixed number of (two) senators per state. Thus, the conscious decision to distribute political power - at least in part - in a manner unrelated to the size of a state stabilized a spatial distribution that diverged from the default, POLO form. Many other subsidiary measures ensure that boundaries of states are frozen (e.g., only the federal government maintains an army).

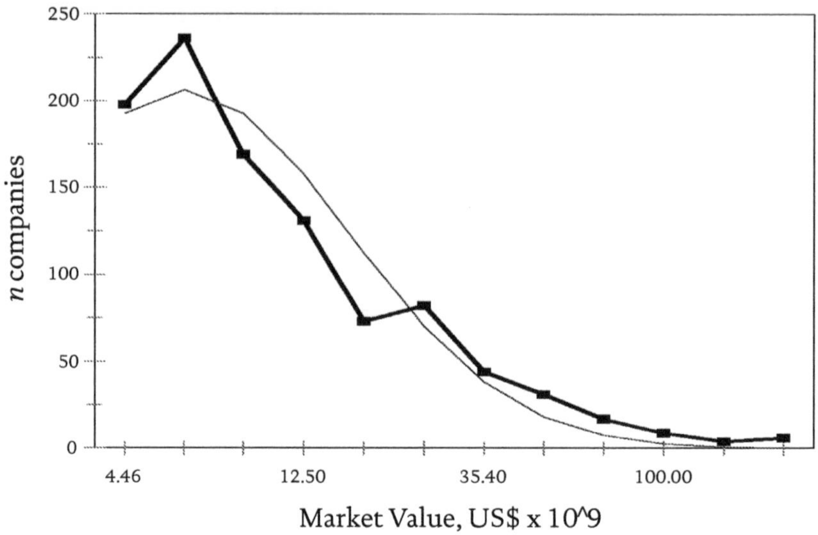

Figure 3 *Distribution of wealth (■) and lognormal fit (-) among 1000 largest multinational companies (from The Global 1000, Business Week, 7 July 1997).*

Applications of Resource Distributions

Given the ubiquity of the POLO distribution, scientists have proposed a range of practical applications. The most benign uses involve ecological indicators (Patrick, 1963; Gray, 1981; Nelson, 1987; Hill & Hamer, 1998; Kevan, 1999). More worrisome applications have also emerged. Zipf (1949) showed that departure from a POLO distribution of human populations among cities preceded the US Civil War and World Wars I and II. He stopped short of contending that such a deviation could predict, or even justify, a tendency to war or conflict, but the implications were apparent. More recently, however, the POLO distribution has been used to derive and defend social prescriptions of unequal wealth (Preston, 1950) and private vs. public transport systems (Bak, 1997), and these uses are cause for ethical concern. Indeed, justifiable concern would seem to extend beyond these explicit statements. Although few scientists go as far, many are willing to make forceful statements of universality of POLO patterns (and laws thereof) and implicitly suggest that there is no escape from such patterns.

The Naturalistic Fallacy: 'Is' Implies 'Ought'

An Historical Perspective

Menand (2001) provides an excellent overview of the way in which science was used to formulate and justify moral positions in the 19th century. In 1800, probability theory and statistics converged in what was known as the *Law of Errors*, made famous by Pierre-Simon Laplace. Astronomers had sought a means of finding the true location of celestial objects, given the variation among scientific observations. The method of least squares provided the answer they sought, and Laplace saw no reason to limit this approach to physical data. He proposed that the Law of Errors could be applied to any phenomenon that varied, including human behavior.

The real breakthrough in applying the normal distribution to social data was made by Adolph Quetelet (1846), who began his seminal work with the assertion that, "Man is born, grows up and dies according to certain laws which have never been studied." Quetelet discovered remarkable regularities in the French murder rate (categorized by the method of killing), and he contended that the mean was the true value, with higher or lower rates being "errors." The Law of Errors, he maintained, provided the means for deriving social laws that were every bit as determinate as the law of gravity. Indeed, he called his new science, "social physics."

The appeal of statistics was that it revealed an order beneath what had appeared to be randomness. Even if individuals were unpredictable, their aggregate behavior conformed to natural law. Later in the 19th century, this desire for order - a hope that appears to motivate modern scientists, even those developing chaos theory today - played a substantial role in the acceptance of evolution. Darwin proposed natural selection as an orderly process that sorted out the chance events of biological change. The idea that the world was self-regulating conferred upon statistical and morphological patterns a 'cosmic seal of approval' that provided the moral foundation for the political doctrines of individualism and *laissez fair* capitalism. Free markets, like unfettered natural systems, would not devolve into anarchy, as some worried, but through the pursuit of individual self-interest aggregate efficiency would emerge. And this optimal pattern would not just be understandable in terms of science, it would be morally virtuous.

This leap from science to ethics was perhaps most powerfully expressed by Thomas Henry Huxley, an abolitionist who argued that slavery was unnecessary. White people did not need a social institution to do that which natural law made inevitable. He maintained that the "laws of social gravitation" would prevent blacks from ascending to stations beyond their natural limits. Although some people objected to this racist characterization, the greater concern was with regard to the philosophical implications of natural law. Such scientific expressions were seen as denying free will, reducing human behavior to deterministic formulae.

The conceptual move from science (a descriptive enterprise which endeavors to divulge the nature of physical existence) to ethics (a normative undertaking which aspires to reveal the nature of virtue and duty) was the fundamental transition that has been most recently termed the naturalistic fallacy. However, the more fundamental problem - that one cannot validly derive a conclusion about 'what ought to be' from a premise concerning 'what is' - was first identified in the 18th century by David Hume. For example, from the assertion that a ring is made of gold it does not follow that the ring is valuable. We must also know that gold is valuable, and arguing for the validity of this 'missing' premise is precisely the problem.

From 'Is' to 'Ought' in Complexity

Within the field of complexity, the leap from 'is' to 'ought' can be traced back to Preston (1950). He suggested that although most people (the poor) would want to see more equality, extreme taxation to achieve this would lead to a static, zero-income society. Preston argued that,

"*Perhaps, indeed, we ought not to try to change it [the lognormal distribution of wealth]. It is not always wise to try to change the laws of nature. Success does not usually attend such attempts, and if success is only to be expected when all life has ceased and nothing moves any longer in the economic system, we should hardly welcome the success. We might do far better to try to understand the law, and the causes that bring it about, than to try to change it out of indignation and without understanding.*"

More recently, Bak, *et al.* (1988) explained that the process of 'self-organized criticality' was the mechanism accounting for power distributions. Bak (1997) then moved from description to prescription:

"*Maybe Greenspan and Marx are wrong. The most robust state for an economy could be the decentralized self-organized critical state of capitalistic economics, with fluctuations of all sizes and durations. The fluctuations of prices and economic activity may be a nuisance (in particular if it hits you), but that is the best we can do! The self-organized critical state with all its fluctuations is not the best possible state, but it is the best state that is dynamically achievable.*"

Bak's statements epitomize, in a nutshell, the forceful beliefs which have led the International Monetary Fund (IMF) and World Bank to coerce many countries into economic programs which have maintained or increased inequality, human suffering and poverty (UN-DESA, 2005). This conceptual leap shows how complexity theory can be used to endorse a normative position. Such a move is particularly worrisome in this case, given that self-organized criticality is known to lead to chaotic termination of large parts of the system (e.g., mass extinctions) as well as to dynamic stability. Scientists working with POLO distributions are not the only ones to be using complexity as a means of deriving normativity.

Holling, *et al.* (2000) and Holling (2001) have explicitly taken the leap from 'is' to 'ought' in research concerning economic, ecological, and social systems. Using the model of an adaptive cycle, Holling (2001) not only describes the complex feedback loop of human systems, but consistently refers to this process in value-laden terms (e.g., "positive change," "creativity," "invigorated," and "safe") and describes alternative pathways as being "degraded" and "maladaptive." Holling, *et al.*'s (2000) report to the MacArthur Foundation refers to dynamics other than the adaptive cycle as being pathological, and they argue that "bad regional policy and management can typically be corrected." The form of this correction is a matter of enticing or coercing these failed societies to conform to Holling's law of growth. Most notably, one of the three criteria that Holling (2001) advances for a theoretical framework is that it "be dynamic and *prescriptive*, not static and *descriptive*" (emphasis added).

Thus, workers in the field of complex economic and social systems are deriving ethical norms from mathematical models. Economic development and wealth distribution are not simply understood in terms of the adaptive cycle and POLO distributions, but the fundamental processes have a modern-day 'cosmic seal of approval' conferred by conformation to natural law. But does it truly follow that we should consider them to be 'the best achievable' outcomes, as Holling and Bak imply?

Natural Laws and Ethical Principles

Whether the move from description to prescription is defensible depends on the resolution of positivist versus transcendentalist ethics. One of the leading scientific advocates of positivism is E. O. Wilson, who refers to this approach as empirical ethics. In 1975, he argued in his landmark book *Sociobiology* (Wilson, 1975) that "ethics should be biologized." Wilson contended that ethics is nothing more than a biological phenomenon arising from complex epigenetic rules. He developed this line of thinking even further in *Consilience* (Wilson, 1998), wherein he maintains that the natural sciences will provide the foundation for all human interpretations of the world, including art, spirituality, and morality. Ethics is merely "conduct favored consistently enough throughout a society to be expressed as a code of principles." As such, ethics can be dismissed as a social crutch of diminishing relevance, a modern-day 'god of the gaps'[4].

However, even if the empiricists are correct - and there is plenty of reason to contend otherwise - this line of argument may not lead to positivistic ethics. That is, one might contend that the epigenetic rules of human nature actually compel a transcendental ethics. The human mind has arguably evolved beyond the limits of materialistic explanation and reductionist experience. As such, we are now free to think and act outside of the rules that constrain pre-conscious organisms.

There are other philosophical concerns with deriving the moral from the empirical. In *Principia Ethica,* Moore (1903) built on Hume's earlier analysis of the is-ought problem and both named and clearly explicated what we know as the *naturalistic fallacy*. Using the argument that one cannot logically leap from the factual 'is' to the normative 'ought', Moore showed that one cannot extract ethical principles from science. This posi-

4. Interestingly, even Wilson (1998) seems ultimately unable to maintain a consistent ethical empiricism when the issues are intimately related to that which he most passionately values - biological diversity. In *Biophilia* (Wilson, 1986), he outlines the extent and rate of environmental degradation and then makes an ethical plea for humans to reverse the present course. His moral argument is primarily expressed in terms of human psychological and social well-being, a form of instrumental valuing that stretches his earlier notion of empirical ethics. Furthermore, he strongly alludes to the possibility of intrinsic value, although this line of ethical argument never fully develops.

tion is grounded in Kant's (1890) contention that to act morally is to act freely, and materialistic determinism precludes genuinely ethical behavior. Indeed, many and perhaps most modern philosophers advocate some form of transcendental ethics in which morality may be informed by, but is not reducible to, material facts.

Thus, the natural laws that lead to particular distribution patterns are not 'good' simply because they exist. This naturalistic fallacy is particularly tempting in the context of contemporary environmental concerns. Deep ecology (Sessions, 1995; Katz, *et al.,* 2000) and other systems of environmental philosophy ascribe intrinsic value to nature. Hence, it is a small, but problematic, step to conclude that what is natural - a POLO distribution, for instance - must be good. In the 19th century, the study of natural history was linked to Christian values via the connection between the creation and the Creator (Barber, 1980). Using this religious endorsement of nature as being a manifestation of the divine, social Darwinists applied natural selection to justify how economic survival of the fittest reflected the will of God. Today, the linkage of nature to virtue is not explicitly religious (although spiritual arguments abound in this context; see Devall & Sessions, 1985) but the normative value of being 'natural' still serves as a foundation on which socio-economic prescriptions derived from Darwinism and complex systems are advocated.

However, even the staunch ethical transcendentalist must take seriously at least the central notion of the ethical empiricist: What 'is' (or more accurately, what 'can be') is relevant to what 'ought' to be. In short, Kant's precept that 'ought implies can' suggests that if we are incapable of performing an act then we cannot be morally compelled to do so. Although the empiricist is unable to derive what we ought to do, s/he might be able to reveal what we cannot be obligated to do. If we are unable to escape from a POLO distribution of resources, then there is no ethical argument that such a pattern is wrong or that we should work towards some better arrangement.

The Naturalistic Constraint: 'Ought' Implies 'Can'

Can We Escape Natural 'Law'?

Natural laws are more properly called 'forces' or 'tendencies' in a human context[5]. These laws describe the mechanisms that push or pull a system toward a particular condition or outcome, but forces can be counteracted by other forces, their effects neutralized or even reversed[6]. However, the notion of opposing forces is overlooked in the search

5. The word 'law' is used for natural (physical and biological) and human contexts. A human-made law is meant as a rule that cannot be broken without social consequences. Of course, whether one ought to break the law is open to question. Henry David Thoreau's *Civil disobedience* (1849) is an impassioned argument for the intentional violation of law when obedience would be immoral (in Thoreau's case he refused to pay his taxes which were being used to wage an unjust war).
6. The second law of thermodynamics has been implicated in a 'dispensation' of a natural law. Living organisms and the biosphere in general 'seem' to violate the

for a universal, or at least unified, understanding of complexity. Bak (1997) contends that, "Self-organized criticality is a law of nature for which there is no dispensation," and this sentiment pervades Holling's (2001) work on adaptive cycles.

Such assertions can result in ethical fatalism. In a relatively benign context, Bak (1997) justifies the irregularity of traffic by arguing that, "the critical state, with [traffic] jams of all sizes, is the most efficient system. The system has self-organized to the critical state with the highest throughput of cars." However, it is not clear that we can equate efficiency (by any definition) with what is desirable, let alone what is good[7]. Many people may be happier with a lesser throughput if that meant they were not even occasionally caught in massive traffic jams. We can choose to engineer highways or use mass transit systems that allow predictable, even if less efficient, flows of traffic.

Consider a better known force and the absurd implications of acceding to physical law. The force of gravity dictates that our 'natural' posture is a prone position. However, most of the time we counteract this force by applying energy to sit, stand, or walk. Indeed, we blatantly defy gravity and steady state conditions by flying in machines that are heavier than air. Yet, even when flying we haven't 'escaped' gravity; we are simply applying forces that counter that of gravity. Not only do we counteract the law of gravity, but we have no ethical compunctions in doing so and might well contend a life spent lying flat would be an immoral waste of human potential. Indeed, there are many examples of natural systems which push themselves away from a POLO distribution through the development of appropriate rules. Such qualitative changes towards and away from the POLO have happened many times in evolution when organisms have evolved from indefinite growth (i.e., no bounds to growth and competition) to limited growth and vice-versa. After having published papers showing the prevalence of the lognormal, Preston (1980) made a special effort to highlight that there are also exceptions.

But keeping away from the lognormal is not without cost. There is a 'price' to pay for defying gravity or maintaining a resource distribution other than the POLO. However, in many cases we are free to decide if the benefits justify the costs[8]. We can choose to deviate from a POLO distri-

law that all closed natural systems should increase in entropy with time. In fact, however, the apparent escape from natural law is achieved because organisms are energy dissipative structures (Prigogine, 1980) rather than closed systems, and thus are able to increase their complexity by utilizing external sources of energy.

7. An attraction of economic deregulation is the claim that competition increases efficiency which reduces prices and improves everyone's purchasing power. However, this benefit is not evenly spread, as lower prices are achieved at the expense of lower wages and higher unemployment.

8. To complicate matters, costs and benefits are scale dependent, in space, time, number and type of people who pay and who benefit. This lack of clarity as to the outcome of decisions is one of the reasons why economists have often tried to simplify human behavior to that of rational, omniscient agents, leading to unrealistic conclusions (Gell-Mann, 1994).

bution of wealth through expending 'social energy' in the form of laws, taxes, morals, gossip, religion, or other such resources (Sen, 1973, 1987; Boehm, 1993). Furthermore, social resources such as education, transportation, and medicine have been variously disconnected from a POLO distribution of wealth. During the 20th century it became possible, in at least some countries, to learn, move about, and remain healthy independent - or nearly so - of personal finances. However, recent changes in the societal values of many industrial nations has led to a diminishment of 'social energy' and a reversion to more POLO distributions (US Census Bureau, 1997; O'Connor, 1995; UN-DESA, 2005)[9].

Should We Escape the Natural Law of POLO?

The law of complex system structure (resource attraction) tells us that in the absence of countermanding forces, purely selfish agents acting in explicit competition for resources will lead to an increasingly unequal (more POLO) distribution of wealth (Halloy, 1998; Barabási & Albert, 1999; Halloy & Whigham, 2004). Of course, we do not mean to suggest that capitalism is a passive economic system; even the most capitalistic economies constrain greed to some extent. But given that we can deviate from a POLO distribution of resources, should we?

In that we are considering the ethical distribution of resources, the language and insights of justice would seem most appropriate to consider. There are various approaches to understanding justice, but Rawls's (1971) theory is particularly compelling and comprehensible. In essence, he contends that to decide what is fair, one should act in a manner consistent with not knowing one's own status in life. Behind this 'veil of ignorance' a person will select that course of action that is most just, because his/her own place in the consequent world is equally likely to be that of any one of the affected individuals. It seems that most people, if placed behind the veil, would not choose a POLO distribution of wealth, food or medicine given the overwhelming odds of finding themselves among the desperately poor, hungry, and sick of the world.

We would hasten to note that the 'natural' or POLO distribution is not inherently or necessarily immoral for all resources. Many non-vital resources reflecting human wants, rather than human needs, can be justly allocated in this manner. For example, given the range of aesthetic sensibilities, it may be entirely fair for a few people to own most of the rap music CDs, while many people own few, and most people own none.

So why do many people seem to accept, even prefer, a POLO distribution even for vital resources? We can only touch on this important and

9. This trend is not unique. Past civilizations and empires have repeated the cycle from more egalitarian origins to inequality (Osborn, 1953). Such reversions are set into motion when the level of poverty (including numbers and intensity) is such that poor people are no longer a resource (typically labor) for the rich, but become a burden to them and material production. This condition typically precedes a period of productive stagnation, inflation, environmental degradation, social upheaval and other adverse socio-economic effects.

difficult question, which deserves a treatise in itself. However, in the words of Mary Wollstonecraft, a 19th century feminist philosopher, "Most of the evils of life arise from a desire of present enjoyment that outruns itself." Many people who may well have sought long and worked hard to secure their place in the right tail of the POLO distribution[10], rationalize their position - and the weakening of social investment that would foster a more egalitarian distribution - based on the beliefs that: 1) unlimited material gain is a source of happiness, 2) one's own wealth is unrelated to another's poverty, and 3) higher productivity provides a trickle-down of benefits to all sectors of society[11].

Perhaps the most compelling explanation of our apparent preference for POLO distributions of resources is rooted in the work of the ex-

10. Many of those who have not 'made it to the top', cling to an irrational hope for ascendency which drives our consumerism as well as our tremendous appetite for games of chance such as lotteries. The underlying motives of many aspiring to an egalitarian society may of course be just as selfish. In effect, an egalitarian redistribution of wealth stands to benefit a large number of people who are below the median wealth level, which suggests a strong utilitarian basis. Understanding these motives is relevant; as mentioned earlier, empiricism can and should inform, but not define, our ethics.

11. In modern society, an underlying justification behind the drive to wealth polarization is that free markets will power economic growth and hence increase living standards of the poor as well as the rich. Since 1984, New Zealand has tested this hypothesis to a unique degree with free-market reforms. The ultimate goal was to increase economic growth and reduce poverty. However, from 1984 to 1996 the lowest 10% of the income distribution have lost 8.7% of their spending power, while the top 10% have increased their share (Chaterjee in Dalziel, 1999; see also Collins, 1987). Furthermore, New Zealand's GDP growth rate has diverged strongly from the Australian growth rate (to which it was almost equal prior to 1984) and has been almost static for 8 years (Dalziel, 1999). Neighboring Australia acted as a rough 'control' to this experiment, by not applying such radical free market reforms while being subject to similar external market forces. Under IMF coaxing, New Zealand's free-market experiment has been repeated in various degrees and forms around the world (The Ecologist, 1993; Beder, 2003), and as with New Zealand, the increase in total poverty and inequality is less than encouraging (UN-DESA, 2005). Freeing the market implies letting a system drive itself with simple, unconstrained, internal interactions, and should therefore lead toward an increase in POLO distribution of wealth. Although it is claimed to lead (but hasn't) to better national, economic performance, it surely leads to an increase in both poverty (increasing the number of poor and decreasing their level of income) and wealth (the number of ultra-rich) (Sachs, 1999; Stiglitz, 2002). Preston's (1950) fear of a zero-degree economy under perfect equality may be true in theory (although absolute equality is theoretically unachievable), but the above information suggests that totally unconstrained markets may also tend toward a zero-growth economy. Collins (1987) asked the question in regard to the New Zealand free-market experiment "can a society be dynamic and diverse but not unequal? how? If not, which should we prefer and to what extent?" As in most complex systems, our best management strategies will probably emerge from following a middle path between such extremes, with both pure POLO and perfectly uniform distributions being untenable.

istentialist philosophers such as Arthur Schopenhauer, Søren Kierkegaard, Friedrich Nietzsche, Edmund Husserl, Martin Heidegger, Albert Camus and Jean-Paul Sartre (Kaufman, 1988; Cooper, 1999; Marino, 2004). The central and relevant concept that they offered in the context of understanding the human affinity for allowing ourselves to become the subjects of external laws, forces and dictates lies in the contention that western culture has fallen into a slave morality. Many people have abrogated personal responsibility to the authority of the Church or science. But these institutions and their 'truths' are human constructs, not objective, eternal, mind-independent realities. Existentialists contend that while we may be constrained by our condition (these philosophers differ in terms of the extent to which they believe we are limited in this regard), we are entirely free to choose how we respond to the world. They call upon humans to become masters of their own condition - to act from internal integrity and authenticity rather than from obedience to the external tyrants of religion or reason.

Acquiescing to the rule of law (whether it be purportedly that of logic, God, government, or society) is dehumanizing, but it is also profoundly comforting. As a slave of moral or scientific masters, we needn't struggle with perennial questions, take responsibility for creating value, commit to autonomous action, or construct an ethical life. Accepting that we make the world - neither creating reality by whim nor discovering preexisting 'truths' - is a frightening prospect. Rather than being cast adrift, most of us cling desperately to false gods, illusionary certainties, religious potentates, and scientific sovereigns. And POLO promises to become a modern, authoritative buoy. Although many people may well have struggled to secure a place in the right tail of the POLO distribution, the existentialists would contend that they have not wrestled with the essential question of what it means to be a 'winner' of a game of pawns whose moves are dictated by vacuous rules imposed by arbitrary authorities. The superficial and illusory victory of material wealth comes at staggering moral and spiritual cost.

Conclusions

As communication technologies, transportation systems, and ecological interdependencies continue to unite people around the world, it may be reasonable to propose that the greatest moral challenge facing society in the 21st century will be the just distribution of resources on a planetary scale. Along with this expansion of the human network has come the emergence of radically new ways of understanding complex systems such as global human society. These promising approaches to perceiving how human and ecological systems 'are' can be classified in the broad category of the complexity sciences. But as exciting as these lines of investigation might be, they are not immune from the errors of earlier forms of scientific inquiry. Using the naturalistic fallacy, social Darwinists provided a rationale for favoring the wealthy more than a century ago. And several authors in contemporary complexity science have made a similar, dangerous error in reasoning.

Although complexity encompasses far more than Preston's patterns, Bak's self-organized criticality, or Holling's adaptive cycles, these programs represent worrying efforts to jump from complexity theory to normative claims. We know of no counter-examples (i.e., subdisciplines whose investigators have explicitly avoided - or warned against - using their scientific findings as a basis for ethical claims). As such, we maintain that there are sufficient grounds for concern that some contemporary scientists, in the area of complexity, are providing a fundamentally flawed justifications for the unequal distribution of resources by misrepresenting natural laws.

No scientist would doubt that natural laws and forces are a reality, but nor should any of us make the mistake of reducing the world to these terms. The reality of human freedom and social will are as evident and compelling in complex systems as any material form or mathematical dictate. By avoiding the wicked irony in which complexity is simplified to the terms of physical determinism, we come to understand that although the attraction toward the POLO is a natural law, the resultant pattern is not inevitable. Human societies have the choice of letting the attraction lead to unequal distributions, or investing in efforts to achieve a fair allocation of resources.

References

Arita, M. (2005). "Scale-freeness and biological networks," *Journal of Biochemistry*, ISSN ,138: 1-4.

Bak P. (1997). *How nature works. The science of self-organized criticality*, Oxford, UK: Oxford University Press, ISBN 0387947914.

Bak, P. and Chen, K. (1991). "Self-organized criticality," *Scientific American*, ISSN 0036-8733, 264: 26-33.

Bak, P., Tang, C. and Wiesenfeld, K. (1988). "Self-organized criticality," *Physical Review A*, ISSN 1094-1622, 38: 364-374.

Barabási, A. L. and Albert, R. (1999). "Emergence of scaling in random networks," *Science*, ISSN 1095-9203, 286: 509-512.

Barber, L. (1980). *The heyday of natural history, 1820-1870*, New York, NY: Doubleday, ISBN 022401448X.

Beder, S. (2003). *Power play: The hidden costs of deregulating and privatizing power*, Auckland, NZ: Scribe, ISBN 0908011970.

Boehm, C. (1993). "Egalitarian society and reverse dominance hierarchy," *Current Anthropology*, ISSN 0011-3204, 34: 227-254.

Brown, J. H., Marquet, P. A. and Taper, M. L. (1993). "Evolution of body size: Consequences of an energetic definition of fitness," *American Naturalist*, ISSN 0003-0147, 142: 573-584.

Brown, J. H., West, G. B. and Enquist, B. J. (2000). "Scaling in biology: Patterns and processes, causes and consequences," in J. H. Brown and G. B. West (eds.), *Scaling in biology*, Oxford, UK: Oxford University Press, ISBN 0195131428, pp. 1-24.

Collins, S. (1987). *Rogernomics: Is there a better way?* Wellington, NZ: Pitman, ISBN 0908575750.

Cooper, D. E. (1999). *Existentialism: A reconstruction*, Oxford, UK: Blackwell, ISBN 0631213236.

Dalziel, P. (1999). "Benefits failed to materialise," *Otago Daily Times*, 20th July, p.13.

Devall, B. and Sessions, G. (1985). *Deep ecology*, Salt Lake City: Peregrine Smith, ISBN 0879052473.

Eeckhout, J. (2004). "Gibrat's law for (all) cities," *American Economic Review*, ISSN 0002-8282, 94: 1429-1451.

Fechner, G. T. (1860). *Elemente der psychophysik*, Leipzig, Germany: Breitkopf und Härtel.

Fechner, G. T. (1897). *Kollektivmasslehre*, Leipzig, Germany: Engelmann.

Galton, F. (1879). "The geometric mean, in vital and social statistics," *Proceedings of the Royal Society*, 29: 365-367.

Gell-Mann, M. (1994). *The quark and the jaguar: Adventures in the simple and the complex*, New York, NY: W. H. Freeman, ISBN 0716727250.

Gray, J. S. (1981). "Detecting pollution induced changes in communities using the log-normal distribution of individuals among species," *Marine Pollution Bulletin*, ISSN 0025-326X, 12:173-176.

Groth, B. H. A. (1914). "The golden mean in the inheritance of size," *Science*, ISSN 1095-9203, 39: 581-584.

Halloy, S. R. P. (1998). "A theoretical framework for abundance distributions in complex systems," *Complexity International*, ISSN 1320-0682, 6:12.

Halloy, S. R. P. (1999). "The dynamic contribution of new crops to the agricultural economy: Is it predictable?" in J. Janick (ed.), *Perspectives on new crops and new uses*, Alexandria, VA: ASHS Press, ISBN 0961502703, pp. 53-59.

Halloy, S. R. P. and Barratt, B. (1999). "Patterns of species abundance as indicators of ecosystem status," in A. Holt, K. Dickinson and G. W. Kearsley (eds.), *Environmental indicators*, Dunedin, NZ: Environmental Policy and Management Research Centre, University of Otago, p. 87.

Halloy, S. R. P. and Whigham, P. A. (2004). "The lognormal as universal descriptor of unconstrained complex systems: a unifying theory for complexity," in R. Stonier, H. Qing-Long and L. Wei (eds.), *Proceedings of the 7th Asia-Pacific complex systems conference*, Cairns, QLD, Australia, pp. 309-320.

Hamada, H. (2004). "A generative model of income distribution 2: Inequality of the iterated investment game," *Journal of Mathematical Sociology*, ISSN 0022-250X, 28: 1-24.

Hammond, C. S. C. (1964). *Hammond's world atlas and gazetteer*, Maplewood, NJ: Hammond.

Harte, J. (2004). "The abundance and distribution of Species at multiple spatial scales," http://www.santafe.edu/events/abstract/211.

Hill, J. K. and Hamer, K. C. (1998). "Using species abundance models as indicators of habitat disturbance in tropical forests," *Journal of Applied Ecology*, ISSN 0021-8901, 35: 458-460.

Holling, C. S. (2001). "Understanding the complexity of economic, ecological, and social systems," *Ecosystems*, ISSN 1432-9840, 4: 390-405.

Holling, C. S, Folke, C., Gunderson, L. and Mäler K. G. (2000). "Resilience of ecosystems, economic systems and institutions," final report to the John D. And Catherine T. MacArthur Foundation, http://resalliance.org/file_download.php/macarthur_report.pdf?URL_ID=2116&filename=10467276131macarthur_report.pdf&filetype=application%2Fpdf&filesize=110191&name=macarthur_report.pdf&location=user-S/.

Hutchinson, G. E. (1953). "The concept of pattern in ecology," *Proceedings of the Academy of Natural Sciences of Philadelphia*, ISSN 0097-3157, 105: 1-12.

Kant, I. (1890). *Critique of pure reason*, Amherst, NY: Prometheus, ISBN

0879755962 (1990).

Kapteyn, J. C. (1903). *Skew frequency curves in biology and statistics*, Groningen, The Netherlands: Noordhoff, Astronomical Laboratory.

Kapteyn, J. C. (1916). "Skew frequency curves in biology and statistics," *Receuil des Travaux Botaniques Néerlandais*, Groningen, XIII: 105-158.

Katz, E., Light, A. and Rotherberg, D. (eds.). (2000). *Beneath the surface: Critical essays in the philosophy of deep ecology*, Cambridge, MA: MIT Press, ISBN 0613911342.

Kaufman, W. A. (ed.). (1988). *Existentialism: From Dostoevsky to Sartre.* New York, NY: Plume, ISBN 0452009308.

Kevan, P. G. (1999). "Pollinators as bioindicators of the state of the environment: Species, activity and diversity," *Agriculture, Ecosystems and Environment*, ISSN 01678809, 74: 373-393.

Li, L., Alderson, D., Tanaka, R., Doyle, J. C. and Willinger, W. (2004). "Towards a theory of scale-free graphs: Definition, properties, and implications (extended version)," http://arxiv.org/PS_cache/cond-mat/pdf/0501/0501169.pdf.

Limpert, E., Stahel, W. A. and Abbt, M. (2001). "Log-normal distributions across the sciences: Keys and clues," *BioScience*, ISSN 0006-3568, 51: 341-352, http://stat.ethz.ch/~stahel/lognormal/bioscience.pdf.

MacArthur, R. H. (1960). "On the relative abundance of species," *American Naturalist*, ISSN 0003-0147, 94: 25-36.

Marino, G. (ed.). (2004). *Basic writings of existentialism*, New York, NY: Modern Library, ISBN 0375759891.

May, R. M. (1975). "Patterns of species abundance and diversity," in M. L. Cody and J. M. Diamond (eds.), *Ecology and evolution of communities*, Cambridge, MA: Harvard University Press, ISBN 0674224442, pp. 81-120.

May, R. M. (1981). "Patterns in multi-species communities," in R. M. May (ed.), *Theoretical ecology: Principles and applications*, 2nd edition, London, UK: Blackwell Scientific Publications, ISBN 0721662056, pp. 197-227.

McAlister, D. (1879). "The law of the geometric mean," *Proceedings of the Royal Society*, 29: 367-376.

Menand, L. (2001). *The metaphysical club: A story of ideas in America*, New York, NY: Farrar, Straus & Giroux, ISBN 0374528497.

Mitzenmacher, M. (2004). "A brief history of generative models for power law and lognormal distributions," *Internet Mathematics*, ISSN 1542-7951, 1: 226B251.

Moore, G. E. (1903). *Principia ethica*, New York, NY: Prometheus, ISBN 0879754982 (1988).

Nelson, W. G. (1987). "An evaluation of deviation from the log-normal distribution among species as a pollution indicator in marine benthic communities," *Journal of Experimental Marine Biology and Ecology*, ISSN 0022-0981, 113: 181-206.

Newman, M. E. J. (2005). "Power laws, Pareto distributions and Zipf's law," http://arxiv.org/pdf/cond-mat/0412004.

Nobuhara, H. and Numata, N. (1954). "Essentials of the law of geometrical progression: Studies on the fundamental structure of biological populations. III," *Bulletin of the Society of Plant Ecology (Japan)*, 3: 180-185.

O'Connor, J. (1995). *Monitoring environmental progress*, Washington, DC: World Bank, ISBN 0821333658.

Osborn, F. (1953). The limits of the Earth, Boston, MA: Little, Brown and Co., ASIN B0007DEZWE.

Pareto, V. (1897). *Cours d'economie politique*, republished as *Manual of political*

economy, Augustus M. Kelley Publishers, ISBN 0678008817 (1969).
Patrick, R. (1963). "Structure of diatom communities under varying ecological conditions," *Annals of the New York Academy of Sciences*, ISSN 0077-8923, 108: 353-358.
Perline, R. (2005). "Strong, weak and false inverse power laws," *Statistical Science*, ISSN 0883-4237, 20: 68-88.
Preston, F. W. (1948). "The commonness, and rarity, of species," *Ecology*, ISSN 0012-9658, 29: 254-283.
Preston, F. W. (1950). "Gas laws and wealth laws," *Scientific Monthly*, ISSN 0096-3771, 71: 309-311.
Preston, F. W. (1980). "Noncanonical distributions of commonness and rarity," *Ecology*, ISSN 0012-9658, 61: 88-97.
Prigogine, I. (1980). *From being to becoming: Time and complexity in the physical sciences*, San Francisco, CA: W.H.Freeman, ISBN 0716711087.
Quetelet, A. (1846). *Théorie des probabilités*, Paris.
Raunkiaer, C. (1918). "Recherches statistiques sur les formations végétales," *Biologiske Meddeleser Kongelige Danske Videnskabernes Selskab*, 1: 1-80.
Rawls, J. (1971). *A theory of justice*, Cambridge, MA: Belknap, ISBN 0674000781.
Sachs, J. (1999). "Helping the world's poorest," *The Economist*, ISSN 0013-0613, 352: 17-20.
Sen, A. K. (1973). *On economic inequality*, Oxford, UK: Clarendon Press, ISBN 0198281935.
Sen, A. K. (1987). *On ethics and economics*, New York, NY: Basil Blackwell, ISBN 0631164014.
Sessions, G. (ed.). (1995). *Deep ecology for the 21st century*, Boston, MA: Shambhala, ISBN 1570620490.
Shiode, N. and Batty, M. (2005). "Power law distributions in real and virtual worlds," http://www.isoc.org/inet2000/cdproceedings/2a/2a_2.htm.
Soares, D. J. B., Tsallis, C., Mariz, A. M. and da Silva, L. R. (2005). "Preferential attachment growth model and nonextensive statistical mechanics," *Europhysics Letters*, 70: 70-76.
Stanley, M. H. R., Buldyrev, S. V., Havlin, S., Mantegna, R. N., Salinger, M. A. and Stanley, H. E. (1995). "Zipf plots and the size distribution of firms," *Applied Economics Letters*, ISSN 1350-4851, 49: 453-457.
Stiglitz, J. E. (2002). *Globalization and its discontents*, New York, NY: W. W. Norton, ISBN 0393324397
Sugihara, G. (1980). "Minimal community structure: An explanation of species abundance patterns," *American Naturalist*, ISSN 0003-0147, 116: 770-787.
The Ecologist. (1993). *Whose common future? - Reclaiming the commons*, London: Earthscan, ISBN 0865712778.
Thoreau, H. D. (1849). "Civil disobedience," in J. W. Krutch (ed.), *Walden and other writings by Henry David Thoreau*, reprinted, New York: Bantam, ISBN 1586632116 (1982).
Thompson, D. A. W. (1942). *On growth and form*, 2 vols. Cambridge MA: Cambridge University Press, 0486671356 (1992).
Times Books. (1993). *The Times atlas of the World - Comprehensive edition*, London, UK: Times Books, Harper Collins, ISBN 0723004927.
UN-DESA. (2005). *Report on the World social situation: The inequality predicament*, United Nations Division for Social Policy and Development of the Department of Economic and Social Affairs, ISBN 9211302315.
US Census Bureau. (1997). http://www.census.gov/hhes/income/income97/in97dis.html.

Weber, H. (1834). *De pulsa resorptione auditu et tactu. Annotationes anatomicae et physiologicae*, Leipzig, Germany: Koehler.

Wilson, E. O. (1975). *Sociobiology: The new synthesis, Cambridge*, MA: Belknap, ISBN 0674002350 (2000).

Wilson, E. O. (1986). *Biophilia*, Cambridge, MA: Harvard University Press, ISBN 0674074424.

Wilson, E. O. (1998). *Consilience: The unity of knowledge*, New York, NY: Alfred A. Knopff, ISBN 0679450777.

Wright, I. (2005). "The social architecture of capitalism," *Physica A: Statistical Mechanics and its Applications*, ISSN 0378-4371, 346: 589-620.

Zipf, G. K. (1949). *Human behavior and the principle of least effort: An introduction to human ecology*, reprinted (1972), New York, NY: Hafner Publishing Company, ASIN B0007FMHHW.

8. What is Complexity Science? Towards the End of Ethics and Law Parading as Justice

Matthew Abraham

Complexity theory figures prominently in recent debates about ethics, justice, contingency, and law. At the heart of such debates, radical reconfigurations of sacred concepts such as rights, principles, and entitlements have led to the re-examination of ethical and legal discourse's provenance and authority. The concept of complexity ("large-scale, nonlinear interaction"), a term used to describe the interaction of a large number of nodes or agents in a dynamic environment, has been used to discuss possible structural resonances between the brain, natural language, artificial intelligence, deconstruction, and the legitimation of knowledge in modern society. Such conceptualizations are in stark contrast to the rule-based descriptions of complexity that impose the rigidness of principled behavior on the nodes that cannot account for the contingency of environmental conditions and localized adaptations.

How can we begin to deploy a theory of complexity to interrogate the symbolic parameters and Habermasian consensus-based limitations that structure any ethical and legal system, and begin to engage in a Lyotardian, Derridean, and Benjaminian interrogation of "ethicality" and "legality" through such a theory of complexity? This article will quibble with the premises and articulation of just such a question and argue that Jean-François Lyotard's *The Postmodern Condition*, Jacques Derrida's "Force of law: The mystical foundations of authority," and Walter Benjamin's "Critique of violence" offered theories of complexity long before complexity studies came into vogue. All three of these works recognized the inability of a system of knowledge, legality, or ethicality to transcend the "conditions of its possibility" or to "get outside of itself " to view its systemic effect upon a society. Drawing on Niklas Luhman's appropriation of Humberto Maturana's concept of autopoeisis, we are left to contemplate the ways in which ethical and legal systems "emerge as the by-product of the system's attempt to preserve its own reproduction from the ravages of moral infection" (Rasch, 2000).

Through an examination of these works, in conjunction with many ongoing discussions about complexity theory, we can begin to resolve a conversational absence between the organizational sciences and the humanities. Ultimately, whether we labor in the organizational sciences or the humanities, we are led to the following troubling question: "Are we doomed to system immanence due to autopoeisis in examining ethical and legal systems?" If the answer is "yes," how can a theory of complexity lead us to an awareness of the process through which this process of autopoeisis occurs, and how can one lead it to "embarrass" itself in legal and ethical discourses? Lyotard, Derrida, and Benjamin offer key answers to these crucial questions.

The "terms" of the Habermasian–Lyotardian debate are by now legion and are in little need of recapitulation. Jürgen Habermas's commitment to a consensus-based discourse ethics stands in stark contrast to Jean-François Lyotard's insistence on a dissensus-generated notion of justice. While Habermas offers us a "theory of communicative action," Lyotard conducts a sort of intellectual guerilla warfare on the "system," which he describes as being fueled by the terror of the legitimation game. While Habermas recognizes the development of a "legitimation crisis" within contemporary society due to a variety of discourses competing for the status of "truth" "authority," and "the story to end all stories" – leading to a loss of faith in all three – Lyotard views the production of a radical notion of illegitimacy as being the result of living within a heterogeneous society that he comes to celebrate as bearing witness to the predicament of living within the "agonistics of a network" that we have come to call the postmodern condition, which can be described through a theory of complexity. While Habermas holds tightly to a rule-centered and discourse-governed "reason," Lyotard seems to commit a heresy in his advocacy of the production of a systems failure that will lead to the renunciation of the grand narrative and an embrace of our *petit recits* ("local" narratives). William Rasch views Habermas's formulation of a theory of communicative action as a response to societal complexity in that consensus becomes the price of admission into a human community – comprised of individual subjects with perhaps wildly differing notions of the good and the right – that must develop social norms to live in harmony (Rasch, 2000: 32).

Lyotard's *The Postmodern Condition: A Report on Knowledge* characterizes the postmodern condition as being principled by the "inventor's paralogy" and not the "expert's homology" (Lyotard, 1977: xxv). Paralogy "refers to the uncodifiable moves we make when we communicate with others, and ontologically, the term describes the unpredictable, elusive, and tenuous decisions or strategies we employ when we actually put language to work" (Kent, 1993: 3). Simply put, paralogy is communicative guesswork because it denies the possibility of a logical system that could possibly predict how language interaction will take place. Lyotard states that paralogy "is a move (the importance of which is often not recognized until later) played in the pragmatics of knowledge" (Lyotard, 1977: 61). Paralogy can be viewed as a strategy, positioned outside of systems of logical legitimation, that seek to "subsume" logic and technical skill. Indeed, paralogy claims its "guerilla tactics" by virtue of its marginality in relation to logic. Paralogy is antilogic. It is this power of antilogic that Lyotard describes in the following:

"*It is necessary to posit the existence of a power that destabilizes the capacity for explanation, manifested in the promulgation of new norms for understanding or, if one prefers, in a proposal to establish new rules circumscrib-*

ing a new field of research for the language of science. This, in the context of scientific discussion, is the same process Thom calls morphogensesis. It is not without rules (there are classes of catastrophes), but it is always locally determined. Applied to scientific discussion and place in a temporal framework, this property implies that "discoveries" are unpredictable. In terms of the idea of transparency, it is a factor that generates blind spots and defers consent" (Lyotard, 1977: 61).

Lyotard envisions the "science and knowledge of today as a search, not for consensus, but very precisely for 'instabilities,' as a practice of paralogism, in which the point is not to reach agreement but to undermine from within the very framework in which the previous 'normal science' had been constructed" (Jameson in Lyotard, 1977: xix). Paralogy provides a tool of epistemological subversion. It is a continually new way to play the legitimation game and keeps the dominant legitimation process continually guessing. Paralogy is the suppressed other in this game. Thomas Kent argues that "without the suppressed other, no logical construct can exist because every logical construct, such as a discursive argument, demands that we ignore or suppress elements that lie outside the construct" (Kent, 1993: 4). Complexity science moves us toward the celebration of the paralogical, allowing for an understanding of the dynamic interactions that characterize social systems.

Paul Cilliers (1998) highlights the ways in which complexity is being put to work in the analysis of contemporary social problems. Through distributed representation, Cilliers circumvents the shortcomings of the rule-based understanding of complexity, because he is able to demonstrate that distributed representation is not representation at all but rather the recognition of localized contingency. Drawing on Lyotard's *The Postmodern Condition*, Cilliers (1998: 119-20) asserts that postmodern societies meet all of the 10 criteria for complex systems:

1. Complex systems are comprised of a large number of elements;
2. The elements in a complex system interact dynamically;
3. The level of interaction is fairly rich;
4. Interactions are nonlinear;
5. The interactions have a fairly short range;
6. There are loops in the interconnections;
7. Complex systems are open systems;
8. Complex systems operate under conditions far from equilibrium;
9. Complex systems have histories;
10. Individual elements are ignorant of the behavior of the whole system in which they are embedded.

By analogy:

1. Postmodern societies have millions of agents operating within them at any one time;
2. These agents fulfill roles in a number of dynamic and multiple roles (teacher, consumer, parent, child, etc.);
3. In a postmodern society, the level of interaction between agents and between agents and mechanisms of the societal system are extremely rich and diverse. Examples include economic transactions and market consumption;
4. Social relationships in postmodern society are nonlinear. It is within these asymmetrical power relationships that people operate as teachers, students, consumers, and citizens. Cilliers is careful to note that he is not in anyway supporting the exploitation that results from such asymmetric relationships. He is simply stating that individuals, to use Lyotard's words, must enter "the agonistics of the network" to disturb these asymmetric relationships;
5. Individuals interact on local levels. Although local levels influence other local levels, there is no "metalevel controlling the flow of information" (Cilliers, 1998: 121);
6. All interpretations are local, contingent, and provisional. In this situation, paralogy and dissensus rather than homology prevail;
7. Open systems such as the social interact with other open systems such as the ecological;
8. Social disequilibrium characterizes the postmodern condition;
9. Although the concept of history is dismissed as a grand narrative in the postmodern, local narratives tell the histories of individuals and groups;
10. It is impossible for an individual to have a complete understanding of the operations of the entire social system in which they live and interact.

Cilliers uses his analogy between complex systems and postmodern societies to dismiss the notion that postmodernism sanctions an "anything goes mentality" in which relativism reigns supreme. Instead, he asserts, postmodernism leads us to new ethical horizons and commitments. He draws on Lyotard to emphasize this point:

"The breaking up of the grand Narratives ... leads to what some authors analyze in terms of the dissolution of the social bond and the disintegration of social aggregates into a mass of individual atoms thrown into the absurdity of Brownian motion. Nothing of this kind is happening: this point of view, it seems to me, is haunted by the paradisaic representation of a lost "organic society" (Lyotard, 1977: 15).

As Cilliers states, "A careful reading of Lyotard shows that his understanding of the individual is formulated in such a way as to counter the idea of fragmentation and isolation that could result from a dismissal of the grand narrative" (Cilliers, 1998: 115). He goes on to argue that individuals constitute part of a vast social scene in which each enters into an "agonistic network" in which discourses compete for legitimacy. Within this framework, paralogy and dissensus rather than homology and consensus "supply the system with that increased performativity it forever demands and consumes" (Lyotard, 1977: 115). Cilliers compares paralogy to self-organized criticality by which "networks diversify their internal structure maximally" (Cilliers, 1998: 117). Ultimately, Cilliers is most intrigued by the Lyotardian concept of justice that is within the postmodern condition. Through the work of Drucilla Cornell and Jacques Derrida, he outlines four criteria for "responsible judgment" in the wake of postmodernism and complexity (Cilliers, 1998: 139-40):

1. Respect otherness and difference as values in themselves;
2. Gather as much information on the issue as possible, notwithstanding the fact that it is impossible to gather all the information;
3. Consider as many of the possible consequences of the judgment, notwithstanding the fact that it is impossible to consider all the consequences;
4. Make sure that it is possible to revise the judgment as soon as it becomes clear that it has flaws.

These four criteria could very well be called a "postmodern ethic" or "postmodern attitude."

Cornell, quoting the work of Nicholas Luhmann, writes that ethics "emerges as the by-product of the system's attempt to preserve its own reproduction from the ravages of moral infection" (Rasch & Wolf, 2000: 99). The conditions of ethics' possibility emerges from the rule-governed structure that sets the parameters of any social system. Indeed, the impossibility of the ethical allows us to speak of ethics. The approbation surrounding justice operates in exactly the same way. Cornell writes: "Justice, in like manner, marks the limits of law, particularly in a modern legal system, demanded by the legal system itself" (Rasch & Wolf, 2000: 102).

Because it is dependent on the rules of law that condition what "can and cannot be," justice is impossible outside these very conditions of legality. The reliance of lawyers and judges on the concept of *stare decisis* ("let the decision stand") precedent, leads to the creation of a system that can be described as complex, because within this system there are a huge number of interacting actors and components. Luhmann finds that a system can be described as complex when there are more possibilities within it than can be realized. Robert Gibbs (2000) writes:

"Luhmann's theory focuses on what sort of communications law makes possible. Law, as a mere code, but as a way of organizing interpretations and communications, sorts materials. Law is a process of making texts relevant to the present: it directs the communications to a wide range of materials, stretching back even millennia. But unlike historical discourse, law makes explicit the need for materials to be relevant in the present ... Luhmann accentuates how legal reasoning negotiates with precedence, making the legal tradition alive in the present case. Luhmann quips that law serves the continuation of communication by other means, parodying the axiom of war as the continuation of politics by other means. Law directs us to reexamine the previous record, to discover disagreements in the past that can be interpreted in our present. But the value of law for society is precisely its ability to generate many conflicts: to formalize and communicate just what we disagree about. Law not only produces the conflicts, but also the complexity with which to treat the conflicts. The promise of legal judgment provides the space in which dissent flourishes, in which factions can be tolerated. We judge through law in order to allow conflict to become more complex, outside the realm of force. But law marshals the complexities of past and present" (Gibbs, 2000: 215).

Law generates its own complexity by presenting us with a maze of possibilities, of which only one can be realized. Within this maze, the impossibility of justice – one of the possibilities – provides the backdrop against which legal authority parades. Clearly, legal authority involves a paradox that can be best described through complexity theory.

Jacques Derrida's "The force of law: The mystical foundations of authority" (1992) confronts us with the consequences of living within the limits of representation that structure any system of language or thought. This essay questions the inauguration of a discourse that has come to be known as "the Law" through Derrida's close reading of Walter Benjamin's "Critique of violence." For Derrida, this work, written in the midst of the Nazi German state's ascendance to world power, grapples with the question of the state's justifications for its legal mandates. Samuel Weber writes, "Derrida finds, or rather rediscovers in [Benjamin's 'Critique of violence'] precisely that '*différentielle contamination*' that has traversed and troubled his own writings at least from his student days on" (Weber, 1991: 1183).

The gravest atrocities in the course of human history have been sanctioned in the "name" of the law, the nominalization of the law; and in his close reading of Benjamin, Derrida comes to describe deconstruction, law, and justice in a way that complexity theorists cannot fail to appreciate. He writes that "deconstruction occurs in the interval between the deconstructibility of *droit* (legal authority) and the indeconstructibility of justice." Through this description of deconstruction in relation to its larger interaction with *droit* and justice, Derrida presents us with a theory of complexity in the realm of ethics and law.

It is the threatened violence that the state can unleash as one stands before the law with which Derrida (1992) brings us to terms in an address given before the Cardozo Law School in 1991. For Derrida, Walter's Benjamin's "Critique of violence" exercised a deconstructive practice (and, as I would argue, a theory of complexity) prior to the inauguration of deconstruction (or complexity science). Although the law and the law's origins cannot be grasped, the violence that accompanies the law continually manifests itself. Indeed, "a totally non-violent resolution of conflicts can never lead to a legal contract" (Derrida, 1992: 288). A contract, "however peacefully entered into by the parties, leads finally to possible violence. It confers on both parties the right to take recourse to violence in some form against the other, should he break the agreement" (Derrida, 1992: 288). There is an implicit necessity for violence within the law.

Benjamin writes:

"*The question that concerns us is, what light is thrown on the nature of violence by the fact that such a criterion or distinction [between sanctioned and unsanctioned violence] can be applied to it at all, or, in other words, what is the meaning of this distinction?*" (Benjamin & Demetz, 1978: 279).

For Benjamin, this distinction is crucial to the operation of the state. For him, the "tendency of modern law is to divest the individual at least as a legal subject, of all violence, even that directed only toward natural ends" (Benjamin & Demetz, 1978: 283). The individual relinquishes their right to commit violence in order to inaugurate the legitimating power of the state to commit sanctioned violence in the name of justice. Benjamin (Benjamin & Demetz, 1978: 287) argues that violence must be either law-making or law-preserving; if it is neither, "it forfeits all validity." Jonathan Boyarin (1991: 1193) writes that "Benjamin's achievement, by breaking the conventional dichotomy between violence *per se* and non-violence and introducing a hypothetical distinction inside violence, was more to reveal the full extent of mythical violence and to identify an alternative." Despite drawing this distinction within violence, Benjamin (Benjamin & Demetz, 1978: 286) writes that "in the law of violence over life and death more than in any other legal act law reaffirms itself."

Benjamin finds that strike law and military law present the legal structure of the state with a paradox, insofar as both enact a violence that the state protects; while at the same time such state-protected violence, if taken to extremes, could threaten the authority of the state. Thus, lawmaking depends on violence to protect the state's authority and power.

"*For the function of violence in lawmaking is twofold, in the sense that lawmaking pursues as its end, with violence as the means, what is to be established as law, but at the moment of instatement does not dismiss violence; rather, at this very moment of lawmaking; it specifically establishes as law not an end unalloyed by violence but one necessarily and intimately bound to it, under the title of power. Lawmaking is power making, and, to that extent, an immediate manifestation of violence. Justice is the principle of divine end making, power the principle of all mythical lawmaking*" (Benjamin & Demetz, 1978: 295).

"*The state fears this violence simply for its lawmaking character, being obliged to acknowledge it as lawmaking whenever external powers force it to concede them the right to conduct warfare, and classes the right to strike*" (Benjamin & Demetz, 1978: 284).

"*This tendency of law has also played a part in the concession of the right to strike, which contradicts the interests of the state. It grants this right because it forestalls violent actions the state is afraid to oppose*" (Benjamin & Demetz, 1978: 290).

However, the state must fear the consequences of the right to strike being taken too far, insofar as the very foundations of the state can be rocked by proletariat revolution. Benjamin relies on Georges Sorel to make this point:

"*The general strike clearly announces its indifference toward material gain through conquest by declaring its intention to abolish the state; the state was really ... the basis of the existence of the ruling group, who in all their enterprises benefit from the burden borne by the public*" (Sorel, 1919: 250).

Derrida elaborates on Sorel's observation by extending it to the context of interpretative reading in the following:

"*For there is something in the general strike, and thus of the revolutionary situation in every reading that founds something new and that remains unreadable in regard to established canons and norms of reading, that is to say the present state of reading or of what figures the State, with a capital S, in the state of possible reading*" (Derrida, 1992).

Benjamin writes that military violence inherently possesses "a lawmaking character" if one concludes it is "primordial and paradigmatic of all violence used for natural ends" (Benjamin & Demetz, 1978: 283). He finds the law of military conscription to be a law preserving violence.

The police represent both lawmaking and law-preserving violence insofar as they represent the state's admission that the law will not always yield decisions or solutions that individuals will accept as fair and just; therefore, the presence of the police force enables the state to authorize itself in two ways: through statute and through the promise of force. Benjamin writes:

"*The ignominy of such authority, which is felt by few simply because its ordinances suffice only seldom for the crudest acts, but are therefore allowed to rampage all the more blindly in the most vulnerable areas and against thinkers, from whom the state is not protected by law-– this ignominy lies in the fact that in this authority the separation of lawmaking and law-preserving violence is suspended. If the first is used to prove its worth in victory, the second is subject to the restriction that it may not set itself new ends. Police vio-*

lence is emancipated from both conditions. It is lawmaking, for its characteristic function is not the promulgation of law but the assertion of legal claims for any decree, and law-preserving because it is at the disposal of these ends. The assertion that the ends of police violence are always identical or even connected to those of general law is entirely untrue. Rather, the 'law' of the police really marks the point at which the state, whether from impotence or because of the immanent connections within any legal system, can no longer guarantee through the legal system the empirical ends that it desires at any price to sustain" (Benjamin & Demetz, 1978: 286-7).

Benjamin paradoxically quips that the monarch's foot soldiers are vastly superior to the police of the democratic state, because the monarch symbolizes the unification of legislative and executive rule, whereas in a democracy the separation between the executive and legislative branches only possesses the pretense of a separation of powers and interests, when in actuality the gravest atrocities are committed through each of them.

"*Unlike law, which acknowledges in the 'decision' determined by place and time a metaphysical category that gives it a claim to critical evaluation, a consideration of the police institution encounters nothing essential at all. Its power is formless, like its civilized states. And though the police may, in particulars, everywhere appear the same, it cannot finally be denied that their spirit is less devastating where they represent, in absolute monarchy, the power of a ruler in which legislative and executive supremacy are united, than in democracies where their existence, elevated by so much relation, bears witness to the greatest conceivable degeneration of violence*" (Benjamin & Demetz, 1978: 287).

Within the police forces of democracies, Benjamin recognizes the potential for the evolution of such evil incarnates as Hitler's SS and Stalin's Gulag. With this possibility in mind, Derrida's "The force of law: The mystical foundations of authority" deserves a more thorough treatment.

Douglas Litowitz (1997: 91) identifies five central themes that emerge from Derrida:

1. Deconstruction is not politically nihilistic (on the contrary, it recognizes an unceasing call to do justice to the other at all costs).
2. There is a distinction between law and justice in that justice is not deconstructible while law can be deconstructed.
3. Deconstruction reminds us that law can never reach a stage of complete justice, since justice is transcendent and never wholly imminent.
4. Justice takes the form of an experience of three aporias: *epokhe* ["suspension"] of the rule, "the ghost of the undecidable," and "the urgency that obstructs the horizon of knowledge."

5. Justice requires a commitment to traditional emancipator ideals and the recognition of marginalized groups.

First, Derrida wishes to defend deconstruction against characterizations of it as nihilistic and also hopes to convince his readers (audience) that deconstruction can enable justice. Secondly, Derrida wishes to draw a sharp distinction between law and justice by asserting that law is infinitely deconstructible while justice is not.

"*Derrida seems to think that justice is outside the law; it is a relation or debt from one person to another, an irreducible and incalculable duty to act without thought of repayment. Derrida thinks of justice as something that exceeds the law and perhaps even contradicts the law in extreme cases. Justice is deemed an "experience that we are not able to experience" and involves aporia*" (Litowitz, 1997: 92).

Third, Derrida finds that, because justice cannot be fully comprehended nor deconstructed, it never reaches completion; the law will never reach a complete stage of justice. Litowitz writes, "Derrida feels that justice cannot be fully present and can only be experienced as something other than itself. That is, its presence is always deferred, always to come" (Litowitz, 1997: 95).

Fourth, justice is experienced as three aporias. *Epokhe* of the rule "arises because a judge must follow the law (in the form of legal precedent) yet must also decide each case on its own terms and must be free to overturn or reject (or distinguish) the precedents which impinge upon him or her" (Litowitz, 1997: 95). The second aporia arises from the observation that "a legal case can be decided in favor of either party, depending on the precedents" (Litowitz, 1997: 96). The third aporia arises because justice must be administered immediately but to satisfy the infinite demands of justice one would need infinite time and knowledge (Litowitz, 1997: 96).

Fifth, as part of the responsibility to answering the Levinasian call of the Other, Derrida situates deconstruction as a practice that can be used to "reinterpret the very foundations of law such as they had previously been calculated or delimited" (Derrida, 1992: 97). All five of these themes require a radical interrogation of the legal institution.

Benjamin (Benjamin & Demetz, 1978: 288) writes that "when the consciousness of the latent presence of violence in a legal institution disappears the institution falls into decay." For Derrida, Benjamin understands that this decay is met with the creation of a conservative violence that is emblematic of the founding violence of the institution.

"*This repression – and droit, the juridical institution, is essentially repressive from this [Benjamin's] point of view – never ceases to weaken the founding violence that it represents. And so it destroys itself in the course of this cycle. For here Benjamin to some extent recognizes this law of iterability that insures that the founding violence is constantly represented in a conservative violence that always repeats the tradition of its origin and that ultimately keeps nothing but a foundation destined from the start to be repeated, conserved, reinstituted. Benjamin says that founding violence is "represented" (repräsentiert) in conservative violence*" (Derrida, 1992: 55).

Derrida (1992: 55) claims that the "founding or conserving violence of *droit* constitutes an oscillation in which the violence that conserves must constantly give itself up to the repression of hostile counter-violences."

"*The critique of violence is the philosophy of its history – the "philosophy" of this history, because only the idea of its development makes possible a critical, discriminating, and decisive approach to its temporal data. A gaze directed only at what is close at hand can at most perceive a dialectical rising and falling in the lawmaking and law-preserving formation of violence. The law governing their oscillation rests on the circumstance that all law-preserving violence, in its duration, indirectly weakens the lawmaking violence represented by it, through the suppression of hostile counter-violence. This lasts until either new forces or those earlier suppressed triumph over the hitherto lawmaking violence and thus found a new law, destined in its turn to decay. On the breaking of this cycle maintained by mythical forms of law, on the suspension of law with all the forces on which it depends as they depend on it, finally therefore on the abolition of state power, a new historical epoch is founded*" (Derrida, 1992: 299-300).

It is with the abolition of state power that the possibility of nonviolent agreement becomes realizable.

Benjamin (Benjamin & Demetz, 1978: 289) writes that "nonviolent agreement is possible wherever a civilized outlook allows the use of unalloyed means of aggression." This unalloyed means of aggression is located wherever "there is a sphere of human agreement that is nonviolent to the extent that is wholly inaccessible to violence: the proper sphere of 'understanding,' language" (Benjamin & Demetz, 1978: 289).

"*The distinction between means and violence is once again very much a question of the violence of language, but also of the advent of nonviolence through a certain language. Does the essence of language consist in signs, considered as means of communication as re-presentation, or in a manifestation that no longer arises, or not yet, from communication through signs, from communication in general, that is, from the means/end structure?*" (Benjamin & Demetz, 1978: 49).

The representational capacity of language instantiates the logocentrism and the "metaphysics of presence" that Derridean deconstruction seeks to dismantle. The logic of the Holocaust arose from an implicit faith in the representational capacity of the sign that parades as a correspondence to the real.

For Derrida, the Final Solution embodied an instance in human history when the limits of our belief in representation were taken to a horrific conclusion. Derrida locates four pitfalls that led to the Final Solution as they reside within Benjamin's "Critique of violence":

1. *The radicalization of evil linked to the fall into the language of communication, representation, information (and from this point of view, Nazism has been the most pervasive figure of media violence and of political exploitation of the modern techniques of communicative language, of industrial language and of the language of industry, of scientific objectification to which is linked the logic of the conventional sign and of formalizing registration).*
2. *The totalitarian radicalization of a logic of the state (and our text ["Critique of violence"] is indeed a condemnation of the state, even of the revolution that replaces a state by another state, which is also valid for other totalitarianisms – and already we see prefigured the question of the Historikerstreit).*
3. *The radical but also fatal corruption of parliamentary and representative democracy through a modern police that is inseparable from it, that becomes the true legislative power and whose phantom commands the totality of the political space. From this point of view, the Final Solution is both a historico-political decision by the state and a decision by the police, the civil and military police, without anyone ever being able to discern the one from the other and to assign the true responsibilities to any one decision whatsoever.*
4. *A radicalization and total extension of the mythical, of mythical violence, both in its sacrificial founding moment and its most conservative moment. And this mythological dimension, that is at once Greek and aestheticizing, this mythological dimension also responds to a certain violence of state law, of its police and its technics of right totally dissociated from justice, as the conceptual generality propitious to the mass structure of opposition to the consideration of singularity and uniqueness. (Derrida, 1992: 59)*

There is an ávenir for justice and there is no justice except to the degree that some event is possible which, as event, exceeds calculation, rules, programs, anticipations and so forth. Justice as the experience of absolute alterity is unpresentable, but it is the chance of the event and the condition of history. (Derrida, 1992: 27)

Humankind's attempts to understand and symbolize the world in which they live seem ultimately futile in the context of a universe that is beyond comprehension. However, as Nietzsche recognized, any attempt to categorize or interpret the universe is an expression of the will to power. Levinas grapples with this paradox:

"Is not this silent world, that is, this pure spectacle, accessible to true knowledge? Who can punish the exercise of the freedom of knowing? Or, more exactly how can the spontaneity of the freedom that is manifested in certitude be called in question? Is not truth correlative with a freedom that is this side of justice, since it is the freedom of a being that is alone?" (Levinas, 1969: 90).

Our freedom and willingness to understand and symbolize the limits of our comprehension must be met with the utmost skepticism. The practice of the hermeneutics of suspicion requires eternal vigilance by intellectuals inside and outside of the academy. Walter Benjamin was a visionary in his own right, because of the many ways in which he foresaw many of the politico-ethical problems that were so tragically witnessed during the twentieth century. He warned us that governmentality must be rabidly and continually scrutinized if we are to create a world in which deconstruction will ever have a possibility of yielding justice. Complexity theory provides an essential understanding of how legal discourse maintains its authority under the masquerade of justice.

References

Benjamin, W. & Demetz, P. (eds) (1978) *Reflections: Essays, Aphorisms, Autobiographical Writings*, New York: Schocken Books.

Boyarin, J. (1991) "Walter Benjamin: Justice, right, and the critique of violence," *Cardozo Law Review*, 13(4, December): 1193.

Cilliers, P. (1998) *Postmodernism and Complexity: A Study of Complex Systems*, New York: Routledge.

Cornell, D. (1992) *The Philosophy of the Limit*, New York: Routledge.

Derrida, J. (1992) "The force of law: The mystical foundations of authority," in D. Cornell, D. G. Carlson, & M. Rosenfeld (eds), *Deconstruction and the Possibility of Justice*, Routledge: New York.

Gibbs, R. (2000) *Why Ethics? Signs of Responsibilities*, Princeton, NJ: Princeton University Press.

Kent, T. (1993) *Paralogic Rhetoric: A Theory of Communicative Interaction*, Cranbury, NJ: Bucknell University Press.

Levinas, E. (1969) *Totality and Infinity*, Pittsburgh, PA: Duquesne Press.

Litowitz, D. (1997) *Postmodern Philosophy of Law: Rorty, Nietzsche, Lyotard, Derrida, and Foucault*, Lawrence: University Press of Kansas.

Lyotard, J.-F. (1977) *The Postmodern Condition: A Report on Knowledge*, Minneapolis, MN: University of Minnesota Press.

Rasch, W. (2000) *Niklas Luhmann's Modernity: The Paradoxes of Differentiation*, Stanford, CA: Stanford University Press.

Rasch, W. & Wolf, C. (eds) (2000) Observing Complexity: Systems Theory and Postmodernity, Minneapolis, MN: University of Minnesota Press.

Sorel, G. (1919) *Réflections sur la violence*, J. Jennings (ed.), Cambridge, UK: Cambridge University Press, 1999.

Weber, S. (1991) "Deconstruction before the name: Some preliminary remarks on deconstruction and violence," *Cardozo Law Review*, 13(4, December): 1183.

9. Complexity and TINA

Pedro L. Sotolongo
Complexity Cathedra, Instituto de Filosofia de la Habana, CUB

Introduction

The recent Complexity and Philosophy Workshop that was held in Rio (November, 2004), was an alternative to the 'developed' region of the World, i.e., it was an event held in the so-called 'Third World'; in Brazil, a country that is showing that alternative ways of thinking, and most significantly, alternative (to the prevailing one) ways of development *are* possible. As such, it seems proper enough to address the topic of "Complexity and TINA" – TINA being the acronym in English for the quite often heard statement – especially since the 90s – "There Is No Alternative": TINA to a free market economy, TINA to neoliberal globalization, TINA to Capitalism and more recently – after 9.11.01 – well you guessed it, TINA to open military aggressions against terrorism.

But Complexity and TINA give shape to a true *contradictio in adjectum*, that is, a contradiction by itself, in its own formulation. Complexity theory has shown us that every stage of real open systems – as our modern societies are – has historical limits, that they are always 'in the making', and so for them TINA does not hold, but SARA (Some Alternatives Remain Always) does. If this is truly the case, as Complexity Theory would suggest, then why are we being made to believe – and most of us indeed have come to believe – in the TINA argument-plot? Why does the TINA argument-plot often seem so plausible to many, including many noticeable (and noticed) Complexity Theory scholars (most probably even to the majority of those who gathered at the Rio workshop)?

This paradoxical situation is worthy of being addressed, particularly given the apparent inability – whether from the right or from the left of the political spectrum – of much current social Complexity thinking and social Complexity modelling to deal appropriately with SARA instead of TINA. When it comes to the study of complex social systems that go beyond separate organizational systems (business-type separate social organizations and/or enterprises) and their interactions in a changing global social environment SARA is always true. Such systems are not reducible to mostly businesslike and/or managerial (i.e., linear) ones. This inability is preventing social Complexity thinking – and social Complexity people and their Journals, Meetings and actions – to make a more substantial and serious contribution to many of the more pressing global issues of today's world. Such contributions remain, however, much needed.

As a possibility: what contribution could the social Complexity community (from across the entire political spectrum) make by showing convincingly that even the most stubborn advocates of TINA-to-the-western-way-of-life have brought upon themselves (and upon others) their own SARA – and the worst SARA: fundamentalist terrorism. And, moreover, that one of the responses to the now often proclaimed TINA–to-open-military-aggressions-against-terrorism is the feeding of reasons to adversaries advocating TINA-to-terrorism: the narrow minded cause is co-creating an equally narrow-minded response.

Let's give a closer look to the situation just outlined.

Two 'Thought-Action' 'Barriers'

Two main reasons are found underlying this undesirable state of affairs:

1. Not taking into consideration the historical limits of global-scale social complex phenomena (e.g., the historical limits of a free market economy, of neoliberal globalization, of unrestrained capitalism (sic), of open military aggressions); thus entailing a 'thought-action' 'barrier' inherent to the TINA-only axiomatic base, which is commonplace for most social Complexity specialists from the right – and even the center – of the political spectrum, when dealing with the global problems and issues of our time.

This Complexity-TINA-type treatment of such key problems and issues (in spite of its unfaithfulness to – and inconsistency with – a true Complexity approach) is nevertheless widely considered in the social Complexity community as 'scientifically correct' (because it is supposedly 'non-ideological') and 'politically correct' (because it does not deal with any alternatives to the social status quo).

2. Taking into consideration only the historical limits of the aforementioned global-scale social phenomena without due consideration of their complex global-organizational systemic nature and of the possibilities that Complexity, and the technosciences, have provided and still provide to move away and keep away (at least for whole periods of time) from those historical limits; thus entailing a 'thought-action' 'barrier' inherent to the SARA-only "axiomatic base" which is commonplace for most critical social thinkers from the left of the political spectrum when dealing (more often than not without involvement of global-organizational system Complexity considerations) with problems and issues of global social changes.

This Complexity-lacking-SARA-type treatment of such global social problems and issues (because of its mostly 'rhetoric' pathos and the absence in it of any true Complexity heuristic) makes it much easier for Complexity scholars to label it as 'scientifically incorrect' (because it is 'ideological'); in addition to 'politically incorrect' (because it deals with alternatives to the social status quo). Although, it should also be said, this last labelling would most probably occur one way or another ideologically (and precisely by those, paradoxically enough, that advocate only 'non-ideological' thinking). Such circumstances, at the same time, and equally paradoxically, help considering such a treatment as 'radical' and even 'revolutionary' by those that put it forward, and by their left-wing followers, thus wrongly reassuring themselves to be 'on the right critical path'.

We shall now examine in more detail some of the circumstances through which the above labelled 'thought-action' 'barriers' are constituted.

Circumstances Leading to the TINA-Only Axiomatic Base

Five circumstances are, among others, most commonly involved in the process through which the TINA-only axiomatic base (whether consciously or often unconsciously) is constituted:

1. Not paying attention (not taking into consideration) the importance of 'small' phenomena;
2. Not paying attention (not taking into consideration) 'the different' on its own merits and in its diversity;
3. Considering only the triggering (enabling) effects of the premises and omitting their restraining (disabling) effects;
4. Considering only probabilities and not possibilities;
5. A bias towards 'objectivity' and 'universality'.

These circumstances, feeding upon each other, lead unmistakably to the consideration of only distinctionally closed (social) systems; that is, those in which nothing 'of–the–other–sort' can ever emerge. Which is equivalent – whether one wishes (or likes) it or not – to *de facto* validation of the (social) status-quo.

Let us take those TINA-only-constituting-circumstances one by one:

Not Paying Attention (Not Taking Into Consideration) the Importance of 'Small' Phenomena

This is a circumstance that is present in – and thus affects – much actual social research, including much Complexity social research and modelling. It refers to the overwhelming attention given to the so-called 'macro'-social processes and phenomena (that is, those that are taken to be decisive in triggering historical events – social and political changes). The downside is that there is little or no attention given to the (small-scale) events and processes that stem from the everyday contributions of ordinary people, day in and day out, that in the long run lead to those (large-scale) social and political changes.

More often than not those 'small' phenomena are overlooked, as if they did not exist and/or as if they did not lead to any significant effects. In reality, it is these 'small' social phenomena that turn into, more often than not, the time limits and the historical character of the 'big' phenomena, of any prevailing social reality.

It is worth noticing that 'small', in reference to this particular circumstance, means not only one or another 'small' quantifiable phenomenon in time (low frequency phenomena) and not only one or another 'small' quantifiable phenomenon in a spatial-geometrical sense (phenomena of restricted social, geographical or demographical extension), i.e., those that are 'small' in a scalar way; but it means also those which are 'small' in a qualitative, relative way to the privileged 'macro'-social processes and phenomena. Thus, the Third – under-industrialized – World (the 'South') as a whole, and its events and processes, *are* taken as 'small' phenomena (relative to the First – industrialized – World: the North) in many social inquiries, including many social Complexity inquiries. These 'small' phenomena are regarded as anomalies that will be 'removed' with further rigorous TINA-type thinking.

What is then taken as a 'small' social phenomenon is then often labelled with a negative label or resonance, sometimes even caricaturized. Special terms are coined – so that they can be supposedly be dealt with 'rigorously' – in a 'single-stroke' manner; terms which (inadvertently?) wash over much of the Complexity of those social realities: *underdevelopment, underdeveloped, periphery, peripheral, 'third-worldly', 'southern'*, etc.

There are also some other phenomena that, while belonging to the 'big' – industrialized – First World, are often considered as 'small' in much social research, and thus they receive their own negatively-charged (and equally Complexity reducing) terminology too: *marginal groups, ethnic minorities, immigrants, below-poverty-level-income-persons*, the *political left*. The reverse side of this coin is of course the existence of developed – First-World – enclaves in some underdeveloped societies. Paradoxically enough, these realities are often also considered as 'small' (although 'of-the-right-type') phenomena when characterizing those societies.

'Small' in this sense, thus, is what is considered not worthy of being included in the researched reality; what is undervalued in it, not granted credit, remaining 'in the shadow'; what is invisible or not seen, in a word, what 'does not exist' for much social research.

What is the ontological equivalent of a treatment of social affairs that is not faithful to the bottom-up emergence of variety? What is the ontological 'price' paid by neglecting 'small' phenomena? Answering this question leads us into the next of the aforementioned TINA-only-constituting-circumstances.

Not Paying Attention (Not Taking Into Consideration) 'The Different' On Its Own Merits and In Its Diversity

The ontological consequence of not taking into consideration the importance of 'small' things is no other than overlooking that social terrain from where 'what-is-different from the prevailing social realities' could emerge. It leaves no social space left for coming into being by its own merits of anything socially qualitatively different from those social prevailing realities. Thus, in an *a priori* fashion, no social space is conceded for alternative, diverse, 'different from the status quo' events and processes to occur and uprise.

The under-developed social realities as well as those 'southern-in-the North' social realities (such as marginal groups, ethnic minorities, etc.), are thus (for the second type of TINA-only-constituting-circumstances) what-ought-not-to-exist; being considered as residual and infertile realities not worthy of being taken seriously into consideration unless they are regarded as social negativities, i.e., as sources of social disorder and disintegration. From this particular perspective these sources of disorder and disintegration can only be avoided by promoting those 'undesirable' social realities to the status of developed, centered, 'first-worldy', 'northern' realities, i.e., the only way to 'correct' the infertile realities of the 'south' is to either ignore them or try to rebuild them through the eyes of the 'north'. This view conveniently overlooks that the reality of the 'north' is very much dependent on the maintenance of the 'south' as is!

Again, through this second class of TINA-only-constituting-circumstances those 'undesirable' ('southern') social realities can find a proper place in social research only by turning themselves into the realities of the prevailing social type ('northern'), i.e., those that are characteristic of the developed, industrialized, centers of power (and in a way that only God should know, since it is seldom explicitly argued by the advocates of this kind of treatment). In other words, by making them become precisely what-they-are-not. The result of this cognitive procedure, which in my opinion is certainly not faithful to the range of emergent phenomena allowed by social Complexity, is the elimination of the potential for any social qualitative diversity; for any social truth or real alternatives to the existing social status-quo of the dominant 'north'. The only social emergences that are admitted are those within the range of "more-of-the-same (prevailing social realities)". In other words, TINA!

The two TINA-only-constituting-circumstances already examined reinforce – and are reinforced by – the next one to be examined.

Considering Only the Triggering Effects of the Premises and Omitting Their Restraining Effects

It is not possible to deal with one or another phenomenon without first bringing forth some starting premises. That is the nature of any human cognitive enterprise – to begin the process of thought and examination we must first identify/construct the initial boundaries to the process. Thus, when studying social phenomena we often assume, for example, a given set of (always present) social interests, social aims and social ideological commitments. Such commitments are characteristic of one or another person, social group, social class, type of society and/or historical context. There is nothing wrong with that (assuming, of course, that the right assumptions were made in the first place; which is, of course, a major assumption on its own). But even if the social assumptions made prove to be right, that is only the beginning. Those social interests, aims and ideological commitments must be taken both in their *social enabling capacities* (that is, in terms of what they socially favour, agree with, make possible, impulse, push forward, foster, praise – whether rightly or wrongly – or, encourage) and in their *social constraining effects* (that is, in terms of what they socially unfavor, fight with, make impossible, restrain, push backward, hinder, criticize – whether rightly or wrongly – or, prevent from emerging). In this way we see that social assumptions enable as much as they disable, and we must wonder as to the possibilities that might be realized if the potential realities disabled/suppressed by the prevailing social assumptions were allowed to flourish.

More often than not, in much social research (including much social Complexity research), one of these two 'sides' of the inquiry is omitted in favour of the other. And the omitted 'side' (the disabled side) is strongly linked with the researcher's own social interests, aims and ideological commitments exemplified by the social group, social class, type of society and/or historical context, in which the researcher 'exists'.

This is, of course, one of the ways ideology sneaks 'through the back door' into science (even into supposed non-ideological research). Thus, in much social Complexity work, when the Complexity specialist shares the assumed premises for a given person, social group, social class, type of society and/or historical context, their social *enabling potentialities* are explicitly considered, developed and assessed. However, no equivalent explicit consideration and assessment is made in relation to their *restraining potentialities.*

Granted that such one-sided inquiry can be made in regard to *any* person, social group, social class, type of society and/or historical context involved in the studied social phenomena. But we are now primarily concerned with TINA-only–constituting-circumstances and because of that we will point at this moment only to the case when the researcher's social interests, aims and ideological commitments coincide or converge with those of the empowered person, social group, social class, type of society and/or historical context in the studied reality (with those at the 'helm' of the researched social reality). In such cases, such one-sided use of the social premises leads to not taking into due consideration the restraining effects of those same social premises.

And it leads to that TINA-only treatment of the researched reality as a direct result of the fact that the prevailing consideration of only the triggering effects of the assumed social premises favours the recognition of emergent phenomena (of the 'more-of-the same' type, which of course is a direct consequence of the previous TINA-only–constituting-circumstance examined above) in detriment of the recognition of phenomena related to persistent unresolved social problems; persistent social problems that remain unsolved precisely because of the restraining effects of the starting social assumptions made, which are systematically omitted from the analysis.

Thus, this type of TINA-only-constituting-circumstance (which is not simultaneously faithful to both the enabling and constraining effects of much social dynamics) results in a distortion of the studied social reality. This distortion reinforces the previously examined types. This is because due consideration to the restraining effects of the assumed social premises is not given, which seriously restricts the scope of the covered social phenomena and, thus, introduces a bias towards considering 'the probable' ones and ignoring 'the possible' ones.

Considering Only Probabilities and Not Possibilities

This bias towards probabilities helps shape the result – already pointed to above – of the TINA-only-constituting circumstances considering only distinctionally closed (social) systems; those in which nothing 'of–the-other–sort' can emerge.

To consider any system as distinctionally closed means considering it as:

- *Synchronic* (everything in it occurs along a common, unique and linear time);
- *Causally connected* (any event that occurs in such systems is causally connected with all other events);
- *Isolated* from its environment (not exchanging anything with it).

However, real social systems do not exhibit any of these traits. Many events in them are not synchronic but *concurrent* (exhibit independent time series); social systems are not (completely) causally connected, not all events in them are causally connected with all others; and social systems are open systems, permanently and overwhelmingly exchanging mass, energy, information and meaning with their natural and social surroundings.

The logic-of-probabilities works well in distinctionally closed systems. It does so because it refers to series of event occurrences determined on their own and that can be accurately specified. But it does not work so well in distinctionally open systems. The modern notion of 'probability' arose within a definite epistemic atmosphere – that of the linear paradigm – and around specific and definite (although not necessarily simple) problems; but it is not the only one feasible. A logic-of-possibilities arises in another epistemic atmosphere and around different problems; it refers to the becoming of events not clearly determined and not sharply specified on their own (as, for example, network events).

A logic-of-possibilities intends to capture the shaping of the 'conditions of possibility' for one or the other phenomenon to happen. It is not biased in favour of 'the probable' events, but is in favour of any 'possible' one. Thus it is better suited to apprehend the emergence of 'anything-of-the-other-sort' with respect to the prevailing realities, thus, helping not to lose sight of the time limits – historical ones – of such systems.

The bias towards probabilities and not possibilities of much of the social Complexity studies (not faithful to the presence of a whole range of alternatives inherent in social Complexity) ignores all these circumstances and enhances its TINA-only-constituting-circumstances. But it is not its only bias.

A Bias Towards 'Objectivity' and 'Universality'

This bias stems as a resulting effect from all the other TINA-only-constituting circumstances examined here, i.e., not taking into consideration 'small' phenomena; not paying attention to 'the different' for its own merits, considering only the triggering effects of the adopted social premises while omitting their restraining effects; giving preference to probabilities and not possibilities. Little by little – even if inadvertently – the researcher constructs a rationalized version of the historical social realities under scrutiny that makes invisible that historical character and turns it into a kind of transcendent universal (acontextual) reality. This turns the assumed social interests, aims and commitments into a justified – and justifiable – social status quo (because you do not argue against what is 'universal').

What is 'universal' and *ahistorical* does not posses time limits, it is historically limitless and thus there is no use in trying to consider any such thing. It is only worthy of being studied in its 'objective' existence. Thus, through this theoretical 'hocus pocus' three things are achieved:

- What is no more than a rationalized, objectified, universalized version of what are really historical social phenomena, becomes what is considered 'scientifically right' to study;
- What is really worthy of being studied – the historical time limits of those social phenomena – becomes what is considered 'scientifically wrong' to study (because you ought not to study what does not exist); furthermore, to study them is considered 'politically wrong' because it challenges taking the social status-quo as unchangeable;
- It takes all 'social critical values' out of true social science. Social value judgements are a burden for social scientific knowledge.

Sooner rather than later, such reductionist, standardizing, homogenizing, cognitive procedures, lead to not taking into consideration the historical limits of global-scale social complex phenomena (the historical limits of a free market economy, of neoliberal globalization, of capitalism, of open military aggressions). Because of the reflexivity of social systems (well known to – and characterized by – contemporary second-order epistemology) such an approach begins to create its own social reality (through the circulation of its ideas in society, and their social diffusion), i.e., ideas that do not initially reflect social reality, can become to define social reality – social reality becomes our idea of what (the dominant party thinks) social reality *should* be.

The social result: the TINA-only-axiomatic-base of much social research. This includes much of the social Complexity research coming from the right and the center of the political spectrum when dealing with the global problems and issues of our time. This is a considerable 'barrier' for any critical social thinking and conduct.

But we pointed above to a second class of 'thought-action' 'barrier'. This time one common to many critical social thinkers. Let us now consider it.

Circumstances Leading to the SARA-Only Axiomatic Base

Two circumstances are, among others, commonly responsible for the process through which, what we have termed the 'SARA-only axiomatic base', is constituted:

1. Considering any 'Complexity approach', and any application of the contemporary technosciences, as tools intrinsically belonging to the social adversaries (to the empowered social forces), or;
2. Considering, in the case of the 'Complexity approach', that it entails nothing new that has not been already attained by previous critical social approaches (this is often the case for researchers belonging to a 'dialectical' approach).

Let us examine them.

Considering Complexity Thinking and the Technosciences Only as Tools of the Adversaries

It is a well known fact that the contemporary technosciences (those that pose their central problems in relation to techniques, using technical instruments, and with the aim of attaining technical solutions to those problems) began to dominate after the second World War; and that Complexity thinking emerged as a recognizable and recognized field of research in the late eighties of the 20th Century. In both cases, the mainstream of those achievements did not 'run' through left-wing critical thinker's territory. In fact, those achievements were primarily used in ways that here and there strengthened the empowered social forces of that period, which had (and still have) no real interest in fostering, even less in provoking, any radical change in the existing social realities. Indeed, some of the applications of contemporary technosciences and of a Complexity approach (by the social forces interested in temporarily postponing those phenomena from approaching such historical limits) have in some respect allowed them to do so. These circumstances have led many critical social thinkers and researchers from the left of the political spectrum to a unilateral – and wrong – appraisal about the technosciences and the Complexity approach, as though they were inherently social-status-quo-defending cognitive and practical tools.

So, although it is commonplace from the side of these critical social thinkers (due to their inherent pathos in favour of social change) to take into consideration the historical limits of global scale social phenomena, more often than not, these historical limits are taken into consideration by them without due attention to the complex global-organizational systemic nature of the alluded social phenomena and to the real possibilities provided by the use of a Complexity approach, and by the technosciences for the social-change-fostering social forces. This results in a lack of global-organizational systemic and Complexity considerations in such critical social research and, furthermore, in a (relative) downgrading of its real heuristic 'cutting edge'.

But the above mentioned circumstances surrounding the historical emergence of the technosciences and of the Complexity approach, in no way mean that the technosciences and the Complexity approach are 'owned' by the empowered social forces. On the contrary, Complexity thinking and the contemporary technosciences are up for grabs, as any scientific achievement sooner or later is – and, as such, can be used with all their heuristic potential by anyone. Of course, they will be put to use in accordance with – and for the social aims – of those that implement them. But that is as it has always been, and as it will always be. And thus, in the case of their implementation by critical social thinkers, they can and would be put to use accordingly to their social critical points of view and in favour of their social aims fostering – and not hindering – social change and making it happen.

Thus, to consider contemporary scientific achievements as only social-status-quo-defending cognitive and practical tools has turned into a 'thought-action' 'barrier' inherent to the SARA-only axiomatic base. This is true for most critical social thinkers from the left of the political spectrum, when dealing with problems and issues of global social changes; and one that hinders a more realistic appraisal of the feasibility of such advocated changes in the global social phenomena concerned. In addition, it hinders the use of those powerful cognitive and practical tools for purposes of fostering and working in favour of the desired and proclaimed needed social changes that such critical social thinkers – justly – advocate.

In the case of the critical social researchers belonging to the dialectical tradition there is still another circumstance that reinforces the ones mentioned, which we will next consider.

Considering Complexity Thinking as a Mere Reformulation of Dialectics

Since ancient times the dialectical current of thought has claimed the priority of movement over rest; of transformations over standing still; of dynamics over statics. It has tried, from Heraclitus to Hegel and Marx, to apprehend in philosophical terms – through the well known dialectical regularities – the behavior of nature and society. Complexity thinking is convergent with this line of thought; it also characterizes the world in dynamical terms, shows us the inevitability of spontaneous fluctuations and perturbations which disequilibrate any system, thus leading it to bifurcations and to the emergence of qualitative new patterns of order, without (necessarily) the intervention of external forces.

But 'convergence' is by no means 'identity'. Dialectical thinking has developed predominantly on philosophical grounds; Complexity thinking, while presenting undoubtedly philosophical implications, has originated in scientific grounds. It is not, by any means, a re-formulation of (which adds nothing new to) dialectics. On the contrary, it is a new scientific formulation which, in some important cases, specifies some of the traditional dialectical philosophical formulations which, thus, achieve concretization from the facts gathered by contemporary science.

So, instead of the inability of many critical social thinkers to recognize the renewed cognitive 'cutting edge' of Complexity thinking, it should be emphasized that this new approach to the world and to knowledge is capable of supplying them with fresh flesh and bones for social critical inquiry. Paradoxically enough, whilst contemporary scientific thinking is arriving, on its own, to conclusions that converge and support what the dialectical tradition has for centuries advocated in abstract philosophical terms, many critical social thinkers from this tradition are turning their back on the new Complexity thinking: dialectical in deeds – in the name of their old tradition – dialectical in words.

Where Are We Heading To: A Worldwide 'Disneyland' or a Worldwide 'Titanic' Type Catastrophe?

The above characterized circumstances that have brought about the mentioned 'thought- action' 'barriers', need to be put in relation to our present global world situation in order to more properly assess their significance.

There is not much possibility to do this in a proper way in a short paper like this. But the present world situation is so dramatic, that even a few references to some empirical data would suffice:

- Nearly 1.3 billion human beings presently live in extreme poverty;
- Year by year extreme poverty rises at the rate of 28 million human beings;
- There are presently more than 800 million hungry human beings;
- More than 120 million children do not go to elementary school (1 from every 5);
- 82 children, out of each 1000 born alive, die before reaching the age of five;
- 11 million children die yearly from curable or avoidable diseases;
- Entire nations in black Africa are menaced with extinction from the AIDS pandemic;
- In 2003, 3 million human beings, overwhelmingly in the Third World, died from AIDS;
- The Third World has paid already, during the last 13 years, US$ 4.1 billion in servicing their external debt, with the net result being that

now this debt has grown from US$ 1.4 billion to US$ 2.6 billion. So, these countries have paid three times what they owed and their present debt is now double what it was originally;

- In 2003 the developed industrialized countries gave US$ 68 billion as Official Development Help to Third World countries; while these countries gave them as payments for their debt US$ 436 billion;
- While all these dramatic situations grow year by year, US$ 900 billion are spent yearly on military expenses, and in the recent U.S. election campaign, approximately US$ 1.5 billion were expended (trying to get the vote of only a little over half of the electorate);
- A single man – Bill Gates – owns a personal fortune that amounts to the GNP of nearly 50 of the less developed countries.

So, we could ask ourselves: Which countries are financing the luxurious way of living of the industrialized developed countries? For how much longer will the masses of the poor countries endure their poverty while financing the wealth of a world minority? For how long will they put up with it without rebelling in one way or another? Will that rebelling be accepted or will it be repressed by the developed states? And the 'question-of-questions': How far away are we from the historical limits of such an unjust, unethical distribution of wealth?

Complexity thinking might help us realize that the present world has already accumulated such substantial social asymmetries that, although we cannot predict it, we can nonetheless foresee the possibility of the present world situation approaching a bifurcation point. In this world context, the TINA-only axiomatic base of much social Complexity thinking, without any consideration of global alternatives – whether consciously or unconsciously – leads, and in fact *is* already leading us, to not being able to prevent the possibility (more and more easily foreseen through the empirical data brought forward above) of a planetary ('Titanic' type) catastrophe. Whereas the SARA-only axiomatic base of much critical social thinking, without any significant Complexity global-organizational system concepts and ideas – whether consciously or unconsciously – leads, and in fact *is* already leading us, to nowhere worthy of being termed as real contributions to facilitate specific contemporary roads (stemming from contemporary circumstances and realities and not from rhetorical abstractions) towards the much needed and urgent social advances and changes.

Thus, the social Complexity community – from both ends of the political spectrum – should not be so pleased with the 'Complexity world' that they are now collectively co-constructing. On the contrary, a real Complexity SARA-type treatment of current global social problems and issues (stemming from every part – right, center, left – of the political spectrum) is urgently needed – together with Complexity SARA-type journals, Complexity SARA-type scientific meetings, and other Complexity SARA-type publications, events and initiatives.

Following further the 'Titanic' metaphor used, Complexity scholars should be urged to come out from their comfortable and separate social Complexity cabins (from which the surrounding ocean, together with the height and strength of its waves and the proximity of its icebergs cannot be sensed) and go out, if not to be at the helm, at least to step into their common upper deck (to be able to envisage where we *all* are heading) and thus to be able to advise (and whenever possible, convince) those at the helm to forsake TINA by showing them SARA.

10. Using Complexity Science in Organization Studies: A Case for Loose Application

Jacco van Uden
Tilburg University, Department of Social Cultural Sciences, The Netherlands

Introduction

According to some authors, using concepts like 'self-organization', 'nonlinearity' or 'emergence' in the field of organization is not without danger. In 1999 Bill McKelvey signalled that complexity science applied to management had all the earmarks of becoming a management consulting fad (McKelvey, 1999). Therefore, together with Steve Maguire, McKelvey made an attempt to differentiate between sound attempts to apply complexity theory on the one hand and reckless ones on the other. As guest editors to a special issue of *Emergence: A Journal of Complexity Issues and Management*, Maguire and McKelvey aimed "to give readers a broad overview of the general quality of complexity applications to CEO problems and to test how vulnerable to faddism they are" (Maguire & McKelvey, 1999: 19). This special issue contained fifty reviews of over thirty books on the relationship between complexity and management.

Like the reviewers whose findings they summarized, Maguire and McKelvey were troubled by the lax treatment of complexity science. "Not unexpectedly, the complexity gurus are most upset with how complexity science terms are loosely, if not metaphorically, defined and tossed in managerial discourse - one goes as far as to suggest that the book offers many insights for managers, but one should simply black out all references to complexity science... [O]ther reviewers worry about 'loose definition' and applications, 'oversimplification', 'incorrect use of concepts', 'superficial' treatments, lack of research, and missing the computational modeling underlay of complexity science" (p. 55). According to Maguire and McKelvey, sloppy applications make dead ends. "From what the reviews indicate, as well as our own reading of the trade books, the New Science is well on its way toward short-lived faddism unless serious research shows that there is more than metaphor to chaos theory and complexity science applications" (p. 57). Maguire and McKelvey are not alone here. Mathews, *et al.* (1999: 20), for instance, write that "it is our belief that the complexity sciences offer an approach to the understanding of social change that has more than metaphorical or analogical benefits." And like Johnson and Burton (1994: 320), who believe that the metaphorical use of complexity science is "problematic", Fuller and Moran assert that "moving beyond metaphor" is of critical importance because when "there is no grounding of these analogies in [the world of small firms], there is no evidence that complexity has validity in describing or explaining empirical observation" (1999: 50).

The reason why Maguire and McKelvey believe that the relationship between organizations and complexity theory is not just metaphorical becomes clear when we think through the authors' reference to "the particular complex systems that are human organizations" (1999: 26). When organizations are seen as a particular kind of complex system, a theory of complex systems simply is a theory of organizations. Logically then the ideas and concepts from the science of complexity apply to organization. According to Maguire and McKelvey complexity science thus gets to the bottom of organizations and allows us to know these phenomena as they really are, in and by themselves. By understanding the relationship between complexity theory and organizations this way, Maguire and McKelvey firmly locate themselves in the 'complexity school' that says that with complex systems, differences between allegedly different phenomena are skin-deep only: if you scratch the surface, you will find that phenomena as diverse as ecologies and brains work in the same sort of way. The reason for this is as simple as it is important: different phenomena "work in the same sort of way because they are the same sort of system" as Byrne (1998: 53) phrases it. As far as Meiss (2000) is concerned the idea that a single science of complexity could account for the behavior of seemingly unrelated phenomena is one of its "most exciting aspects." Complexity science "has applications to a wide variety of fields, from mathematics, physics, biology, and chemistry, to engineering, economics, and medicine" and thus "brings researchers from many disciplines together with a common language" (Meiss, 2000). Complexity scientists are not the first group of researchers to look for a shared scientific language however.

The Ideal of a Common Language

Young (1990) points out that several attempts have been made to reduce the number of academic discourses and find a single language to discuss both social and natural phenomena. "Comte thought that sociology could find the laws of social dynamics. Laplace held that modern science was powerful enough to predict social phenomena with precision were all initial states and all laws of interaction known; he assumed linearity. Pareto said: 'My wish is to construct a system of sociology on the model of celestial mechanics, physics, and chemistry'." Probably the most rigorous and systematic attempt to arrive at a scientific *lingua franca* was made during the Macy conferences, held shortly after World War II. The idea behind the Macy conferences was to "bring together a group of researchers working on the forefront of their fields to forge a new interdisciplinary paradigm that became known, retrospectively, as cybernetics. Christened by Norbert Wiener, cybernetics was conceived as a science that would develop a common explanatory framework to talk about animals, machines, and humans" (Hayles, 2000: 246). According to Bale, the "introduction of cybernetics as an interdisciplinary field led to considerable enthusiasm among the scientists who attended the Macy conferences. Many ... believed the ideas offered were sufficiently deep, yet acceptably overarching, that out of them might come a vocabulary suitable as a unifying conceptual framework for the biological and social sciences" (Bale, 1995). The hope of finding a single discourse for a wide range of phenomena rested on the assumption that deep down these various phenomena were essentially all "information processors that encoded and decoded messages, exacerbated or corrected their actions through feedback loops, and demonstrated causality" (Hayles, 2000: 246). There are interesting parallels between cybernetics and complexity science as introduced above.

Cybernetics was based on the idea that machines and living individuals are in essence information processors. Complexity theorists on their turn believe that when you look carefully you will find that earthquakes, ecologies, brains, cultures, economies, languages or organizations and so on are really all complex systems. This assumption allows for the following situation to be imagined. A student of the particular complex system that is an ecosystem carries out research and 'uploads' his findings into a database where knowledge about complex systems is stored. That piece of knowledge then becomes available to other students of complex systems, for instance to the student of the particular complex system that is the brain. If the latter researcher is not so much interested in the superficial 'brainness' of her research object as she is in its underlying, more fundamental 'complex systemness', she can make use of the knowledge that the student of the particular complex systems that are ecosystems is willing to share with his fellow complex systems researchers. Under what I will call the Grand Complexity Project, complexity scientists working in seemingly different (academic) areas are free to exchange research findings. According to Maguire and McKelvey this includes students of complex systems who work in the faculty of Management and Organization. "Yes, we said 'exchange'. As that field within the social sciences that devotes itself to the study of 'organization' and 'organizing', it would be surprising if management and organization studies had nothing to contribute to complexity scholars as they turn their attention to systems composed of human agents" (1999: 24). Byrne also foresees an interdisciplinary exchanging of ideas and he too maintains that all complexity scientists are created equal. He insists that even though complexity research as we know it has its roots in the hard sciences, the insights offered by social scientists are no less valuable: "in no sense whatsoever is the project of applying the ideas of complexity theory to the social driven by any sort of physics envy... It is true that chaos/complexity emerges from the experimental mathematics and thermodynamics, and has been particularly developed in physical chemistry and evolutionary biology. But, and it is a big but, once the social sciences get going, then other fields of inquiry will have a lot to learn from them" (Byrne, 1998: 17).

There are important conditions that need to be met before students of complex systems can actually profit from each other's research. For one thing, the exchangeability of research findings requires that complexity researchers mind their language. For the results of research on the particular complex system that is a human organization to make any sense to the student of the particular complex system that is an ecosystem, it is necessary that these research findings do not contain references to the form in which the phenomenon expresses itself in daily life. That is, if a student of human organization were to include in the research findings to be uploaded concepts like strategy, culture, leadership, technological developments or budgets the data would be useless to ecologists. Typical features of 'organizationess' must be removed before knowledge can be offered to the community of complexity researchers. Under the Grand Complexity Project it is assumed this can be done. As argued above, the presumption that underlies this Project is that the differences that normally encourage us to distinguish between organizations, economies, languages or coral reefs can be ignored in favour of a focus on the deeper complex systemness. But if not by means of the language they normally talk about organizations (strategy, culture, leadership, technological developments, budgets) how are students of organization to communicate their research findings instead? Here, too, there are important parallels between complexity science and cybernetics. Norbert Wiener left no room for misunderstanding as to what he thought was the appropriate unifying conceptual framework for discussing various sorts of information processors: "cybernetics is mathematics or it is nothing" (Wiener quoted in Hayles, 2000: 148-9). Participants of the Grand Complexity Project share this preference for the language of the hard sciences. As Byrne (1998: 55) sees it, complexity science can only deal with "mathematically formalized accounts of reality." Mathematics offers 'lean', abstract descriptions and as such it has a natural advantage over any language that offers 'thick' descriptions that do depend on references to the earthly appearance of organizations. Knowledge that is embedded in spatio-temporal specifics does not lend itself to being processed by computers, and according to Maguire and McKelvey this poses a real problem: "complexity science and computational modeling go hand in hand" (1999: 56). To overcome the problem of descriptive thickness and to do away with context, Maguire and McKelvey propose alternative concepts to describe organizational life. As they see it organizational concepts such as "conversation elements, people, departments" should be redefined as "stochastically idiosyncratic agents" because the "platform assumptions [of such agents] are more easily tolerated in computational models" (1999: 56). The advantages of abstracted concepts to the Grand Complexity Project are obvious: whereas concepts like 'staff meetings', 'managers' or 'human resource management department' mean nothing to the student of the particular complex system that is the brain, she can make use of the work of the student of organization who publishes on the behavior of 'stochastically idiosyncratic agents'.

The War on Faddism

Having explored the main assumptions underlying the Grand Complexity Project we can now turn to the question of how complexity science research is to be conducted. According to Maguire and McKelvey the fact that complexity science is still a "nascent" field has significant implications for researchers who seek to make sense of organizations using the science of complexity. "If complexity researchers themselves have not yet reached a consensus on how to integrate and synthesize all the ideas and concepts they bring together from diverse disciplines it would appear unwise for organizational researches to charge ahead with the building of models or entire organization theories and management philosophies, drawing a few concepts from one discipline and a few from another based on whatever is convenient or has a nice metaphorical ring to it" (1999: 56, 25). While there is no reason to assume that such mass consensus among complexity scientists could ever be reached - like any other field of research, the science of complexity consists of large numbers of schools that 'fight' one another over issues as fundamental as what complexity really is, over the way in which the complex is to be studied, over how complexity concepts should be interpreted, and so on (see e.g., Van Uden, 2004: 56-64) - Maguire and McKelvey maintain that in the absence of such consensus, attempts to apply complexity to management are bound to fail: "the record is clear over the past several decades - management ideas that do not become legitimized by resting on a foundation of quality research are quickly replaced by the next fad coming down the pike" (1999: 19). Below I will argue there is no compelling evidence to support Maguire and McKelvey's claim that only a rigorous approach can make complexity applied to management and organization 'work'.

According to Maguire and McKelvey one of the main characteristics of quality research is the avoidance of loose or metaphorical application of concepts. In their eyes faddism is inevitable unless we commit ourselves to "the tough work of rigorous definition and operationalization" (1999: 25). This claim is fully consistent with how the authors see the relationship between organization and complexity science, namely that organizations are complex systems. Within this view it makes perfect sense to maintain that when we apply concepts like 'emergence', 'nonlinearity', 'strange attractors' or 'fitness landscapes' to organization, we do not use those words metaphorically: when used correctly, when we are loyal to the original meaning of these concepts, they mirror the true nature of organizations. If, on the other hand, we use these concepts incorrectly, if we simplify their meaning or treat them superficially, these concepts no longer correspond with the reality of the particular complex systems that are human organizations. In the latter case 'the word' no longer matches 'the world' which means the complexity concepts have lost all value.

I see no reason to accept this scenario where loose application necessarily leads to meaninglessness. There is nothing that indicates that students of organization tend to take great interest in the original meaning of the concepts they borrow from new sources, let alone in applying them rigorously. But rather than this leading to poor research quality it can be argued that because students of organization apply new concepts loosely rather than rigorously that such applications turn out to be successful. An analysis of the role metaphor plays in scientific inquiry makes clear that only 'tailor-made' ideas and concepts - ideas and concepts that have been borrowed and reshaped to fit their new environment - stand a good chance of making a valuable contribution in the realm of organization.

Metaphors Gone Literal

The warning that applications of complexity science must go 'beyond metaphor' bears testimony to a distinct philosophical position in which metaphorical descriptions are categorically separated from literal descriptions that represent a phenomenon as is. According to this line of reason, you either grab the subject matter by its very essence, in which case 'you tell it like it is', or you discuss it in an inappropriate language, which makes you a poet or a liar. Other philosophical positions lead to different appreciations of metaphor. Gareth Morgan (1980, 1983, 1986) thinks of metaphor as a lens and argues that with every other metaphor we 'look through', we disclose more and more of the same organization. Others have radicalized Morgan's conclusion that the difference between literal and metaphorical is problematic and claim that a fundamental difference between the two simply does not exist (see e.g., Letiche & Van Uden, 1998). These authors deny that the world is already 'out there', that language has the capacity to objectively describe that world and that we, as 're-searchers' or 'dis-coverers' are thus in the business of finding the particular language that 're-presents' that world accurately. Instead, language is said to 'produce what it denotes'. Chia, for instance, writes that "language and words are not wrappings in which things are packed for the commerce of those who write and speak. Social entities, events and things do not first pre-exist and then suffer descriptive distortion through language. Instead language actively configures such entities and events in the very act of representing... [W]e do not just 'write about' our objects/subjects of analysis. Rather we bring these objects into existence through representational acts of writing" (Chia, 1996: 72). This has far-reaching consequences for how metaphor is valued. Instead of accepting an ontological distinction between right (literal) and wrong (metaphorical) descriptions, it is now assumed that if there is a line that separates between different kinds of descriptions, it is the line that separates standard, old, embedded, ingrained, normal descriptions on the one hand from new, speculative, surprising, odd ones on the other. As Nietzsche (1872: 50-1) describes the situation, "the only intrinsic difference here is the difference between custom and novelty, frequency and rarity." If the difference between metaphorical and literal is a difference of degree rather than of kind, it follows that one can only discriminate between routine and new understandings of organization. To illustrate this point, I will show that the development of the concept of organizational culture is best characterized as a process in which metaphors slowly developed into literal descriptions. The process that underlies the 'career' of organizational culture will then be applied to complexity science and it will be argued that here also success will ultimately be determined by the extent to which new ideas and concepts contribute to existing organizational research programmes.

The Culture Example

According to Edgar Schein (1992: 490) "the fields of organizational psychology and sociology have developed a variety of useful concepts for understanding individual behavior in organizations and the ways in which organizations structure themselves. But the dynamic of why and how they grow, change, sometimes fail, and - perhaps most important of all - do things that don't seem to make any sense continues to eludes us". Here is where Schein inserts culture: "the concept of organizational culture holds promise for illuminating this difficult area. I will try to show that a deeper understanding of cultural issues in organizations is necessary not only to decipher what goes on in them but, even more important, to identify what may be priority issues for leaders and leadership". According to Shafritz and Ott (1992: 481-3), the organizational culture school accused the dominant research groups of that time of "using the wrong tools (or 'lenses') to look at the organizational elements to understand and predict organizational behavior" and therefore proposed new concepts to make sense of organizational life. Today these 'odd' culture concepts (myths, heroes, rituals, sagas, etc.) are part of the 'official' language of organizations and a satisfactory answer to the question "can you tell me about the culture of your organization?" is considered by most as real and valid a description of that organization as is the answer to a question like "can you tell me about the structure of your organization?" Slowly the speculative concept of culture has become a concept we believe says something meaningful about organizations. What is interesting to observe is that in this process the dependency on the field in which the concept of culture was originally developed is diminished. Initially, when organizational culture was still an underdeveloped concept, its theorization heavily relied on the literature that inspired the pioneers of organizational culture in the first place. Schein for instance aimed to "provide a clear, workable definition of organizational culture that takes into account the accumulated insights of anthropologists, sociologists, and psychologists" (1992: 491). Armed with the insights of these anthropologists, sociologists, and psychologists, authors like Schein faced the challenge of using the culture discourse in a way that made sense. What if these students of organization would have committed themselves to Maguire and McKelvey's guidelines for quality research? Let us presume for the sake of the argument that at some point all anthropologists, sociologists, and psychologists reached overall consensus on the meaning of culture. Even in this imaginary situation it is not clear why students of organization would be interested in such a widely agreed upon definition of culture. What is important for them is that the concept of culture 'works' in the realm of organization and above all this requires that concepts like 'culture', 'heroes', 'sagas', 'rituals', 'artefacts' or 'norms' bear relevance to existing work on organizations. Schein valued culture for the ways in which this concept contributed to his research on leadership in organizations. Other researchers have appreciated the notion of culture for different reasons; as an example, for its capacity to say something useful about dysfunctional bureaucracies. Or because knowledge of myths adds something to our understanding of inter-departmental communication. Or because insight into cultural norms and values makes clear why people are reluctant to file a sexual harassment complaint against a colleague. The point being made is here is that students of organization value the concept of culture first and foremost for what it has to add to standing research. As a result of the process of 'subjecting' culture to existing organizational research projects the link between the meaning of culture in an organizational context and its 'original' meaning has grown weaker over the years.

Much in accordance with how anthropologists, sociologists, and psychologists used the concept, Schein understood culture as a 'conceptual tool' to describe certain phenomena. Here culture was still seen as descriptive in its intent: "a dynamic model of culture will be especially useful in improving our understanding of how human systems evolve over time" (Schein, 1992: 491). But Jeffcutt shows that in the course of time the meaning of organizational culture diversified. He argues that "a number of contradictory discourses have become constructed and embellished in order to give coherence to the differing positions in the interpretation of organizational culture." One of those discourses he describes as "culture as a corporate (managerial) possession, (a.k.a. 'Corporate Culture')." This management-centric twist instigated a departure from the descriptive nature of the concept of culture. "A remarkable turnaround or critical inversion appears to have taken place [in organization studies] in which the manipulation of the non-rational has become the latest tool of managerial control" (Jeffcutt, 1993: 25-6; 1989: 23). This collapse of the 'true' meaning of culture was accompanied by a growing disinterest in the original writings on culture by anthropologists, sociologists and psychologists. As the concept of culture nestled itself in its new habitat, and 'organizational culture' became a phenomenon in its own right, organizational researchers - and this can be concluded from the literature references they make - turned to books and articles that dealt specifically with organizational culture. Over the years organizational culture thus emancipated itself from 'normal' culture. The case of culture is not unique in this respect. Students of organization have borrowed the concept of 'strategy' from war studies the concept of strategy. In the field of organization studies strategy, like culture, has come to be (re-) interpreted in many different ways (see Mintzberg, *et al.*, 1998). And in the very same way that students of organizational culture 'got the idea' at some point and lost interest in the work of anthropologists, students of organizational strategy eventually decided to disregard new theories of strategy developed back in the field of war studies. As a result the military associations of the concept of strategy were washed out, and strategy became an organizational phenomenon in its own right: business strategy.

The process Maguire and McKelvey (1999: 25) refer to as the "importing and translating" of ideas and concepts can be summarized as follows. On an ongoing basis students of organization explore other disciplines, borrow ideas and concepts they believe can make a contribution of some kind, and try to fit these newly adopted ideas and concepts into existing organizational research programmes. Concepts that prove successful in making the contribution to organizational research become part of normal organizational discourse and lose association with the sources from which they were derived. There is no reason to expect that students of organization will approach the science of complexity in a way that is fundamentally different from how they welcomed culture studies. In the same way that researchers linked novel cultural concepts to (already) normal(ized) organizational phenomena such as leadership or structure, complexity concepts are, and will continue to be, tested on their capacity to fit into research on organizations as we know them. Surprisingly enough Maguire and McKelvey actually seem to support this practice. In their eyes, complexity science will end up a meaningless hype unless "serious research ... shows that CEOs using the New Science produce more competitively advanced firms than CEOs who do not" (1999: 57). Maguire and McKelvey's norm for successful "importing and translating" is clearly not of an 'intra-complexity science' nature but is instead defined in very specific managerial terms: does complexity science help us build competitively advanced firms or does it not? Other students of organization value complexity for very different reasons. Letiche (2000: 555), for example, hopes to find that complexity theory can "provide us with ethnographies of emergent action" while Irvin uses insights from the science of chaos and complexity to scrutinize "one of the main features of the conventional credo of organizational management: the 'ethic of self-preservation'" (2002: 359). Svyantek and Brown have reasons to believe that the representation methods associated with complex adaptive systems studies, and in particular its theories of the way in which agents interact with each other so as to form coherent, self-reinforcing clusters, may be useful in developing a better understanding of the US automobile industry (2001: 42-3). And having first pointed out that theories of complexity have already been used in studies on 'organizational transformation', on 'corporate strategy', and on 'organization design', Styhre, Ingelgard and Roth set out to examine "the notion of knowledge from the nonlinear, non-reductionistic perspective put forth by complexity theory" (2000: 51). So far from success depending on rigorous application, it is in these very specific organizational terms that complexity science must live up to the expectations.

Applying Complexity Theory to Organization: Concluding Remarks

In this article the argument has been developed that the discourse of organization cannot be replaced with the supposedly 'deeper' discourse of complexity. From the claim that rigorous application is not the key to successful use of complexity science in the field of organization, it should not be concluded that complexity science has no 'intrinsic' value however. The reason why students of organization turn to ideas, words or even entire models developed in other disciplines is because they expect that these ideas, words and models have something to add to existing research projects. A borrowed concept is deemed valuable because it offers something that was not already available to the student of organization prior to the 'appropriation'. There are many ways in which a new concept can make a contribution to existing research. A new concept may enable us to call into existence a phenomenon that was inexpressible in our pre-existing language, for instance. Alternatively, we may borrow a concept to undermine the foundation of a dominant theory we seek to destroy. Or we hope to find that the concept combines existing organizational themes in a way that was inconceivable in the absence of it. In any of these cases the concept must connect to an existing research project. At the same time, it is important that the concept is not fully consumed by that project. The value that the borrowed concepts add is that they make a difference, one way or the other. Given that we turn to the use of new concepts because they add something to an existing discussion, it makes no sense to kill its potential by sterilizing it - by arguing for example that the concept we take from complexity science is really but a synonym for an already existing organizational concept. The successful import and translation of 'alien' concepts thus boils down to a balancing act: for the concept to be relevant, it needs to fit into existing research programmes. At the same time, if we want to preserve the potential we appreciated it for in the first place we must be careful not to let the concept be fully assimilated into and taken over by existing body of knowledge.

References

Bale, L. (1995) "Gregory Bateson, cybernetics, and the social/behavioral sciences," *Cybernetics & Human Knowing: A Journal of Second Order Cybernetics, Autopoiesis and Cyber-Semiotics*, 3(1), available at http://www.narberthpa.com/Bale/lsbale_dop/cybernet.htm, downloaded 5th August, 2005.

Byrne, D. (1998). *Complexity and the social sciences: An introduction*, London, UK: Routledge, ISBN 0415162963.

Chia, R. (1995). "From modern to postmodern organizational analysis," *Organization Studies*, ISSN 01708406, 16(5): 579-604.

Fuller, T. and Moran, P. (2000). "Moving beyond metaphor," *Emergence: A Journal of Complexity Issues and Management*, ISSN 15213250, 2(1): 50-71.

Hayles, N. K. (2000). "Making the cut," in W. Rasch and C. Wolfe (eds.), *Observing complexity*, Minnesota: University of Minnesota Press, ISBN 081663298, pp. 137-162.

Irvin, L. (2002). "Ethics in organizations: A chaos perspective," *Journal of Organizational Change Management*, ISSN 09534814, 15(4): 359-381.

Jeffcutt, P. (1989). *Persistence and change in an organization culture*, unpublished PhD thesis, University of Manchester.

Jeffcutt, P. (1993). "From interpretation to representation," in J. Hassard and M. Parker (eds.), *Postmodernism and organizations*, London, UK: Sage, ISBN 0803988796, pp. 25-48.

Johnson, J. and Burton, B. (1994). "Chaos and complexity theory for management: Caveat emptor," *Journal of Management Inquiry*, ISSN 10564926, 3(4): 320-328.

Letiche, H. and Van Uden, J. (1998). "Answers to a discussion note: On the metaphor of metaphor," *Organization Studies*, ISSN 01708406, 19(6): 1029-1033.

Letiche, H. (2000). "Phenomenal complexity theory as informed by Bergson,"

Journal of Organizational Change Management, ISSN 0953-4814, 13(6): 545-557.

Maguire, S. and McKelvey, B. (1999). "Complexity and management: Moving from fad to firm foundations," *Emergence: A Journal of Complexity Issues and Management*, ISSN 1521-3250, 1(2): 19-61.

McKelvey, B. (1999). "Complexity theory in organization science: Seizing the promise or becoming a fad," *Emergence: A Journal of Complexity Issues and Management*, ISSN 1521-3250, 1(1): 5-33.

Meiss, J. (2000). " Sci.Nonlinear FAQ," available at http://amath.colorado.edu/faculty/jdm/faq.html, downloaded 5th August, 2005.

Mathews, M., White, M. and Long, R. (1999). "Why study the complexity sciences in the social sciences," *Human Relations*, ISSN 0018-7267, 52(4): 439-642

Mintzberg, H., Ahlstrand, B. and Lampel, J. (1998). *Strategy safari: A guided tour through the wilds of strategic management*, New York, NY: The Free Press, ISBN 0684847434.

Morgan, G. (1980). "Paradigms, metaphors, and puzzle solving in organization theory," *Administrative Science Quarterly*, ISSN 0001-8392, 25(4): 605-622.

Morgan, G. (1983). "More on metaphor: Why we cannot control tropes in administrative science," *Administrative Science Quarterly*, ISSN 0001-8392, 28(4): 601-607.

Morgan, G. (1986). *Images of organization*, Newbury Park, CA: Sage Publications, ISBN 0761906320.

Nietzsche, F. (1872). "Philosophenbuch," translated in D. Breazeale (1979). *Philosophy and truth: Selections from Nietzsches notebooks of the early 1870s*, Sussex, New Jersey: Humanities Press, ISBN 1573925322, pp. 50-1.

Schein, E. (1992). "Defining organizational culture," in J. Shafritz and J. Ott (eds.), *Classics of organization theory*, Pacific Grove, CA: Brooks/Cole Publishing Company, ISBN 0534173047, pp. 490-502.

Shafritz, J. and Ott, J. (1992). *Classics of organization theory*, Pacific Grove, CA: Brooks/Cole Publishing Company, ISBN 0534173047.

Styhre, A., Ingelgard, A. and Roth, J. (2000). "A nonreductionist view of knowledge: Product development in the pharmaceutical industry," *Emergence: A Journal of Complexity Issues and Management*, ISSN 1521-3250, 2(3): 51-67.

Svyantek, D. and Brown, L. (2001). "Stability in the American automobile industry," *Emergence: A Journal of Complexity Issues and Management*, ISSN 1521-3250, 3(3): 42-57.

Van Uden, J. (2004). *Organization and complexity: Using complexity science to theorize organizational aliveness*, Boca Raton, FL: Dissertation.com, ISBN 1581122225.

Young, T. R. (1990). "Chaos and the drama of social change: A metaphysic for postmodern science," distributed as part of Transforming Sociology Series of The Red Feather Institute.

11. Knowledge, Complexity and Understanding

Paul Cilliers
University of Stellenbosch, ZAF

"The strange thing about television is that it doesn't tell you everything. It shows you everything about life on earth, but the mysteries remain. Perhaps it is in the nature of television."

Thomas Jerome Newton in *The Man Who Fell to Earth*

During most events concerned with knowledge management, someone starts a presentation by saying that they will not revisit the problem of the distinction between knowledge and data. Usually a sigh goes through the audience, seemingly signifying relief. But why relief? Is it because they will not be bored with an issue that has been resolved already, or because they are glad that they will not be confronted with these thorns again? I suspect that they want to believe that the first reason is the case, but that in fact it is the second.

In what follows I therefore want to problematize the notion of "knowledge." I will argue that when talking about the management of "knowledge," whether by humans or computers, there is a danger of getting caught in the objectivist/subjectivist (or fundamentalist/relativist) dichotomy. The nature of the problem changes if one acknowledges the complex, interactive nature of knowledge. These arguments, presented from a philosophical perspective, should have less influence on the practical techniques employed in implementing knowledge management systems than on the claims made about what is actually achieved by these systems.

The issues around knowledge – *what* we can know about the world, *how* we know it, what the *status* of our experiences is – have been central to philosophical reflection for ages. Answers to these questions, admittedly oversimplified here, have traditionally taken one of two forms. On the one hand there is the belief that the world can be made rationally transparent, that with enough hard work knowledge about the world can be made objective. Thinkers like Descartes and Habermas are often framed as being responsible for this kind of attitude. It goes under numerous names including positivism, modernism, objectivism, rationalism, and epistemological fundamentalism. On the other hand, there is the belief that knowledge is only possible from a personal or cultural-specific perspective, and that it can therefore never be objective or universal. This position is ascribed, correctly or not, to numerous thinkers in the more recent past like Kuhn, Rorty, and Derrida, and its many names include relativism, idealism, postmodernism, perspectivism, and flapdoodle.

Relativism is not a position that can be maintained consistently[1], and of course the thinkers mentioned above have far more sophisticated positions than portrayed in this bipolar caricature. There are also recent thinkers who attempt to move beyond the fundamentalist/relativist dichotomy[2], but it seems to me that when it comes to the technological applications of theories of knowledge, there is an implicit reversion to one of these traditional positions. For those who want to computerize knowledge, knowledge has to be objective. It must be possible to gather, store, and manipulate knowledge without the intervention of a subject. The critics of formalized knowledge, on the other hand, usually fall back on arguments based on subjective or culture-specific perspectives to show that it is not possible, that we cannot talk about knowledge independently of the knowing subject.

I am of the opinion that a shouting match between these two positions will not get us much further. The first thing we have to do is to acknowledge the complexity of the problem with which we are dealing. This will unfortunately not lead us out of the woods, but it should enable a discussion that is more fruitful than the objectivist/subjectivist debate.

Complexity and Understanding

An understanding of knowledge as constituted within a complex system of interactions[3] would, on the one hand, deny that knowledge can be seen as atomized "facts" that have objective meaning. Knowledge comes to be in a dynamic network of interactions, a network that does not have distinctive borders. On the other hand, this perspective would also deny that knowledge is something purely subjective, mainly because one cannot conceive of the subject as something *prior* to the "network of knowledge," but rather as something constituted *within* that network. The argument from complexity thus wants to move beyond the objectivist/subjectivist dichotomy. The dialectical relationship between knowledge and the system within which it is constituted has to be acknowledged. The two do not exist independently, thus making it impossible to first sort out the system (or context), and then identify the knowledge within the system. This codetermination also means that knowledge and the system within which it is constituted are in continual transformation. What appears to be uncontroversial at one point may not remain so for long.

---

1. If relativism is maintained consistently, it becomes an absolute position. From this one can see that a relativist is nothing but a disappointed fundamentalist. However, this should not lead one to conclude that everything that is called postmodern leads to this weak position. Lyotard's seminal work, *The Postmodern Condition* (1984), is subtitled *A Report on Knowledge*. He is primarily concerned with the structure and form of different kinds of knowledge, not with relativism. An informed reading of Derrida will also show that deconstruction does not imply relativism at all. For a penetrating philosophical study of the problem, see *Against Relativism* (Norris, 1997).
2. The critical realism of Bhaskar (1986) is a good example.
3. Complex systems are discussed in detail in Cilliers (1998).

The points above are just a restatement of the claim that complex systems have a history, and that they cannot be conceived of without taking their context into account. The burning question at this stage is whether it is possible to *do* that formally or computationally. Can we incorporate the context and the history of a system into its description, thereby making it possible to extract knowledge from it? This is certainly possible (and very useful) in the case of relatively simple systems, but with complex systems there are a number of problems. These problems are, at least to my mind, not of a metaphysical but of a practical nature.

The first problem has to do with the nonlinear nature of the interactions in a complex system. From this it can be argued (see Cilliers, 1998: 9-10 and Richardson *et al.*, 2000) that complexity is incompressible. There is no accurate (or, rather, perfect) representation of the system that is simpler than the system itself. In building representations of open systems, we are forced to leave things out, and since the effects of these omissions are nonlinear, we cannot predict their magnitude. This is not an argument claiming that reasonable representations should not be constructed, but rather one that the unavoidable limitations of the representations should be acknowledged.

This problem – which can be called the problem of boundaries[4] – is compounded by the dynamic nature of the interactions in a complex system. The system is constituted by rich interaction, but since there is an abundance of direct and indirect feedback paths, the interactions are constantly changing. Any activity in the system reverberates throughout the system, and can have effects that are very difficult to predict; once again as a result of the large number of nonlinear interactions. I do not claim that these dynamics cannot be modeled. It could be possible for richly connected network models to be constructed. However, as soon as these networks become sizable, they become extremely difficult to train. It also becomes rather hard to figure out what is actually happening in them. This is no surprise if one grants the argument that a model of a complex system will have to be as complex as the system itself. Reduction of complexity always leads to distortion.

What are the implications of the arguments from complexity for our understanding of the distinction between data and knowledge? In the first place, it problematizes any notion that data can be transformed into knowledge through a pure, mechanical, and objective process. However, it also problematizes any notion that would see the two as totally different things. There are facts that exist independently of the observer of those facts, but the facts do not have their meaning written on their faces. Meaning only comes to be in the process of interaction. Knowledge is interpreted data. This leads us to the next big question: What is involved in interpretation, and who (or what) can do it?

4. The problem of boundaries is discussed in more detail in Cilliers (2001).

Knowledge and the Subject

The function of knowledge management seems to be either to supplement the efforts of a human subject who has to deal with more data than is possible, or to free the subject up for other activities (perhaps to do some thinking for a change). Both these functions presuppose that the human subject can manipulate knowledge. This realization leads to questions in two directions. One could debate the efficiency of human strategies to deal with knowledge and then attempt to develop them in new directions. This important issue will not be pursued further here.

There is another, perhaps philosophically more basic, question, and that has to do with how the human subject deals with knowledge at all. Given the complexities of the issue, how does the subject come to forms of understanding, and what is the status of knowledge as understood by a specific subject? This has been pursued by many philosophers, especially in the discipline known as hermeneutics. However, I am not aware that this has occurred in any depth in the context of complexity theory[5]. How does one perceive of the subject as something that is not atomistically self-contained, but is constituted through dynamic interaction? Moreover, what is the relationship between such a subject and its understanding of the world? A deeper understanding of what knowledge is, and how to "manage" it, will depend heavily on a better understanding of the subject. This is a field of study with many opportunities.

Apart from calling for renewed effort in this area, I only want to make one important remark. It seems that the development of the subject from something totally incapable of dealing with the world on its own into something that can begin to interpret – and change – its environment is a rather lengthy process. Childhood and adolescence are necessary phases (sometimes the only phases) in human development. In dealing with the complexities of the world there seems to be no substitute for experience (and education). This would lead one to conclude that when we attempt to automate understanding, a learning process will also be inevitable. This argument encourages one to support computing techniques that incorporate learning (like neural networks) rather than techniques that try to abstract the essence of certain facts and manipulate them in terms of purely logical principles. Attempts to develop a better understanding of the subject will not only be helpful in building machines that can manage knowledge, they will also help humans better understand what they do themselves. We should not allow the importance of machines (read computers) in our world to lead to a machine-like understanding of what it is to be human.

---

5. An important contribution was made by reinterpreting action theory from the perspective of complexity (Juarrero, 1999). Some preliminary remarks, more specifically on complexity and the subject, are made in Cilliers & De Villiers (2000).

Implications

In Nicholas Roeg's remarkably visionary film *The Man Who Fell to Earth* (1976), an alien using the name Thomas Jerome Newton (superbly played by David Bowie) tries to understand human culture by watching television, usually a whole bunch of screens at the same time. Despite the immense amount of data available to him, he is not able to understand what is going on directly. It is only through the actual *experience* of political complexities, as they unfold in time, that he begins to understand. By then he is doomed to remain earthbound.

I am convinced that something similar is at stake for all of us. Having access to untold amounts of information does not increase our understanding of what it means. Understanding, and therefore knowledge, follows only after interpretation. Since we hardly understand how humans manage knowledge, we should not oversimplify the problems involved in doing knowledge management computationally. This does not imply that we should not attempt what we can – and certain spectacular advances have been made already – but that we should be careful in the claims we make about our (often still to be finalized) achievements. The perspective from complexity urges that, among others, the following factors should be kept in mind:

- Although systems that filter data enable us to deal with large amounts of it more effectively, we should remember that filtering is a form of compression. We should never trust a filter too much.
- Consequently, when we talk of mechanized knowledge management systems, we can (at present?) only use the word "knowledge" in a very lean sense. There may be wonderful things to come, but at present I do not know of any existing computational systems that can in any way be seen as producing "knowledge." Real breakthroughs are still required before we will have systems that can be distinguished in a fundamental way from database management. Good data management is tremendously valuable, but cannot be a substitute for the *interpretation* of data.
- Since human capabilities in dealing with complex issues are also far from perfect, interpretation is never a merely mechanical process, but one that involves decisions and values. This implies a normative dimension to the "management" of knowledge. Computational systems that assist in knowledge management will not let us escape from this normativity. Interpretation implies a reduction in complexity. The responsibility for the effects of this reduction cannot be shifted away on to a machine.
- The importance of context and history means that there is no substitute for experience. Although different generations will probably place the emphasis differently, the tension between innovation and experience will remain important.

These considerations should assist in developing an understanding of knowledge management that could be called "organic," but perhaps also "ethical."

Acknowledgement

The financial assistance of the National Research Foundation: Social Sciences and Humanities (of South Africa) toward this research is hereby acknowledged. Opinions expressed and conclusions arrived at are those of the author, and are not necessarily to be attributed to the National Research Foundation.

References

Bhaskar, R. (1986). *Scientific Realism and Human Emancipation*, London: Verso.

Cilliers, P. (1998). *Complexity and Postmodernism: Understanding Complex Systems*, London: Routledge.

Cilliers, P. (2001). *Boundaries, Hierarchies and Networks in Complex Systems* (forthcoming).

Cilliers, P. & De Villiers, T. (2000). "The complex 'I,'" in W. Wheeler (ed.), *The Political Subject*, London: Lawrence & Wishart.

Juarrero, A. (1999). *Dynamics in Action: Intentional Behavior as a Complex System*, Cambridge, MA: MIT Press.

Lyotard, J. F. (1984). *The Postmodern Condition: A Report on Knowledge*, Manchester, UK: Manchester University Press.

Norris, C. (1997). *Against Relativism: Philosophy of Science, Deconstruction and Critical Theory*, Oxford, UK: Blackwell.

Richardson, K., Cilliers, P. and Lissack, M. (2000). "Complexity science: A 'grey' science for the 'stuff in between,'" *Proceedings of the First International Conference on Systems Thinking in Management*, Geelong, Australia, 532-7. Also see Chapter 2 of this volume.

12. Complexity Theory and Continental Philosophy Part 1: A Review of Letiche's Phenomenological Complexity Theory

Stephen Sheard
Freelance Research/Writer, UK

Introduction

A marked feature of recent conceptual advances in the area of Complexity theory has been the incorporation of themes related to areas such as phenomenology (incorporating theorists such as Heidegger, and Derrida), and poststructuralism (especially incorporating the theories of Derrida). Particular theorists have evolved: for example Cilliers (1998), Letiche and Boje (2001), and Letiche (2000a). A particular integration has occurred around the thought of Henri Bergson.

This article describes key features of Hugo Letiche's theory of phenomenology, and then moves to an analysis of this theorization. It looks at its theoretical sources of inspiration and grounds the discussion with reference to the work of Derrida, especially his concept of *différance* and its status relative to what is known as the ontic-ontological divide (as defined by Heidegger). Derridean *différance*, as opposed to PCT proper, is then debated relative to the themes introduced in the discussion of the recent theoretical constructions of complexity theory and its variable forms - with emphasis on self-organization theory.

In this paper, the difficulty in the transference of an analogy from the physical to the human/social sciences is considered. Letiche's PCT serves as a touchstone to introduce that debate, and also the significance of the project of creating a theory of complexity which has a pre-epistemic or ontological emphasis.

The article is significant in that it points to the possible limits and nature of a complexity theory of human life and social organization, and shows how a synthetic theorization (PCT) introduces a variety of problems which are key to the understanding of the extension of complexity theories towards the human (individual) and social realm.

A Synthetic Approach to Phenomenology and Complexity: Hugo Letiche's Phenomenal Complexity Theory

Letiche (2000a) links the concepts of phenomenology, with the ideas of Derrida (*différance*) and those of Bergson (specifically the concepts of *Durée*, *Élan Vital* and *Intuition*). He adopts a position which disputes the subject/object divide as a cornerstone of management research methodology. This is termed "dualistic," reflective of a "one sidedly rationalist or subjectivist" perspective. A critical stance is engaged, which draws on Derrida, and represents "a process of *différance* thinking wherein complexity - the interactive dynamics of a multifaceted process leading to emergence - occurs..." (Letiche, 2000a: 547).

Letiche links this theorization to Bergson's distinction between analytic and intuitive consciousness; the former being argued to be closer to 'the object', whereas the latter is argued to be an experiential feature, which is "centred on the experience of the knower" (Letiche, 2000a: 548-549). Letiche equates the lived experience of self-organizing emergent systems with Bergson's concept of *durée.* He suggests that "emergence precedes consciousness, though consciousness - as meaning or language - remains constitutive of human understanding of the world..." (Letiche, 2000a: 550). Consciousness is thereby defined as having a correlation with antecedent emergent qualities of 'self-organization'. It is argued that Bergson's *durée* is characterized by a double-movement - of the experience of the present and the experience of continuity of identity - and that in a state of duration one enters into the object (Letiche, 2000b). Letiche argues that "the intentional structure of emergence is logically prior to *durée* and to intuition. Intuition is the form of cognition able to grasp *durée...*"

Letiche also argues that "CAS (Complex Adaptive Systems) is not just a construction of a 'knowing subject', it is a principle of being..." which is to state the epistemology of CAS in ontological terms of significance. He asks whether emergence is atemporal or contextually bound by time:

"*If emergence is emergent it must be historical, contextual and time bound. Thus, is emergence inside or outside of 'time' and process? If 'outside' (ontology), how could it emerge; if 'inside' (ontic), in what sense would it be knowable? How could the 'subject' gain enough distance from emergence to identify its 'laws'?*" (Letiche, 2000b).

Theorists such as Juarrero (1999) and Letiche have emphasized the potentially complex nature of causality[1]. A key aspect of that process is the aspect of bifurcation relative to the self-organizing system (e.g., Belousov-Zhabotinsky chemical clock)[2]. It is possible to construe the transformation which occurs in a self-organizing complex system in terms of an analogy with a shift of identity in the social/human realm. Letiche argues that Juarrero suggests a complex causality as an alternative to the efficient (Aristo-

---

1. As Letiche notes, a construction of causality which is primarily that of a crudified cause-effect billiard ball style causation, has dominated Western epistemology since Newton. This has tended to overlay the interpretation of the nature of emergent phenomena. It has led to the prevalence of a view of causation which is atemporal and non-recursive.
2. See Wheatley (1994) for a popular account which discusses the role of this experiment and self-organization theory.

telian) causality, and that this may result from a process of 'entrainment'[3]. He seems to envisage Juarrero's conception of entrainment as anterior to the self-organizing processes (which he argues can occur at all levels of existence including human consciousness). Hence, he argues that an alternative conception of complexity, built around an affiliation of these ideas with the concept of '*différance*', is more suited to an account of the role of consciousness in the social sciences:-

"*...consciousness can also be understood - in so far as it is currently possible to understand it - as a product of the looseness of difference. Consciousness can be thought as a result of many slightly different weak signals that all voice in the mind. By bringing these differences into one another, 'mind' emerges. Tightly coordinated links of entrainment are but one sort of hypothesized relationship (for self-organizing systems). Much looser, paradoxical and polyvalent forms of order could be at least as important to the formation of consciousness...*" (Letiche, 2000b: 35).

These looser forms of order are also the basis of his phenomenology of complexity:

"*[T]he social activity of emergence goes on somewhere else (other than organizational studies) - in-between consciousness, circumstance and conceptualization. The in-between exists in the dialogic interactions that link the different aggregation levels of organizing and organization*"[4].

Hence interaction between distinct levels of aggregation is defined as emergent, and the province of nonlinear links - organization and organizing, but a "phenomena of consciousness," (Letiche & Boje, 2001: 19) not an economic fact.

PCT: Boje's Instantiation Within Disney

Before moving towards further discussion of Letiche's concept of PCT, it is worth asking: "how exactly might a practitioner (as opposed to a theorist) derive benefit from a study of PCT?

Boje develops his instantiation of Letiche's PCT in two papers (Boje, 2000, 2001); these collectively argue that Disney "dances on the edge of chaos" as diverse protest groups are having a collective strategic/financial impact. Boje notices that Disney is "a complex enough empire that it manages to offend almost every cultural group"[5]. Boje's case is that Disney is a CAS which is subject to "emergent processes of protest, self-organizing networks of labor and family value monitor groups and operating, seem-

3. Letiche (2000b) of Juarrero (1999).
4. This splits the aggregation levels into three: at the top - coherence/reification/organizational - in the middle - 'emergence' - and below these - 'self-intentional'. See Letiche and Boje (2001: 15).
5. Cultural groups so noted include American Red Indians, pro-family groups, anti-homosexual groups, and workers collectives/unions. See Boje (2000: 562).

ingly on the edge of chaos..." (Boje, 2000: 563).

It is difficult to exactly discern as to how Letiche's PCT is specifically applicable to Disney, as opposed to a more generic theory of the CAS. Letiche (2000a) reviews Boje's paper, noting that: "David Boje has in his research created an artefact, that is in opposition to the implicit norms of PCT's, which stress the role of emergence, consciousness and 'self'. 'Nike' is Boje's organizational, anti-phenomenal, reification." Boje has shown an organization (Nike), which displays principles that are contrary to PCT. This goes beyond Boje's looser application of complexity theories to Disney; rather Letiche construes Disney as exhibiting a "false emergence" as "so many of the participants (are) not allowed to have any influence on the action (of Disney)." Clearly, these comments apply to external stakeholders also. These accounts of Nike as reification, and Disney as false emergence point to a treatment of organization in terms of "its ethical and social significance" (Boje, 2000, 563).

So, what is Letiche saying about Disney? I think he is supplementing PCT's credentials by interpreting these firms in terms of PCT and emphasizing the consciousness of individuals rather than a company orientated or corporatist view. On these accounts Disney exhibits a "false emergence"; that this is "anti-phenomenal". This is actually unlike Boje's Disney - "dancing on the edge of chaos" or anticipating destruction because it is a complex system. Letiche rather emphasizes that Disney does not embody PCT as it constitutes a "false emergence". Boje is clearly drawing on the distinct idea of chaos theory to represent his ethical disapproval of Disney's policies. It can be argued that Letiche's view of Boje's instantiation of his ideas, stresses an aspect of moral and ethical significance relative to emergence within PCT. Emergence is not about remarkable transformation in terms of engaging profits or company success, but from a human perspective in terms of engaging our hearts and minds, according to Letiche. This is a radical contrast to what is suggested (often implied) in many corporatist accounts of complexity theory (e.g., Kelly & Allison, 1998). Disney are (like all firms) attempting to emerge in competitive terms. This can mean accounts of complexity which emphasize competition and the theory of competitive advantage, the evolution of "plateaus of competence over time" for instance. Letiche's definition of complexity sits uneasily with these kinds of definitions in that they may imply, as a feature of the desire for efficiency, a lack of ethicality or concern for individual consciousness.

By examining Boje and Letiche on the example of Disney, it is notable that Letiche adopts a view of complexity theory which is indirectly capable of antagonism to the ideology of profit maximization or efficiency. PCT is therefore, albeit discretely, a form of critique of modern capitalism, in that it espouses humanitarian values quite alien to those which may be a necessary feature of competitive progression of a firm. PCT may be radically opposed to more corporatist views of complexity also, in terms of its links to more radical influences on a phenomenological perspective, containing the seeds of radical critique of orthodoxy - harnessed towards a scientific analogy which is more usually conceived of as a means of promoting organizational (rather than individual) transformation.

Ought reflective practitioners to examine the often dubious extension of a commercially orientated ideology to the detriment of the individual consciousnesses of their employees and related stakeholder groups? Discussion of PCT reveals its radical potential relative to practitioners but more needs to be said about the status of PCT in terms of its philosophical credentials. What are the phenomenological sources that so emphasize the individual in PCT? We have mentioned complex causality and Derrida's *différance* (above). These themes will now be discussed.

PCT: How Does It Relate to Derrida's Concept of *Différance*?

What may be noted is that PCT dwells on the cusp of the social sciences and the sciences; it extends a theory of phenomenology to the close comparison between Complexity theory and human life. PCT extends a particular interpretation of phenomenology, of which a vast array is possible. It also emphasizes a theory (Derrida's *différance*) which has affiliations with areas of radical critique of phenomenology as it was envisaged by Husserl; a founder of phenomenological enterprise, which itself emphasized a non-empirical, transcendent view of phenomenology[6].

We can see an immediate problem with Letiche's PCT; in that he is concerned with creating an interpretation which has the promise of linking with empirical cases in organizational studies (see above, the case of Disney). However, Husserl establishes an initial but not exclusive claim towards phenomenological theory. Schroeder describes phenomenology in terms of a distinction between transcendental and existential phenomenologists. A transcendental phenomenologist - such as Husserl, would 'bracket' perceptions of the world[7], However, it is possible to emphasize the existential - the 'lived experience' - rather than the transcendent[8]. On these accounts, phenomenology (which to Husserl had a Platonic aspect) as "an a priori science of the essences of all possible objects and experiences"

6. Husserl (1913) notes that: "Pure phenomenology is accordingly the theory of the essences of 'pure phenomena', the phenomena of 'pure consciousness' or of a 'pure ego': it does not build on the ground, given by transcendent apperception, of physical and animal, and so of psychophysical nature, it makes no empirical assertions, it propounds no judgements which relate to subjects transcending consciousness: it establishes no truths concerning natural realities, whether physical or psychic..."

7. Husserl (1913) notes that phenomenology: "takes all apperceptions and judgemental assertions which point beyond what is given in adequate, purely immanent intuition, which point beyond the pure stream of consciousness and treats them as purely the experiences they are in themselves: it subjects them to a purely immanent, purely descriptive examination of essence..."

8. Schroeder (2005: 174) indicating that: "Existential phenomenologists embrace the complexity of lived experience, believing that such bracketing is artificial and inimical to clarifying experience as lived. In addition, phenomenologist attitudes, assumptions, and goals all influence the type of phenomenology they pursue."

(Moran & Mooney, 2002)[9] is anything but unitary. Letiche reflects a more existential view, when he notes that such research is a form of embodied practice wherein the inability to side entirely with the 'subject' or the 'object' (the observed) applies[10]. This becomes meaningful in the light of the equation of 'subject'-orientated perspectives with phenomenal experience and those which emphasize the 'object' with a more realist inclined perspective. Letiche follows Merleau Ponty in terms of his perception that phenomenology follows the middle-ground between empiricism and cognitivism (in this latter context we can read 'subject'-orientated experiences as related to idealism). It is also possible elsewhere to see shades of Husserl's own theorizations (transcendental phenomenology) within Letiche's PCT. For instance, when Letiche (and Boje) describe the phenomena of intentional emergent coherence, this has resonance with Husserl's 'intentionality', but also the 'rhizome' (Deleuze):

"*Intentional emergent coherence links the experiential to the complex. Identity is achieved on the intentional level. It is on this level that the self determines activity ... The intentional is never coherent, it is desire, rhizome and inchoate*" (Letiche & Boje, 2001: 23).

Already we can see that these observations are indicating that it is not purely Derrida, and his *différance*, which is emphasized in PCT, but a variety of theorists of continental origin are drawn into a form of synthetic theorization. Nevertheless, despite the sense of a melange or variety of influences, Bergson and Derrida are the predominant theorists unified in this theoretical construct[11].

As Letiche cites Derrida, and especially his concept of *différance*, it would be especially relevant to look at the intersection of Derrida and the phenomenological tradition, relative to Letiche's PCT. Derrida follows, in part in the wake of Levinas, who emphasized the ethical aspect within phenomenology; that it was more important to "respect other people over the urge to know or objectify them" (Schroeder, 2005: 177). One can see the hand of Levinas in the ethical conception of Disney as a 'false emergence' due to its ignoring the individual views of its workers. Schroeder also notes, apart from is distinction between transcendental and existential phenomenologies (above), that "Levinas constitutes a pivot within phenomenology, anticipating key claims of poststructuralism" (Schroeder, 2005: 178). We might tentatively note these influences, that PCT seems closer, generally speaking, to the existential account of phenomenology. However, these

9. Contains extracted writings of Husserl (including *Logical Investigations*) and Heidegger (including *Being and Time*, pp. 1-26.
10. Letiche (2000a) citing Merleau-Ponty (1962).
11. The creative aspect of synthetic theorization is a feature of Letiche's PCT but the admixture of phenomenologists tends to mask differences in their individual perspectives. That might then imply a unity between distinct theorists which was only partial. To some extent this aspect will be examined in the second paper in this series (Sheard, 2006).

points can take us only so far in understanding PCT - which dwells, as Letiche notes, on the issue of a unity of the idea of *différance* with a view of complexity theory.

A view of lived experience as an 'awakening' is common to all phenomenologists, Schroeder notes, and it is the construal of that theme of phenomenology or 'awakening' in terms of the complexity theory state of emergence, which is the novel move of Letiche (2000b). Whether emphasizing emergence as within or outside time (a central concern of phenomenology being the evaluation of perception relative to time states and the status of 'transcendence'), Letiche draws heavily on Derrida; his complex notion of *différance*. This tends to act as a satellite concept which links the disparate origins of Letiche's conceptions of phenomenology with his views on another continental theorist - Bergson. What exactly is *différance* according to Derrida?

"*In a conceptuality adhering to classical strictures 'différance' would be said to designate a constitutive, productive and originary causality, the process of scission which would produce or constitute different things or differences...*" (Derrida, 1982: 9).

Hence, *différance* is 'the possibility of conceptuality'; it is the "non-full, non-simple, structured and differentiating origin of differences. Thus the name 'origin' no longer suits it" (Derrida, 1982: 11). Contemporary theorists have diverse interpretations. Only a few cases can be examined in the space available and particular attention is paid to the interpretation of Lawlor, as being clearer than most. For instance, Lawlor describes *différance* in terms of a construction/interpretation of the evolution of the present in terms of its continued passage; because of this he terms "subjective presence" as something which is continually passing and in relation to which "I must bring myself over and over again" as the "power of representation in its general form." This return of a memory of presence is the 'trace'. As the sense of the present must survive, a need for survival exists; "the sense must survive and that means it must be no longer finite but infinite, no longer immanent but transcendent..." This second aspect of difference is a "graphic possibility (which) can make the sense be different from me as a singularity." The re-evocation of the trace - its being made immanent again - is as it were envisaged as a recreation of what is dead (Lawlor, 2002: 232-233). In contrast, Stellardi notes (when discussing the inter-relationship of *différance* and metaphor) that *différance* is an example of an indecidable - "negatively defined by their inability to tolerate any definition"[12]. *Différance* in its play in and out of reality, can be conceived in terms of a kind of feature expressed as episteme (typified in terms of language), which

12. Stellardi (2000: 108-109) gives an example of an indecidable in terms of binary oppositions, for instance as a "bundle of signified's disposed within the space of an opposition (e.g., signifier/signified) that has been barred", wherein "A notion (*une marque*) is systematically kept out of a symbolic space it is supposed to indicate."

is seeking to express an insight into the heart of the nature of the ontological inter-relationship of thought, including the aspect of the inter-relationship of self and interior monologue, with both conceptualization in its evocative and re-evocative states. Hence, it is concerned with an interpretation of the relationship between memory and thought, and the aspect of language articulation as a feature of thought, including the intervention of memory[13].

Différance thereby obliquely addresses the psychological aspect of the nature of interplay of language, and the sphere of formulation of mental acts expressed in terms of word-structures (the temporal dimension). However, it closely engages this insight (in terms of its emphasis on the transcendental nature of experience) by suggesting that it locates in the present, differences between the construction of words within the immanent - or here and now. As language (a metaphoric entailment in certain senses) is engaged in this project, *différance* is described in terms of its lack of confinement, that reinforces a view of language which is scarcely chained by metaphysically subordinated conceptions of the role of metaphor in language formulation (for example the Aristotelian)[14]. These aspects transfer the concern with causality, or origins, of thought to an analysis vitally related to language (and hence socially implicated). The persistence of the trace (of language) may in a certain sense be construed as writing. Hence, in a macro-sense, the issue of the text as a feature of indeterminacy due to an infinite reiteration of its articulation, from the perspective of the separate reader(s)[15].

Différance therefore permits a wide scope of discussion, because the originary process of the thought-memory relationship infers the universality of the 'conception' of difference, as a feature which stands instead of a less dynamically conceived theory of perception. In this sense, *différance* allows a mode of discussion as to the relationship between immanence and transcendence in human thought; related to the issue of past - present and future articulation of language (as thought reflected in the iterability of the 'trace'). This emphasis on language as a feature of intervention between knowledge and Being, dwells on the area explored by Heidegger - the ontico-ontological divide[16].

---

13. See for instance the emphasis on the role of what is not present in alterity, i.e., the past and present - as a feature of *différance* - which is a feature of the re-evocation of the trace which constitutes the basis of a semiotic conditioning of alterity as a feature of identity (Smith, 2005: 36-37).

14. The inter-relationship of Derrida's views of metaphor and difference are discussed in the next paper (Sheard, 2006). Stellardi (2000: 108-109) indicating *différance* as having a notional value for a metaphor (or catachresis), which can be conceived in terms of spacing and temporization.

15. Derrida (1982: 12) notes that: "Retaining at least the framework, if not the content, of this requirement formulated by Saussure, we will designate as *différance* the movement according to which language, or any code, any system of referral in general, is constituted 'historically' as a weave of differences..."

16. See the next paper (Sheard, 2006) in this series for fuller discussion of Heidegger relative to Derrida and Ricoeur.

As Lawlor notes, whereas Derrida initially envisaged *différance* as a form of dialectic, the fully-fledged conception contained a marked deviation from the idea that *différance* was a dialectical conception akin to Hegelian dialectics. This is worth dwelling on, because it lies at the heart of the Derridean conception of *différance* (see Lawlor, 1992). Lawlor further notes that difference is based on the displacement of univocity by equivocity (terming this 'supplementarity'). He also notes that: "dissemination is a genesis based not on temporalization and imagination, but rather on spacing and the aleatory" (of difference). This alludes to the conception of Stellardi, that *différance* can be spatially and temporally conceived - as a neographism - capable of acting as a kind of notional substitute for a word which could act as the basis for metaphorization. Difference in these senses, does not offer closure, because its function is not to seal or hermetically isolate the process of mediation, unlike as Lawlor suggests, in the case of Ricoeur[17]. This aspect of lack of closedness, of splitting itself (if it can be so described as an 'it') at every point including that of origin or causality; inscribes the character of *différance.* This lack of closure, or process of infinitization (iterability of the process of formation of the 'trace'), prevents the sense of dialectical completeness (á la Hegel). It also points to the sense of resemblance between the flow of *différance* and the sense of an open system (CAS), because on this account language is not permitted a completeness in terms of the certainties of an abstract (or philosophical) resolution of the play of determinacy of meaning. These features give the sensation of a possible allegiance with a complexity view of science.

However, such sensed resonances, between the CAS and *différance*, do not necessarily underpin the case represented for PCT. They may well relate to a case for a Derridean view of complexity, however. As said, a theory which entertains a relationship between thought and memory (*différance*, put simply), encompassing within language both transcendence and immanence, would accomplish Letiche's purpose of representing a kind of theoretical basis for a phenomenological view of complexity. Following Lawlor's construction of Derrida's theoretical construct of *différance* (for instance), and the trace coming in and out of being as thought - expressed in terms of a shift between immanence and transcendence (almost but not quite a dialectic of the moment of perception) - would enable one to define

17. See Lawlor (1992: 128) in which he describes the difference between the two theorists as follows:

"The formulae for the Derridean idea is:

Origin + mediation + end + the "+" = idea

...for Derrida, the Idea's 'strange presence' arises in Husserl from the space between the idea's form and content, between totality and infinity...

Whereas, in the case of Ricoeur, the formulae is:

Origin + mediation + end = Idea"

emergence in terms of a co-relationship of these features. On that account each moment of each psyche would be emergent. Emergence would not be a remarkable or unusual idea - or a marvellous transformation of a collective (as in the case of most corporatist views) for instance - but the universal presence of perception and thought in humanity.

A problem exists in that Letiche does not specifically argue the case I have made for a Derridean phenomenology. He does not fully elaborate a connection between Derrida's thought and his own theory of organizational phenomenology as represented in PCT. Rather, he infers the constitutive nature of difference as an organizational phenomenon, through talk of 'weak signals' and the like (Letiche, 2000b). In short, he assumes the validity of Derrida's conception as a feature, which can fulfil the role of a phenomenological account of human ontology. In that respect, he borrows from Derrida's own universalizing tendency in relation to his claims for the concept of *différance*.

A further problem also relates to the definition of emergence. This concept has a more technical meaning as related to complexity theory and the emergence of self-organizing effects that Letiche often uses - especially those represented in the ordering aspects of dissipative systems (e.g., Prigogine & Stengers, 1984). There is a second more obvious sense in which the term can be used and that is in its vernacular or common usage. A term such as 'emergence' tends to sit easily as a pre-theoretical term which may align with a conception like 'difference'; both are connected with origins and the process of becoming. That does not necessarily mean that because a link between Derrida's thought (in the loosest sense) and the term 'emergence' exists, that a theoretical link can be firmly established between the sciences of complexity theory and phenomenology.

Letiche then tends to build on this pre-theoretical affiliation between emergence and *différance*, but does so without fully articulating the theoretical link or scrutinizing the conception of *différance* (which I have tried to do). This means that the account is more interpretative than it ought to be. Clearly the goal is to underpin a (felt) notion of allegiance between complexity theory (generally conceived) and (currently) fashionable continental theorizations[18]. This tends to open Letiche to the charge of exploiting the credentials of vaguely poststructuralist notions, without a full sense of argument as to their extensibility relative to another domain (complexity).

However, before passing early judgement on Letiche's PCT it is important to examine other features of this theorization - firstly, the link he makes between PCT and Bergson, and second, the implications of correlating a 'phenomenological' theory with aspects of complexity theory science. These features will be dealt with in the following sections.

---

18. See for instance, the biting case of Sweetman (1999). See also, Habermas critique of Derrida, for instance discussed in Fleming (1996).

PCT: The Significance of Bergson

Letiche adopts Bergson's *durée*, and follows Bergson in terms of emphasizing the non-identicality of successive moments of perception[19]:

"*Experiential durée is not a succession of states resembling a collection of objects. One instant doesn't replace another - consciousness isn't a drawer wherein moments of time are put and taken out. In duration the one instant does not simply replace the other - as if 'time' was merely a succession of present moments*" (Letiche, 2000a: 522).

Whereas Derrida's *différance* can support a form of psychological theorization as to the nature of memory and its interaction with thought, this dimension is underplayed by Letiche. Letiche rather synthesizes mention of *différance* with *durée*. There are distinctions between the concepts - the death of the trace within Derrida's theory is not about the death of identity of the subject, but an important feature of the temporalization of the trace which Derrida evolves to intimate the transition between transcendental aspects of the trace and the current perception (at the present moment) of the same:

"*...this aspect of différance is crucial, since the form of sense (or something of me, some intuition or action) survives only in writing or in the trace of memory; it is really dead; it is merely a body with a subjectless transcendental field. Here we return to différance's first necessity: the trace must be made present, must be immanent, must be made mundane, must be made close once more, and must be made close once more and must be made the same as me; it must be made to live again. When this return happens, presence that is both objective and subjective is constituted...*" (Lawlor, 2002: 232).

Death here is the hinge of the micro-process of the qualitative process of iteration of the function of the mechanism of consciousness. Conversely, in Bergson's quotation (immediately above) the concern is the identity of being relative to the nature of the moment. If the moment is a uniform feature it is a quantifiable feature (and hence Bergson argues - such a conception is a feature of consciousness at a level of reflection which evades the lived experience of the moment).

The difference is that Bergson's thought tends to emphasize the relationship with the external or extrinsic aspect of time as a function of consciousness. Homogenous time overlays our perception of duration in Bergson's theories. Both Bergson and Derrida were concerned with the phenomena of consciousness, relative to reflexive internal epiphenomena, expressed in the terms of the influence of Levinas, as 'the other'. Derrida being keener on phrasing the issue in terms of the issues related to the origination of thought and its evanescence - followed by its recrudescence. In

19. Bergson (1912: 6) notes that: "A consciousness which could experience two identical moments would be a consciousness without memory. It would die and be born again continually."

contrast, Bergson takes a slice on this process and depicts it as a duality, or tension, as between the reality of the process (*durée*) and its veiling through our cultural accretions of consciousness (especially the tendency to seize time in terms of spatial forms).

Phenomenology and the Issue of Temporality

Bergson's views are concerned with the contrast between the interior and exterior of the psyche relative to the passage of time; this becomes a concern with identity relative to the perception of time. Identity is a discrete feature of a problematic within Letiche's theorization, as Letiche notes that if emergence is an ontic phenomena, (inside time) then, it is not possible to get a fix or 'eagle's eye' perspective on this activity (Letiche, 2000a). In this he follows the metaphysical presumption suggested by Derrida, that ontic phenomena do not permit law-like utterances (Derrida conceives in terms of the term '*Arche*'). However, it has been noted that *différance* fulfils many of the features of an axiological principle or *Arche*, albeit without such explicit claim. If we consider that the physical sciences are realized in an epistemic (or ontic) realm, the only possibility for their alternative definition (within the scope of the ontic/ontological distinction) is if they are described within the frame of an ontological vantage point.

It is fair to make the point that, this idea of a contrast between ontic and ontological realms is one which Heidegger inherits from Kantian discourse, and which Letiche works with also. Derrida's views are tending towards a form of disarrangement of the conceptual underpinnings which constitute Western metaphysics, including these types of distinction which are (despite their axiological primacy in Western thought) nevertheless a form of categorical opposition giving rise to an 'indecidable'; perhaps the ontic-ontological distinction is the root 'indecidable'. This aspect therefore within Derrida, might be treated differently to that outlined in Letiche's adoption of his ideas.

However, to work within the structure of PCT which encompasses both views of Bergson, and Derrida - one can note that Letiche builds on the binary structure of the ontic-ontological divide to make his case. Letiche links emergence (the state - of a scientific extrinsic nature in complexity) to the inner psyche; hence the view that the self can attain a creative realization of emergence in terms of *durée.* Consciousness constitutes an outcome of emergence, which forms a "pre-structure" of the outlay of consciousness (Letiche, 2000a: 550). Emergence, on this account, straddles both ontic and ontological dimensions of the framing of identity and experience:

"The result of self-organizing emergence is perceived in itself and is thus, absolute. Perception is direct, perspectivism or symbolic process/es of representation do not intervene between the observer and the observed. Bergson calls such a cognition 'intuition'. In it consciousness abandons the usual reflexive attitude, wherein the givens of consciousness are treated as if they were objects in space. Consciousness as durée, or pure emergence 'retains what is just

no longer there, and anticipates what isn't just present..." (Letiche, 2000a: 553).

In this context, Letiche's PCT argues that Bergsonian intuition or *durée* becomes synonymous with the pre-ontic state of emergence. It is suggested that emergence becomes synonymous with Bergson's process metaphysics at the level of individual identity. That is, a description of a psychological ego, in which the inference of the ontic states of complexity science are analogously suggested as transposed. Derridean theory (expressed by '*différance*') is analogously suggested as resembling this operation, in that the process of indeterminacy (which *différance* implies) is a feature which avoids direct identification with either absence or presence, and therefore equates well with the feature of *coming into being*, inferred by the position of Bergson. Thereby, on this reading of PCT, *différance* is argued to hover on the ontico-ontological divide at the level of articulation, expressing a phenomenological state discerned by Bergsonian *durée*.

However, one might take critical issue with this amalgam of the separate theories of Bergson and Derrida. To Bergson, the distinction between duration and extensity is the basis for his view that duration is the form of intuition which acts as a focal point for present experience (Evans, 2003). Bergson believes that these two aspects (of perception) are two sides of an absolute. *Durée* in this sense is akin to the primordial perception which Heidegger's process of destruction is set to realize. In my view, it is still a metaphysical conception; albeit conceptualized as the end-point or resolution of metaphysics (as conceived by Derrida in terms of Heidegger's onto-theology). From this perspective, *durée* is a creature of presence - albeit pure presence - revealingly we note Letiche's mention of 'pure' emergence as opposed to simply 'emergence' (above).

Hence, unifying *durée* and *différance* on the issue of conception within the processes which approximate to perception, shifts *différance* towards a consolidated point of psychological life, as exhibited in the experiential crystallization of the moment which is *durée*. *Différance* points in Derrida's formulation towards a moment of essence which is older than Being; a process of scission around a point - not focused towards a point (*durée*). So, although both theories are orientated towards the transient nature of experiential life, *durée* does not so much emphasize the passing of the 'trace' (its ephemerality), but rather its uniqueness within time. On this account *durée* might lend itself to a theorization with a bias towards a clear distinction between the ontic and ontological; but if *durée* is akin to *différance* - as suggested in PCT - then it ought to stand on the cusp of these categorizations. Rather, *durée* represents the ontological defined and lozenged clearly within the ontic (which *différance* proper - as envisaged by Derrida - resists). *Durée*'s point of intensity - its coming into being (albeit continual) - rests in the ongoing present, and the emphasis is not on the process of declension of the seizure of the moment (Derrida). This obliteration of the distinctions of these theorists therefore creates a problem for the synthetic construction of PCT.

Comparative Implications of 'Emergence' in PCT and Complexity Sciences

In summary, we have seen that the disparate views melded in PCT are only loosely a particular philosophical position; rather different views are interpreted into commonality to create an anchor conception for a novel theory of complexity. This is a theory of complexity which emphasizes emergence *per se*, and not strictly the other sciences of complexity. Hence rather than a theory of emergence derived from complexity sciences in any strict sense, we have a translated, or generalized, conception of 'emergence' as a mark of a quasi-phenomenological account.

What do I mean by 'translated' in this context? I think that the term 'emergence' has become a kind of arch-metaphor in complexity theory accounts. It partly owes its success to its being able to hook towards a complex or a simple meaning. The concepts of complexity can be used in strictly scientific senses, implying different axiological prerequisites. When unified with collateral concepts such as *durée* or *différance*, the concept of emergence gains new credentials from the arena of continental philosophy. It might be argued that in this context, the PCT theory was a kind of basis for suggesting the viability of the analogy between complex sciences in the physical and human realms (see Figure 1). By implication PCT is a knowledge claim for the extensibility of the complex sciences outside the epistemic area of the sciences.

However, it would help us to understand these inter-related categories if we could see more clearly the features of emergence as they relate to the physical sciences which categorize complexity sciences[20]. The link between non-equilibrium thermodynamics and nonlinear dynamic systems theory ties chaos and self-organization theory. Dissipative structures in far-from-equilibrium states are subject to nonlinearity (Prigogine & Stengers, 1984). Nonlinear equations may be used to model these states - hence such equations may be used to chart the emergence of complex (order), within a thermodynamic environment that is subject to bifurcations, (the 'phase transition'). Within the science of chemistry this phenomena is exemplified by the Belusov-Zhabotinsky reaction. However, drawing this model as an exemplar of emergence is to read the relationship of the physical sciences in a particular way (and with the accent on self-organizing theory). However, links with the phenomena of nonlinearity should not be underestimated as a feature of the aspect of the drawing out of a complex order in the transformation[21]. The inclination is towards an extension of

20. Mathews, *et al.* (1999) defines a range of theories which comprise the complexity sciences, of which the most important for the emergence phenomena are the areas of nonlinear dynamic systems theory; non-equilibrium thermodynamics, dissipative structures, and chaos theory.

21. Formerly emphasis tended to be on chaos and chaotic effects including attractors, for instance, Butz (1997) stating that "the three major distinctions in chaos theory are between dynamical systems theory, self-organization theory and fractal geometry" (Butz, 1997: 8). The modeling of nonlinear equations within 'phase-space' is a feature which depicts the form of a particular attractor (Capra, 1996:

the self-organizing principle as a feature of universal tendency, towards the extension of epistemology from physical to social science[22]. The project of the Santa Fe Institute underpins this endeavor. However, the attempt to extend the (primarily) epistemic foundations of complexity towards the ontological, meets the issues of philosophical complexity attached to recent controversies as implicated in the 'linguistic' turn (especially Derrida's reactions to Husserl and Heidegger). Also, the seemingly reductionist aspects of some of the sciences (Letiche notes Holland in these terms, for instance) may lead to the notion of an epistemic strait-jacket approach to human sciences (and hence derivative social sciences).

Hence, the fear of an imperialism of complexity, which 'dumbs down' not only the richness of human experience, but also the advances of a philosophical nature which incline to challenge the hegemony of particular philosophical axiologies (such as Derrida's view of western metaphysics). Of course, alternative theorizations attempting to utilize these approaches do exist, and some element will be introduced towards the close of this paper (c.f., Cilliers, 2005).

On this account PCT becomes rather an idiosyncrasy; but what exactly would a post-structural theory of complexity be like given the definition of the same? PCT clearly does act as a kind of indirect validation of complexity sciences relative to their applicability towards the social sphere. Is then the match of his emergence and the emergence of the complexity sciences *per se* an aporia - or form of enigmatic 'blind-spot' - at the heart of Letiche's theorization? I think that whatever else, PCT tends to be self-validating on the axis of the (assumed) analogical veracity of the epistemology of the complexity theory of the physical sciences to the human sciences.

If one considers Derrida's theory on this matter one can see that he dwells on the relations of thought and memory in terms of the continued inter-relationship of these terms, relative to the coming into being of the trace; its 'subjective presence' and its tendency towards infinitude (as a principle of its nature, incipient yet iterative, within the human psyche). At the micro-level of perception, this activity alludes towards the cusp of the ontic-ontological divide. These processes, and their allusion to the synthe-

---

129). An unstable or strange attractor constituting aperiodic oscillations. Kellert (1993: 4) defines chaos theory as a "specialized application of dynamical systems theory." Specific transformations of end states became the province of an area of theorization known as catastrophe theorization (Thom, circa 1970s). Whereas, a range of transitions occurs in self-organizing open systems, it is also possible to conceptualise the theorization of transitional states in terms of catastrophe theory as an alternate description of end state departures or bifurcations.

22. Overlap in areas like chaos and self-organizing effects are amongst the more notable. However, the overlap can also be a cause of ambiguity, and is present in other areas of the complex sciences. For instance, the notion of 'edge of chaos' at which transitions towards order occur builds on the aspect of the presence of non-linearity within human systems (Packard & Langton, quoted in Waldrop, 1992). This correlates with the work of Holland (1998) related to the presence of the complex adaptive system (CAS), which in turn can be related to the contra-Darwinism of Kauffman's self-organizing systems.

sis of diverse theorists by Letiche, are represented as showing emergence at an impasse - whether ontic or ontological is not resolved. Letiche rather problematizes emergence as Derrida problematized the ontic-ontological 'question' posed by Heidegger. However, Letiche does not seem to realize that he has transposed the enigma of ontology and being[23] to that of emergence.

The Promise of the Project of Phenomenology and Complexity

A further question remains to consider: whereas we have spoken of the nature of Derrida's *différance*, and alluded to the differences between the conception and its use in PCT, might *différance* nevertheless prove the ideal project for a phenomenological view of emergence, and hence complexity. In other words, is PCT pointing in the right direction in spirit if not to the letter?

Fink (a theorist of Husserl, influencing Derrida) describes the "phenomenological reduction" as "radically unmotivated." It is not about "the human possibility of knowledge" (expressed as an epistemological feature); that phenomenology may be distinguished from speculative philosophy's concern with origins. Rather (for Fink) it is about "(a) transcendental subjectivity which originally achieved the possession of the world" (Lawlor, 2002: 21)[24]. In this type of view "the world does not present itself as an object over and against a human subject, because the human subject finds itself within the world" (Lawlor, 2002: 15). This aspect of immersion, or existential dimension, is a feature which forms a part of Derrida's case, it could be argued that Derrida's position is to inter-relate both transcendental and psychological egos (Schroeder, 2005: 280-287). In this context identity suffers from a radical impermanence, due to the continuous process of dispersion he terms *différance* (Schroeder, 2005). That instability is a feature which may be described in both temporal and spatial terms. Derrida - with *différance*, is hunting for a relationship between the transcendental and psychological egos at each passing moment. Hence, Derrida's concern with the process, which the underlying network of associations that sustain the process of difference(s) are just as important to the constitution of experience as the subset of features that derive this 'present' moment.

Next we turn to complexity theory and consider the aspects of identity and the transcendental as opposed to psychological egos. If one con-

23 The project of Martin Heidegger.

24. Lawlor (2002: 21) discussing Fink in relation to Derrida. Fink's position is as follows: "Separating and distinguishing [the transcendental ego from the psychological ego] is as false as their direct equation" (PP 155/144). The relation, as Fink says, consists in a "unique identity" (PP 121/115), a "singular identity in difference... [a sameness in being-other" (PP 155/144). The ego's identity consists in a "tension" (PP 123/116); the ego's in other words overlap and coincide. No matter what, for Fink, the relation must be posited as "transcendental," and this means that man's worldly finitude is conceived as a constituted sense and thereby "taken back into the infinite essence of spirit" (PP 1555/145).

siders a particular case of the phenomena of complexity-based emergence, namely, the Bérnard equation, this self-organizing phenomena may also be expressed as a form of attractor (the Lorenz attractor) (Mainzer, 1994: 58-59). The Lorenz attractor contains fractal forms. The 'phase-space' within which such attractors exist is an abstract mathematical space for visualizing time-space data[25]. Nonlinear differential equations express the self-organizing phenomena, which give rise to complex forms of emergent phenomena. Hence, the notion of 'phase-space' permits a mode of understanding, or depicting, of self-organizing phenomena (in a certain respect, it might be termed a mode of 'bracketing' of conceptual phenomena). Returning to the preceding argument related to the nature of Derrida's ideas, with the phenomenology of the kind which Derrida and Fink follow, the sense is to get beyond that process of 'bracketing' (a feature of Husserl's concept of intentionality) and to enter a existential sense in which we are not knowing beings or observers.

For instance I might argue that the point in phase-space was a continued *becoming* - an analogical case of the shift between the transcendent and the immanent (the Derridean theory of *différance*). The common feature is a temporal context but that is extrinsic to both the point in phase-space and the human 'subjective consciousness'. Unlike the human entity, the point in phase-space lacks no element of absorption at the subjective level in the life-world. The phase-space point charts the coming into being of the self-organizing phenomena; in sum it may be said to characterize its form - like the map of a maze charts the distinct paths around the same. In this sense, the phase-space denotes the identity of the self-organizing phenomena, but it does not connote the validity of an analogy with an individual mind (for example) simply by virtue of our interpretation of its shifting from immanence to transcendence with each passing moment.

Can we thereby argue that phase-space (an instance of an epistemic form of complexity theory) comes a step too late for the kind of theorization that Derrida suggests? On this account, a claim for truth within complexity as a science may link less with the existential than the 'transcendent' account of the ego. So, if we look again, and view our conception of the phase-space movement, that construction is capable of resolution in terms of the models of *différance* (or any other phenomenological construction). The issue then becomes, is Derrida's *différance* (for instance) a true insight which at the level of social construction (of the actor) permits a validation of the epistemic insights of a complexity theory model (such as phase-space)? But, here, at least with Derrida, we enter the problematic issue of Derrida's argument being a case which suggests itself as inclining towards the margins of a rationality that can strictly accord a 'yes' or 'no' to a claim for truth status.

On this issue, Cilliers has noted that our knowledge of complexity theory means that "we cannot have perfect knowledge of complex systems" (Cilliers, 2005; see also Cilliers, 1998). But this aspect of the incomplete-

25. It is not, as Goldstein (1997) notes in his contradiction of Kellert, a geometry which supersedes causal explanation.

ness of our categorization of the features of a complex system - including predictabilities - is quite distinct from our knowledge of complexity theory as an issue of principle being indistinct. We can therefore submit complexity theory to an evaluation in terms of its reconciliation relative to 'truth' at an abstract level, one of principle, if not expecting to find complete knowledge at an empirical level. This is somewhat akin, at a basic level of analysis, to one's expectancy that one cannot know the long term parameters of a weather system operating in terms of chaos theory.

Another aspect of argument is that somehow complexity theories have left behind the Newtonian scientific order, and that the sciences are not subject to the same theoretical basis as in prior cases. However, the issue of a potential shift in the theoretic correlates linked to epistemic structures of knowledge traditionally associated with sciences (like the Kantian with the Newtonian) at a philosophical level, does not evade the accountability of complexity theories at a fundamental level of philosophical discourse. It is at this feature which an orientation towards Derrida's *différance* - of complexity - might press against. If the insights of complexity cannot be resolved in contrasting argument relative to fundamental metaphysical structures, features of debate such as the 'performative contradiction' in post-structural theories also arise within the analogous deployment of complexity. Somewhere one gets the sense in all this - as with PCT - that complexity theory *per se* (the science) is the Cinderella at her own ball.

Even if we argue that Derrida operates within the contours of Western metaphysics (or its 'rationality')[26], we still have the problematic issue of whether the ontic-ontological divide does not carry a burden of philosophical irresolution as to the nature of certainty relative to the epistemic aspect of the sciences. Does the adoption of a mix of continental philosophies allow us to do away with the distinction of the epistemic and the ontological? Along the lines perhaps that the ontic (real) and ontological are blurred, or contradictories, that can be set aside and hence so the distinction of ontological and epistemic? I think that maybe this is partly behind PCT, but not stated explicitly. Clearly the physical sciences by traditional standards are perspectival. The frame of interpretation of a complexity theory model, which is suggested as the individual phenomenological case, is distinct to the abstract model which arguably presents a kind of ideal insight. In this respect the critique of Derrida relative to Husserl would operate to argue a case to the effect that complexity sciences had a problematic seizure on 'truth' revealed in any singular sense, once disseminated as texts within 'discourse'[27]. Such models are clearly removed from an experiential phenomenological case in terms of their truth, whatever else. The problem then extrapolates from PCT's dwelling on the cogency of emergence as a feature of phenomenology, to that of the 'truth' claims of complexity

26. See Cilliers, (2005). Note his discussion of the issue of whether we can have perfect knowledge of complex systems, and the status of claims to knowledge relative to Derrida and complexity theories.

27. My reflections here relate to Husserl's *Origin of Geometry*, contained as an appendix in Derrida (1989).

theory relative to ontology, once placed under the lens of philosophy. However, this can be skirted, if we evade the notion of relevant 'truth-claims'. This is partial in Letiche as he does adopt the ontic/ontological divide as a stated problematic. This said, Derrida's resolution is partly to incorporate difference as indecidable and so to situate the onto-ontological divide as a problematic which also obliquely represents an entailment of the influences of Western metaphysics.

Conclusion

In this paper I have critically assessed the admixture of theories which is represented by PCT, and shown some problems with this theorization but also the sense of its promise relative to the field. It is, I suggest, a worthy area of theorization, and Letiche has developed a novel hybrid of ideas. However, I do not think he has done this thoroughly or with full consideration of the implications. I think that these include the difficulty of phrasing an epistemic theory of complexity in terms of theories of philosophy which sit uneasily with the notion of veracity or 'truth', especially Derrida's *différance*. Theorizations of this nature are ineluctably ontologizing the sciences from their own perspectives, and this colonization thereby passes the enclave of complexity sciences and extends to the nature of 'truth claims' of the sciences in more generalizable terms.

A radical phenomenology of human organizations as represented by Letiche's PCT[28], or indeed any such radical phenomenology drawing on theorists such as Derrida/Heidegger - meets certain problems. One is the tendency of phenomenology to concern itself with the nature of the hidden life world of the individual and to erect sophisticated theorization. Second is the issue of the social (and individual) construction of the world and its allegedly subjectivist nature, as opposed to the prior assumptions of objective knowledge being a prerequisite of scientific insight.

More generally, if a phenomenological or related approach were construed to be valid as a feature of human life it would hardly be a science of human life, on our accustomed view. Its nature is to get beyond the epistemologies which equip us and look for a foundational pre-epistemic base. This, I think, is distinct from the complexity science aspiration to seek a global epistemic basis of knowledge which infers the extension of a demonstrable scientific principle as the regulating feature of human life.

Indeed, a danger - though some might say a promise - is present in theorization such as PCT, in that it is tending to leave both phenomenology and complexity theory derived from the physical sciences behind in an aspiration to synthesize these disparate features. A unifying discourse occurs in PCT without the full evolution of a stringent argument as to the nature of the disparate theorists, and this is perhaps a further danger aslant this

28. The issue of definitions - of concepts like 'phenomenology' is stretched in the case of the inclusion of Derrida Schroeder (2005), Derrida's form of post-structuralism as a "philosophy of dispersion," and classes it as in opposition to phenomenology, as also it seems to be in opposition to a certain extent to areas of thought like hermeneutics.

enterprise, which is that in seeking to break the divisions of sciences and philosophical conceptions, insights are drawn which rely on presuppositions of the applicability of the cases of novel theorists, as between areas of analogical transfer.

In the next paper I examine further these ideas, and also look more fully at the role of metaphor in analogical transfer. This will bring into consideration an evaluation of hermeneutics as opposed to Derrida's approaches in relation to the construction of complexity theory issues in the social sciences.

References

Bergson, H. (1912). *The creative mind: An introduction to metaphysics*, New York, NY: Putnam, ISBN 080652363 (2002).

Boje, D. M. (2000). "Phenomenal complexity theory and change at Disney: Response to Letiche," *Journal of Organizational Change Management,* ISSN 0953-4814, 13(6): 558-566.

Boje, D. M. (2001). "Flight of antenarrative in phenomenal complexity theory: Tamara - storytelling organization theory," conference paper at the *25th Conference of Complexity and Consciousness*, Huize Molenaar.

Butz, M. R. (1997). *Chaos and complexity: Implications for psychological theory and practice*, Washington, DC:Taylor and Francis Inc, ISBN 156032418X.

Cairns, D. (1976). *Conversations with Husserl and Fink, (Phaenomenologica S)*, Heidelberg, Germany: Springer, ISBN 9024717930.

Capra, F. (1996). *The web of life: A new scientific understanding of living systems*, London, UK: Harper Collins, ISBN 0385476760 (1997).

Cilliers, P. (1998). *Complexity and postmodernism: Understanding complex systems*, New York, NY:Routledge, ISBN 0415152879.

Cilliers, P. (2005). "Complexity, deconstruction, and relativism," *Theory, Culture and Society*, ISSN 0263-2764, 22(5): 255-267.

Derrida, J. (1982). "White mythology," in J. Derrida, *Margins of philosophy*, A. Bass (trans.), Chicago, IL: University of Chicago Press, ISBN 0226143260 (1984).

Derrida, J. (1989). "Introduction" in E. Husserl, *The origins of geometry*, J. Derrida (trans.), Lincoln, NE: University of Nebraska Press, ISBN 0803265808.

Evans, F. (2003). "The beginnings of post-modernism: Phenomenology and Bergsonism, Derrida and Deleuze," in Lawlor, L. (ed.), *Thinking through French philosophy: The being of the question*, Bloomington, IN: Indiana University Press, ISBN 0253215919.

Fleming, M. (1996). "Working on the philosophical discourse of modernity: Habermas, Foucault and Derrida," *Philosophy Today,* ISSN 0031-8256, 40(1): 169-178.

Goldstein, J. (1997). "Causality and emergence in chaos and complexity theories," in W. Sulis and A. Combs (eds.), *Nonlinear dynamics in human behavior: Studies of nonlinear phenomena in life sciences*, Hackensack, NJ: World Scientific Publishing, ISBN 9810227426.

Holland, J. H. (1998). *Emergence: From chaos to order*, Oxford, UK: Oxford University Press, ISBN 0192862111 (2000).

Husserl, E. (1913). LI VI appendix, II, o.343, Hua XIX/2 765, extract in D. Moran and T. Mooney (eds.) (2002), *The phenomenology reader*, New York, NY: Routledge, ISBN 0 415 22422 5.

Juarrero, A. (1999). *Dynamics in action: Intentional behaviour as a complex system*, Cambridge, MA: MIT Press, ISBN 0262100819.

Kellert, S. (1993). *In the wake of chaos*, Chicago, IL: Chicago University Press, ISBN 0226429741.

Kelly, S, and Allison, M, A. (1998). *The complexity advantage: How the science of complexity can help your business achieve peak performance*, San Francisco, CA: McGraw-Hill, ISBN 0070014000 (1999).

Lawlor, L. (2002). *Derrida and Husserl: The basic problem of phenomenology*, Bloomington, IN: Indiana University Press, ISBN 0253215080.

Lawlor, L. (1992). *Imagination and chance: The difference between the thought of Ricoeur and Derrida*, Albany, NY: State University of New York Press, ISBN 0791412180.

Letiche, H. and Boje, D. (2001). "Phenomenal complexity theory and the politics of organization," *Emergence*, ISSN 1521-3250, 3(4): 5-32.

Letiche, H. (2000a). "Phenomenal complexity theory as informed by Bergson," *Journal of Organizational Change Management*, ISSN 0953-4814, 13(6): 545-57.

Letiche, H. (2000b). "Self-organization, action theory and entrainment," *Emergence*, ISSN 1521-3250, 2(2): 24-37.

Mainzer, K. (1994). *Thinking in complexity: The computational dynamics of matter, mind, and mankind*, Berlin, Germany: Springer-Verlag, ISBN 3540002391 (2003).

Mathews, K. M, White, M. C, and Long, R. G. (1999). "Why study the complexity sciences in the social sciences?" *Human Relations*, ISSN 0018-7267, 52(4): 439-461.

Merleau-Ponty, M. (1962). *The phenomenology of perception*, London, UK: Routledge, ISBN 0415278414 (2002).

Moran, D. and Mooney, T. (2002). The phenomenology reader, London, UK: Routledge, ISBN 0415224225.

Prigogine, I. and Stengers, I. (1984). *Order out of chaos*, London, UK: Harper Collins, ISBN 0394542045.

Schroeder, W. R. (2005). *Continental philosophy: A critical approach*, Oxford, UK: Blackwell, ISBN 1557868816.

Sheard, S. (2006). "Complexity theory and continental philosophy 2: A hermeneutical theory of complexity," accepted for publication in *Emergence: Complexity & Organization*, ISSN 1521-3250.

Stellardi, G. (2000). *Heidegger and Derrida: Imperfect thought*, New York, NY: Humanity books, ISBN 1573928240.

Smith, J. K. (2005). *Jacques Derrida: Live theory*, London, NY: Continuum, ISBN 0826462812.

Sweetman, B. (1999). "Postmodernism, Derrida, and *différance*," *The International Philosophical Quarterly*, ISSN 0019 0365, 39(1): 5-18.

Waldorp, M. M. (1992). *Complexity: The emerging science at the edge of chaos*, New York, NY: Simon and Schuster, ISBN 0671872346.

Wheatley, M. (1994). *Leadership and the new science*, San Francisco, CA: Berret-Koehler, ISBN 1881052443.

13. Complexity Theory and Continental Philosophy Part 2: A Hermeneutical Theory of Complexity

Stephen Sheard
Freelance Research/Writer, UK

Introduction

In the previous article (Sheard, 2005), I evaluated the synthetic theorization of Letiche – PCT (Phenomenological Complexity theory). This theory deploys a mixture of theorists, including both Derrida and Bergson, to argue the value of a phenomenological perspective on complexity. It was felt, on analysis, that phenomenology (so defined) undercut the validity of an epistemological perspective which emphasized the complexity sciences as a feature of activity within time (ontic), and that the emphasis on the ontologization (of complexity sciences) present in PCT, shifted the view of complexity towards an other than scientific basis. It was suggested that the complexity sciences, whatever their novelty, are still a feature of epistemology and created out of the desire for objective knowledge about the world. The problems are brought into sharp relief when the theories of complexity are pre-supposed as current and valid in a social setting (especially the firm) and the leap is then made backwards as it were to the individual present within the firm (and his presumed phenomenological composition or being).

This problem relates in part to the aspect of naturalism which underpins the deployment of complexity relative to its coextension to both individual psyche and the company. It is assumed that complexity endorses a global epistemic basis for a naturalistic case, which extends complexity towards areas of interpretative significance of the psyche (phenomenology), as expressed in philosophical terms. In this respect, the ambit of science to permit a reordering of the social and psychological is extended to the role of the philosophical reinterpretation of the role of the psyche also[1]. However, a phenomenological account might be said to not so much to reduce the higher aspects of cognitive functioning in human beings towards a biological underpinning (reductionism), but, conversely, to infer that a sophisticated human theorization (phenomenology) is extensible to a discrete section of nature (human beings and their group developments). Phenomenology is marked by a desire to seek a pre-epistemic basis of understanding to hu-

1. Khalil notes that "the realm of the social should not be juxtaposed to the natural. The social is simply a variety of the natural. Citing Danto he also notes that "naturalism is polemically defined as repudiating the view that there exists or could exist any entities or events which lie, in principle, beyond the scope of scientific explanation (Danto, 1967: 448)..." in Khalil, (1997: 47).

man life, rather than – I suggest – to extend a series of scientific principles towards human life (complexity theories).

This leads to a problem, in that by pairing complexity and phenomenology, the axiological inferences of authors that Letiche represents as 'phenomenological', are philosophies that have not expressed theories of a universalistic nature – they have rather confined themselves to human-kind, and not to human-kind as a continuum with nature (naturalism). It would be problematic for instance to acquire the actual instances of complexity in biological cases (such as anthill communities) as examples of phenomenology. On this kind of basis, one might at first hand, see theories like that of Bergson as more akin to vitalism (anti-naturalist perspectives)[2]. This leads to a further issue, which is the extent to which complexity theories are necessarily generalizable. It is possible to argue that the complexity sciences can be distinguished in terms of invention from mobilization; that complex phenomena are in this sense local or specific occurrences which do not thereby infer a universalistic case. Hence, the notion that "their regularities are not direct evidence of the workings of fundamental laws but emerge from unexpectedly local interaction, tuning, and balancing"[3].

A non-generalizable theory of complexity, has a problem relative to a phenomenological basis of criticism, in that the direct scientific exemplars cannot be easily mobilized to render a literal equivalence as between the local interaction and the global (at a biological or chemical level for instance). That mobilization becomes a matter of analogical transfer. What is more, such a contingent theory of complexity weakens the inference of a global epistemic unity of a holistic nature, as between distinct levels of nature. In short, analogy in such a perspective as that described by Stengers (2000), is debarred from acting as a kind of 'Trojan horse' by which complexity science advances its claims towards acting as a foundational theory amongst the sciences.

One way of trying to overcome this difficulty – the problem of the universalistic extension of complexity and the dangers of reductionism attending that – is to argue for a feature of special contingency of complexity at a phenomenological level. This leads towards the issue of how to represent that theorization. As analogy is a feature of the suggestion of global epistemic transfer – and analogy takes of the nature of an attempt to extend a principle (complexity) across multiple levels of nature – it can be described as a form of *unificational metaphor* (Khalil, 1998). It is said that the "unifi-

2. Khalil notes that "such theorists (those that assert intentionality), have erected a dichotomy between natural and human phenomena on the supposed grounds that a mechanistic view of behavior and a reductionist view of organization should be restricted to non-human organisms. Similar to vitalism, i.e.) the postulate of the discontinuity between animate and inanimate matter, such anti-naturalist rese*arche*rs have appealed to some extra-natural variables, such as free will and the role of interpretation, in order to establish the uniqueness of human behaviour and organization..." (Khalil, 1997: 48).
3. See Mackenzie, (2005), discussing Stengers (2000).

cational metaphor expresses similarities when they arise by the same law"[4]. Is it possible to deploy metaphoric theories to interpret that process of transference in such a way that the feature of an extensibility of complexity can occur (in some law like fashion) as between distinct levels of activity? Such an open acknowledgement of the role of metaphor would be distinct from the covert expansiveness of an epistemology of complexity towards areas of ontology, which (I have argued, for instance in the case of PCT) embeds tensions of varying kinds – including its failure to acknowledge the issue of the role of metaphor, or its mobility in extending complexity to a global theoretical level. I extend discussion of this issue, at the close of this paper, having first looked at the possibilities for uniting various approaches to metaphor (principally Ricoeur's, and Lakoff and Johnson's 'take' on cognitive science) on the axis of an alternative view of phenomenology to that represented in the previous discussions of PCT[5].

Ricoeur and the Metaphors of Complexity

Ricoeur argues for a tensive theory of metaphor in which he stresses not purely the novelty of the literary metaphor, but its capacity to represent a mode of aesthetic originality which may disclose a fresh, or advancing, epistemic order. He argues that a relational tension exists within a metaphor, in which a "new semantic congruence or pertinence [arises] from the ruins of the literal sense [of the language in question] shattered by semantic incompatibility or absurdity" (Ricoeur, 1979: 151). Ricoeur is following the Kantian conception that metaphor enables a fresh lens to be formed on the world; metaphor is an aspect of our categories of perception as defined by Kantian theories but its import is defined as ontological as well as epistemological[6]. Within metaphor itself a relational tension exists between tenor and vehicle; predicative assimilation involves a tension - the previous incompatibility is identified through the fresh compatibility.

If one extends Ricoeur's theory of metaphor towards metaphors such as 'emergence', one can conceive of the 'relational tension' – which Ricoeur believes is a feature of metaphor – as being a feature which is productive of particular metaphors, such as 'emergence', 'the edge of chaos', etc. These metaphors are then novel variances in prevailing modes of conceiving. This is to suggest that the metaphoric process is emblematic of process-

4. Khalil notes that "For instance, Newton's law of gravitation established a unity between celestial motion and terrestrial acceleration of bodies as they fall downwards. Both disparate events are regulated by the same law of gravity." (Khalil, 1998: 6).

5. I summarize features of the discussion and make fresh points by way of comparison of Letiche's PCT and Ricoeur's perspectives in Appendix 1.

6. Ricoeur defines this in terms of a metaphorical meaning which is a fresh scale of perception. This is a "mode of schematizing a synthetic operation" relative to Kantian theories; in which case it is described as "both a thinking and a seeing. It is a thinking to the extent that it effects a re-structuration of semantic fields; it is transcategorical because it is categorical..." (Ricoeur, 1979: 146-148).

es of change at an epistemological level (in terms of the deployment or mobilization of complexity related concepts). So, for example, the conception 'emergence' – as it applies at distinct levels of application – acquires fresh credence in that it sits tightly between the twin zones of analogical target and source. Hence each distinct level of operation (of the extensibility of the language of complexity) involves a shift from the literal level towards a metaphorical level, which engages a kind of 'semantic shock' productive of the need for interpretation. In phenomenological terms, metaphor invokes a hermeneutic reinterpretation of the value of the words engaged[7], whose dynamism can be described in terms of a tensional situation of split reference (the levels of metaphoric transfer).

This theorization, apart from its debt to Kant, borrows from the ideas of Heidegger, but its emphasis is less on the state of Being, and abstract enquiry relative to that (and rather more specifically on the manifestations of Being in the semantic process). Ricoeur, in this sense, goes in a very different direction than Derrida, although working with a similar source conception. Instead of emphasizing the interminability of the trace – its undecidability and refusal to locate the identity of Being within presence – Ricoeur sees metaphoric vitality as a duality in which Being manifests in the process of schematic reinterpretation of the host or focus of metaphoric extension. Heidegger had essentially reinterpreted the Kantian conception of a timeless being in terms of "being as time... [so as to] ...disclose it as poesis – as an event of creative imagination..." (Kearney, 1991: 52). This move made possible a reconstruction of the ontological in terms of the creative acts of the imagination (for instance)[8]. On this account the creative deployment of the image is a form of reverie, which links the creative imagination with the process of becoming[9]. This is significant, in that Ricoeur's move, if legitimate, might be said to relate a concern with metaphor as alluding to a mode of Being, that acts of analogical transfer are not just acts of consciousness but that they may have special significance (Bruns, 1993: 36).

Considering these issues relative to complexity theory, it is appropriate to ask a range of questions. First, if these theories are correct and extensible to metaphor in general, then what is special about complexity metaphors in particular? The coining of metaphor may be a phenomenological

7. Hermeneutics may be defined as related to Heidegger's phenomenology and discloses a fore-structure of understanding related to ontological structures rather than being purely methodological means of textual interpretation: "It [the hermeneutic circle] is not to be reduced to the level of a vicious circle, or even of a circle which is merely tolerated. In the circle is hidden a positive possibility of the most primordial kind of knowing..." (Gadamer, 1989: 265-307), in Moran and Mooney (2002: 328).

8. Hence, Kearney notes that Ricoeur combines a "psychology of the imaginary" with the semantic theory of metaphor – this is a matter of stressing a creative function for the verbal imagination, which is complemented by a conjunction with "a sensible moment of metaphoric creation." (Kearney, 1991: 151).

9. Kearney notes that Ricoeur is also influenced by Bachelard, *(the Poetics of Reverie)* to whom the image is a trans-subjective construction – which can cut across individual subjective divides. (Kearney, 1991: 91-93).

'act' or a 'mode of Being' as Ricoeur might put it, but is that not to make a quite special occurrence commonplace? The significance of complexity as a unificational metaphor (Khalil, 1998) then comes into play by extending a series of analogical comparisons – one is arguing for the extensibility of complexity from one plane to another (universalistic theory of extensibility of complexity based on naturalism). This is however, quite a distinct operation to a local or contingent theory of complexity as a special occurrence (Stengers, 2000). However, it remains to stress the nature of the significance of the role of metaphor in Ricoeur more clearly. Noted as related to the process of the metaphoric copula and 'relational tension' implied in the metaphoric process:

"*...split-reference [Ricoeur's theory of metaphor], really then mirrors a double dialectic, a dialectic that is operative both at the level of what Jaspers called Existenz... [hence] ...it is the tensive-boundary character of man's being in the world that is in the precondition of metaphoric-symbolic utterances, and second, that it is because of this antecedent anthropological condition that metaphor both speaks and is transparent to human consciousness*" (Olson, 1981: 109).

Olson notes that "metaphors are just the linguistic surface of symbols, and they owe their power to relate the semantic surface to the pre-semantic surface in the depths of human experience to the two dimensional structure of the symbol..." (Olson, 1981). This leads us into a consideration of Ricoeur's views on symbolism.

Contrasting Themes: Symbolism or *Différance* and Complexity Theorization

Apart from his theory of metaphor, which invests metaphor with a potential special significance á la Heidegger's theory of Being (in a strict sense, thereby breaking the Kantian idea of separateness of noumena and phenomena), Ricoeur subdivides forms of symbolical metaphor; he terms them *symbolic*, *oneiric*, *poetic* and *social* imaginations. Kearney (1991) notes that this subdivision was the hallmark of his philosophical enterprise, in which he moved from a "phenomenology of the will" towards a "hermeneutics of the symbol"[10]. In this (later) structure of thought, Ricoeur interprets symbols as being *cosmic* (of cosmic origin); *oneiric* (psychic function of the imagination); or *poetic* (creative imagination)

The theory of metaphor outlined may be said to be a hallmark of Ricoeur's extrapolation of these ideas towards an examination of the role of

10. Kearney states that the significance of this is that Ricoeur is shifting from phenomenology which rests on a reflection of intentionality (following the phenomenological tradition of Husserl and especially, in Ricoeur's case, taking into account the further work of Heidegger), towards, a belief in the primacy of signs and symbols in the process of intuition. (Kearney, 1998: 150).

the imagination relative to linguistic categories. This work came in the 70s with *The Rule of Metaphor*. If we dwell, on our objection to the theory of Ricoeur, in the previous section – the aspect of the almost everydayness of metaphor (as Ricoeur infers it) – his theories of symbolism present a kind of answer to that. On this account, metaphor can attain symbolic prominence, in which case, it can have a significance which correlates (in certain cases) with a cosmic origin (this infers a link between human processes and a causal feature of activity). Might then, a metaphor of special significance (in that it was 'unificational', Khalil, 1998) – in the case of complexity - extending across distinct areas of organic, animal, human and social realms, be conceived as more likely to attain a symbolic stature? In this sense 'unificational' might imply that the generalizability of the complex states was a feature of holistic vitality which confirmed the value of analogical extension, in the first place. In this sense, metaphor itself acquires a kind of *élan vital* as a feature, not of invention, but of the discovery of a correlation between a cosmic (metaphysical) and physical order.

How might the key metaphor 'emergence' sit relative to this idea? Conceptions such as emergence, edge of chaos, CAS, etc. allude to an extension towards areas other than science. These metaphors represent portmanteau ways of grasping the experiments of science in certain areas of research. For instance, the nature of emergence as discussed by theorists of Santa Fe, in which a question is raised as to what constitutes 'truly emergent behavior' and, how is it recognized, is expressed as the key to the problem of 'artificial life'[11]. On this account, emergence is a kind of pre-construct, which anticipates a form of behavior which is recognized as following a given pattern. That pattern indicates reactivity or order, which might express learning (for example). That is the heart of the conception as originally devised. Emergence is a leitmotif in this context, not for human perception, but for that of a simulation which seems to mimic human perception. In essence that might be termed 'mimetic of forms of human conceptualization'. In certain cases, this alludes to the sense that complexity theories are phenomena fitted towards pre-existent categories at a descriptive level. That is a more local than globally epistemic theorization. However, it is a clear part of the extension of incipient meaning in emergence that such a term is extended more generally. Hence, in a certain sense, that metaphor (emergence), once chosen, becomes a seemingly inexorable candidate for future definitions of similar phenomena. But that need not mean that the conceptual framing of the observed phenomena (of complexity) is thereby

11. Waldrop notes as follows, (of Santa Fe theorists on the issue of 'artificial life'): "*Holland remained to be convinced; he had seen too many examples of 'emergent' behavior that was tacitly built into the program from the start... 'But for myself,' he says, 'I couldn't see how you could define 'truly' emergent behavior.' In some sense, everything that happens in the universe, including life itself, is already built into the rules that govern the behavior of quarks. So what is emergence, anyway? And how do you recognize it when you see it? 'That goes to the heart of the problem in artificial life,' he says.*" (Waldrop, 1992: 242).

pure discovery at future levels of application[12].

We could say that the conceptual category of emergence anticipated the phenomena, which would be discerned in a wider complexity theory lens. The approximation in reverse of phenomenology's conception of psychical processes such as emergence, (PCT for instance), and the projection of those towards complexity sciences, is a kind of mirroring of the process whereby tradition (as encapsulated in the more generic formation of the conception of 'emergence' as a feature of language) initially identified complexity theory phenomena in terms of metaphors such as emergence. One might argue that the aspect of naming is the foundation of the possibilities of a symbolism. However, the issue of what significance one ascribes to a particular conception – such as emergence – is a feature of cultural accretion. A sense exists, as Ricoeur notes, in which the issue of the aspect of the universality of a theory (for instance complexity) is traversed in philosophical terms, in that its epistemic extensibility (i.e., the issue of universality) is inferred by its (metaphoric) mobility[13]. Ricoeur's position is that symbols have an inherent quality which embodies their tendency to evolve in layers of complexification, towards more refined forms, and ultimately towards conscious expression in language; mythical narrative is the first great form of that articulation[14].

If one considers the complexity theory metaphors, it is fairly clear that they form a pattern which, in terms of metaphorical parlance, allude to the emergence of order from chaos; the phenomena under observation (complex life) are expressed as metaphors which are redolent of that initial archaic drama as expressed by Ricoeur. A further aspect of the redolence of archaic drama is the parlance of 'chaos'. 'Edge of chaos' for instance might be equated with the aspect of mythical drama, not of the beginning, but of

12. On the theme of discovery and invention, Johnson embellishes the point with clarity around his suggestion that science has its own 'creation myths': "*We are left to wonder to what extent abstractions like fitness landscapes and scales ranging from order to chaos are features of the universe and to what extent they are attractive inventions- more sophisticated than dividing the world into concentric tetrads of mesas and mountain's extending from the pueblo's spirit hold, but projections nevertheless*" (Johnson, 1995: 290).

13. Ricoeur notes: "*How can philosophical reflection be articulated upon the hermeneutics of symbols? I shall say a few words about the question itself. A meditation on symbols occurs at a certain moment of reflection; it answers to a certain situation of philosophy and perhaps even of modern culture. This recourse to the archaic, the nocturnal, and the oneiric, which is also an approach to the birthplace of language, represents an attempt to avoid the difficulties in the problem of a starting point in philosophy*" (Ricoeur, 1974: 287).

14. Ricoeur notes: "*These great narratives, which as was said, put into play space, time and characters woven into story form, have in fact an irreducible function. It is a threefold one. First they place the whole of mankind and its drama under the sign of an exemplary man, an Anthropos, or Adam, who symbolically stands for the concrete universal of human experience. Second, they give to this history an élan, a direction, an orientation, by unfolding it between beginning and end...finally, and more fundamentally, they explore the cleavage in human reality represented by the passage or leap from innocence to guilt*" (Ricoeur, 1974: 293).

the end of time whereas 'chaos' is redolent of that but in that aspect, chaos is allusive of a drama which was prior to our consciousness (or 'awakening') and the Adamic myth which is illustrative of that. Transliterated into the societal realm of the resolution of the imperfection of the social world as set against the market, the emergence sought is a kind of resolution of that imperfection of an exemplary nature[15]. This has an ethical dimension. It represents a kind of resolution of the original sin of Darwinism (which echoes the effect of self-organization theorization in terms of ameliorating the chance orientated stochastic nature of social Darwinism).

Emergence as a metaphor expresses the continuum of time; it is marked by a passage from one state to another within the time or process of *becoming*. An expression of the specific moment of transformation in the scientific realm is analogically re-specified in a societal realm. These features within the complexity sciences themselves are subject to particular variations of classical theorization – for instance that at particular points of non-equilibrium, new forms of order arise (Prigogine, discussed in Capra, 1996: 184)[16]. In this context, irreversible processes are interpreted as having a constructive role. However, these aspects of development are non-psychical at the scientific level; a system such as PCT (loosely) extends them to the human sensibility by virtue of the expression of scientific emergence as a leitmotif of exemplary restoration of self, or awakening in a psychical sense.

An alternative view of phenomenology – *différance* – is resilient against the investiture of the temporal, or the differentiated, with the aspect of Being; it is resilient even to the attempt by Heidegger to gain a foothold on the idea of a primordial state of Being. Derrida, in this sense, follows Nietzsche in that he expresses difference as a free play of forces. This said, he resists the attempt made by authors like Ricoeur, in the wake of Heidegger, to invest metaphor with any particular sacral connotations. His view of metaphor is that it leads to the metaphysical and that this process is one which does not ultimately reflect a transcendence – that is because *différance* is a mode of evading the causal transcendence which authors like Heidegger (to Derrida) persist in believing exists at some causative point (for instance in terms of Heidegger's idea that a pre-Platonic 'conception' of Being – before its dissemination into 'beings' – could 'recover' Western philosophy).

Therefore, the difference (or one difference) between Derrida and Ricoeur, is that Ricoeur believes that "phenomenology will understand a symbol by ritual and myth..." (Ricoeur, 1974: 297). Ricoeur therefore looks to the earliest symbols as they manifest a deeper truth in his eyes – but rather that symbols constitute an indestructible substance "which gives

---

15. Ricoeur notes that: "*This dynamic [that as Ricoeur puts it "myths have never stopped battling each other..."], is animated by a deep seated opposition: on one side are the myths that take the origin of evil back to a catastrophe or primordial conflict prior to man: on the other hand are the myths that take the origin back to man...*" (Ricoeur, 1974: 294).

16. These features relate to the phenomena of irreversibility and the arrow of time.

rise to thought... [therefore] ... I am convinced [Ricoeur] that we must think not *behind* the symbols, but starting from symbols..." In contrast, Derrida is more concerned with the transcendental nature of meaning – its evanescence – within the sensation of time.

How can we summarize these points relative to complexity theory and PCT? There is a sense, in which the theorists of Santa Fe (including Holland, 1998 and Kauffman, 1992) originated metaphors within a scientific area of studies. They perceived phenomena which they sought to place 'ready to hand' hooks of a metaphoric nature towards. For instance, when Kauffman identified a self-organizing principle to supplement Darwinism, the parlance (or nomenclature) of the 'edge of chaos' suggested itself. On this basis of interpretation, one might say that it was fallacious to invest features of scientific discovery with cosmic (or mythic) significance. Are not the metaphors of complexity decorative? This would suggest that they have status as metaphoric creations (being extrapolations or developments of an imaginative nature), which might seem to infer an invented or creative basis of development rather than aspect of scientific discovery.

Clearly, the issue is perhaps, how deep does metaphor go? Can we attach this kind of (interpretative) significance which Ricoeur does? Ricoeur's faith in the value of symbolism is at least proto-Christian, and this infers a kind of moral order underpinning the symbolism which is not necessarily philosophically valid. Are we not trapped in a kind of hermeneutic cultural validation of the significance of metaphor, on this account?

In the next section, I will look at the ideas of Lakoff and Johnson, which provide a further contrast with the deployment of Ricoeur, relative to the interpretation of metaphor.

Lakoff and Johnson: A Perspective on Cognitive Science and Metaphor

Lakoff and Johnson regard semantics "as the study of how linguistic expressions can fit the world directly, without the intervention of human understanding" (Lakoff & Johnson, 1980). This is an alternative to an 'objectivist' theory of metaphor, as pursued by a theorist such as Searle[17]. Their theories are also opposed to a purely subjectivist construction of metaphor (for example Davidson, quoted in Ricoeur, 1979) in that, whereas partly imaginative in function, metaphor is held to be also a feature of reason[18]. Lakoff and Johnson propose an 'experientialist synthesis', in which metaphor is conceived of as uniting reason and imagination. This leads to a conception of a kind of 'imaginative rationality'. It is argued that everyday reasoning involves metaphoric entailments and is therefore imaginative in nature and, conversely, that poetic imagination (novel use of metaphor) is

17. Searle views metaphor as a deviation from literal equivalence and as a fixed range of alternative meanings which can be potentially read into a given metaphor.
18. Davidson argues metaphor has no isolable decipherable meaning, he argues that: "Metaphors mean only what they appear to mean on the surface in their most literal sense..." (Haskell, 1987: 35). However, beyond the surface; subjectively speaking – metaphor has unbounded or unlimited meaning (contrast with Searle).

partially rational in nature. Does this help us? I think it alludes to the recognition of the prevalence of metaphor in rational discourse. However, it does not solve the issue of what measure of status or significance we attach to metaphor of a particular kind – unificational (Khalil, 1998) – as opposed to more mundane or everyday metaphors. If metaphor is so special why is it so commonplace? Why is such an antagonism to metaphor present in the sciences?

Lakoff and Johnson have evolved a theory known as *embodied realism*. The significance of this feature (feeling/embeddedness) in metaphoric theorization is that it lends itself to lived experience, which corresponds to phenomenological theorization. This particular approach towards metaphor tends to follow a distinct line, being more akin to the ideas of Merleau-Ponty, and intuitive/sensation based aspects of phenomenological experience (Merleau-Ponty, 1962). Lakoff and Johnson argue that a strong theory of metaphor "rejects a subject/object dichotomy"[19], and "that we are coupled to the world through our embodied interactions" (Lakoff & Johnson, 1999: 93). Metaphor on this account, is a vital feature of thought, i.e., "can we think of subjective experience and judgement without metaphor?" (Lakoff & Johnson, 1999: 59).

Clearly, both Lakoff and Johnson and Ricoeur share a view of metaphor as of crucial importance. This said, the feature of influence on Lakoff and Johnson, within a philosophical frame is that of Merleu-Ponty (and, also stated, as influenced by Dewey's pragmatism) rather than Ricoeur. Both Ricoeur and Merleau-Ponty have phenomenological credentials, however Lakoff and Johnson in the wake of Merleau-Ponty emphasize the existential dimension of human experience. This can take the form of an emphasis on ontological metaphors, which have many conceptual connections and thereby have an ordering or systematic structure (Medina, 2005: 130). These cohesively link areas of theory (like the role of emergence in complexity), and they may have an orientational nature. So, in the case of emergence – whether in time or out – this can be figuratively depicted (for example the Lorenz attractor or phase-space formations) as a spatial construction.

One mode of reading Lakoff and Johnson's theorization suggests that metaphor is able to project a level of meaning (as in the case of complexity concepts) which is sub-discursive and pre-conscious, exploiting

---

19. My view is that there is a sense in which the subject/object divide is elided by holism (as promoted by complexity theory) but that sense is distinct from the argument of phenomenology as to the dissolution of the subject/object divide; and stems from a different basis of reasoning. This aspect of elision – of the subject/object divide, is a feature of similarity which might be interpreted as being a commonality between the holistic aspect of complexity theories and a particular phenomenological interpretation. These aspects are discussed by Morçöl as phenomenology's questioning of an exo-physical determination of the relationship between subject and object. He argues that complexity theory infers an inter-relationship of subject and object (as relational holism). See also Morçöl (2001: 104-119) in discussion of Makaryk (1993) and Rossler in Holden (1986).

subjective or emotional reactions (Maasen, 2000). On this argument, it is precisely this arational feature of metaphor which allows science to penetrate to a popular state, and allows metaphors such as the 'edge of chaos' to be capable of flexible adaptation to contexts other than their origins (of scientific inception). The feature of relational holism, mentioned in the context of complexity, may apply to metaphors as expressed in terms of the 'imaginative rationality' which Lakoff and Johnson suggest. On this account, metaphor binds distinct levels of conceptual operations – for example the cellular automata may be seen as the heuristic basis for the transference of complexity principles associated with the complex adaptive system. It is discussed in Johnson (1995) as an operative example of the creation of a complex system by virtue of phase-transitions[20]. However, unlike Ricoeur, Lakoff and Johnson do not specifically argue for a heightened or cosmological role for symbolism. That view tends to be more of a supra-discursive nature. Relative to Lakoff and Johnson, complexity theory's conception of emergence would be one primary (orientational) metaphor amongst many. Indeed, on this account, these types of metaphors are reduced to "inferential patterns used in sensorimotor processes that are tied to the body" as representing the neural patterning of primary metaphor (Lakoff & Johnson, 1999: 77).

What special role might complexity hold in the theorization of Lakoff and Johnson from the standpoint of cognitive science? One possibility does exist, and that is to tie these types of theorization as to the nature of metaphor, to the idea of a reinterpretation of cognitive theorization in terms of the dynamics of neural development (interpreted from the angle of behavioral attractors which predicate language emergence as well as bodily behavior at distinct phases of human development). Such a particular interpretation would argue that those motor functions which underpinned learning, and hence language deployment, could be modeled in the spirit of complexity related exemplars (Thelen & Smith, 1994). However, the virtue of this approach would be that it would correlate the significance of metaphor in cognitive sciences with that of motor development as expressed in terms of the complex sciences. This type of combination would emphasize less the phenomenological aspect within Lakoff and Johnson.

Lakoff and Johnson are committed to a theorization which employs a series of theorizations of cognitive science – albeit reinterpreted (as they term it) in terms of a second generation of theories (for example, that conceptual structures arise from our sensorimotor experience and these structures are characterized by image schemas, etc.)[21]. The theorization is

20. See Johnson, (1995: 284-286). Rather interestingly, Johnson notes that Crutchfield (a scientist at Santa Fe), believed that the description 'edge of chaos' was misapplied, by Kauffman, Langton and others - a case of making the facts fit the theory, in that it applied the principles of dynamical systems theory to cellular automata, which Crutchfield believed was mistaken. The case is an interesting one in that it raises the issue of whether scientific discovery is a matter of simply uncovering facts or making "discoveries".

21. It is beyond the point in this article to fully narrate these theorizations, see

argued as non-Cartesian (in contrast to first generation theorizations based on representation). However, it does at least apply an *a priori* adoption of certain philosophical presuppositions as to embodiment (Merleau-Ponty, 1962). As Lakoff and Johnson note: "the distinction [of first and second generation cognitive science] is one of philosophical and methodological assumptions" (Lakoff & Johnson, 1999: 78). One can note clear affinities with Ricoeur's discursive semantic theorizations and Lakoff and Johnson's 'take' on Merleau-Ponty. Ricoeur notes that:

"*I think that it is always through the mediation of structuring operations that one apprehends the fundamental meaning of existence, what Merleau-Ponty called* l'etre sauvage. *Merleau-Ponty sought this* etre sauvage *throughout his philosophical career and consistently criticized its deformation and obfuscation in science. I for my part have always attempted to identify those mediations of language which are not reducible to the dissimulations of scientific objectivity, but which continue to bear witness to creative linguistic potentialities*" (Ricoeur, in Kearney, 2004: 104).

By way of clarification, the latter comment distinguishes Ricoeur from the perspective of Husserl and the quasi-Cartesian perspective. Ricoeur goes on to add that he is looking for an "essential connection between speculative and poetic discourse – evidenced, for example in the whole question of analogy" (Ricouer, in Kearney, 2004: 104). Is this evidence of relativism or anti-scientific thinking? I think Ricoeur's point is the essential aspect of metaphor in discourse to reveal *ontological* truth – 'truth' at an existential level. However, as will be noted in succeeding sections, I do not read Ricoeur as therefore being a relativist regarding the presence (or absence) of underlying truth in ontological matters (what I will refer to as the *arche*)[22].

Phenomenology and Ricoeur's Symbolism Relative to Complexity

I have suggested the applicability of Ricoeur's hermeneutic theory of metaphor (and symbol) to complexity metaphor and especially its mobility between its source scientific realm and the area of phenomenology. However, metaphorical theorization suffers from its own varied and arguable status, in epistemological terms. I have adopted specific views of metaphor and discussed their applicability, but one might argue that these were arbitrary selections. The special claim of trying to develop Ricouer's theorization, as opposed to a contrasting theorization such as Derrida's, is that it allows metaphor to be valued more highly as a feature of the trans-

Lakoff and Johnson, (1999: 77-79) and fuller arguments therein advanced.

22. The issue of Ricoeur's cultural or conceptual relativism or his partiality to a values frame which is from a particularly western perspective (for instance), is further discussed in Medina, (2005: 131+).

formation process of complexity metaphor[23]. This diminishes the idea that metaphor is purely decorative and occurring at the word-substitutive level (Aristotelian theorizations). Also, Ricoeur's theorizations, being phenomenological and hermeneutic – are not caught in the constraints of debate which might mire a pro-Cartesian view.

This admitted, is it possible to form any sort of linkage between Ricoeur and his hermeneutic phenomenology and the other, more seemingly scientifically based theories of metaphor, such as Lakoff and Johnson's? Could this give the phenomenological case a feature of resonance with contemporary cognitive science theorization on metaphor?

The symbol is not a special metaphysical category in Lakoff and Johnson, as it is in Ricoeur, in which case it is the mode of underpinning which stands in stead of a kind of *arche* or source theorization. Correlations may be suggested which tie phenomenology in its form as advanced by Ricoeur, and that suggested by the ideas of Lakoff and Johnson. Rather like Lakoff and Johnson, Ricoeur stresses a role for metaphor in terms of ideational reform, giving metaphor a social dimension (Ricoeur, 1979: 154). He also stresses the emotional aspect of interpretation relative to the formulation and comprehension of metaphor: "feelings, first, accompany and complete imagination in its function of schematization of the new predicative congruence (of metaphor)" (Ricoeur, 1979: 154). That gives a role for feeling which tends to embed the formulation of the metaphoric process (or conversely one's reaction to it) in terms of a non-cognitive mode of thinking. This has affinities with the pre-logical thinking which Ricoeur associates with the cosmological symbol.

Ricoeur believes that the metaphor is a kind of conscious echo of the immanence of the (mythical) origins of an ideal state of Being. That oneiric, or dream imagery, can correlate the cosmic totality or source of mythic symbolism and conscious metaphor[24]. It is possible to find indications of this regressive order within the secondary ('popular') management literature. These act as a kind of equivalent of oneiric or psychic symbolism

---

23. Derrida's views as to metaphor, in *White Mythology,* (Derrida, 1982) tend to represent the metaphoric process in terms of a continual renewal of semantic life by metaphor and then a falling away of metaphoric vitality. It is true that Derrida does import a complex notion of *usure* which might be interpreted to promote the idea of surplus value or additional value to metaphoric creation, but generally speaking the drift of Derrida's viewpoint is to argue that metaphor reinvigorates semantic novelty only to result in a hardening or erosion of that vitality towards metaphysical "normalcy" (Derrida, 1982: 210). Ricoeur adopts an approach whereby metaphors are novel and in time they become trivial, but he also argues that the function of metaphors is epistemic as well as ontological (Medina, 2005: 121-131).

24. Ricoeur defines these regressive features in the following terms: "*The progressive order of symbolism is not exterior to the regressive order of fantasies; the plunge into the archaic mythologies of the unconscious (thereby) brings to the surface new signs of the sacred. The eschatology of consciousness is always a creative repetition of its own archaeology*" (Ricoeur, 1974: 334-335).

(as evinced at a social level), but with cosmic affiliations[25]. On this account, such symbols tie lesser to more conscious states through the unconscious. This has clear links with the pro-cognitive science theorizations of Lakoff and Johnson. Ricoeur is also influenced in his theorization from an encounter with the literature of Freud – hence the development of his idea of the symbol as a *liminal* feature which expresses at a cosmic level, dream symbolism. Such perspectives lend themselves to congruence with the interpretative paradigm in management studies.

If one desired a specifically Christian cultural interpretation of the symbolism in question – the influence of Eliade (e.g., 1971 – in Coupe, 1997: 60) can also be remarked in relation to Ricoeur. Mircea Eliade's views suggest a mythic sensibility as existent at some past state, and a feature which he terms *hierophany*[26]. If we marry this concept with the idea of emergence (á la complexity), and consider our ideas of cosmic symbolism (above), we could say that the Garden of Eden was like a first Complex Adaptive System, in relation to which, Darwinism was a kind of principle of the theme of the Fall, but self-organization a feature of potential redemption. In this vein, the parable of the Garden, or paradise, is a *mythical originary* point, whereas the fall (or loss of innocence) relates to an 'awakening' or emergence. But, also, in terms of cultural accretions, and in the spirit of Ricoeur's own analysis, the creative form of metaphor itself may be viewed as a kind of re-evocation of immanence or mode of discursive hierophany. On this account, metaphor has resonance with the goals of phenomenology, in that it is plausible to suggest that an experiential basis towards the phenomenological quest involves the absorption of the transcendent within the immanent[27].

Ricoeur noted that there was an opposition between reflective thought and mythical thought in *The Symbolism of Evil*, but in essence he transposed that dilemma in connecting myth and metaphor in the layers

---

25. In an earlier study I noticed that mythic *arche*types or themes could be discerned in the management literature, and linked this idea to a consideration of complexity theory. These ideas reinforce the theories developed in relation to Ricoeur here outlined. See Sheard (2001).

26. Eliade's hierophanic state is timeless and resonates with ideas of a pre-Edenic fall. Eliade believed that a restoration into formlessness was necessary if reinvigoration was to be achieved (Eliade, 1971, in Coupe, 1997: 60). He considered Myth as a form of desire to be at one with a "continual present"; ritual can invoke the moment in which order was created out of chaos; the merely individual can become the *arche*typal; the profane is invested (momentarily) with the signs of the sacred. Such moments are 'hierophanies' or special occurrences: "*Any form whatever, by the mere fact that it exists as such and endures, necessarily loses vigor and becomes worn; to recover vigor it must be reabsorbed into formlessness if only for an instant; it must be restored to the primordial unity from which it issued; in other words it must return to 'chaos' (on the cosmic plane).*" Eliade (1971), in Coupe (1997: 60).

27. Statement derived from Evans: "Both the phenomenological reduction and the turn of experience imply the reduction of transcendence to immanence." (Evans, 2003: 19).

of symbols (cosmic/oneiric/poetic/social). This developed a process for the ontologization of the semantic in a discursive narrative, in respect of which, Being was linked with the aspect of metaphoric enaction in a temporal (ontic) context of activity. This move follows in part the prevalence of the view – demonstrated for instance in Nietzsche – that conceptions are metaphoric in their inception, but that it is the forgetting and opposition of this which is a feature of Western metaphysics (Kofman, 1983). Derrida famously developed this aspect of interpretation with his critique of Husserl's 'metaphysics of presence'. Ricoeur therefore follows the tendency of structuralism (and consequentially post-structuralism) towards incorporating the semantic in a consideration of phenomenology, but he does so in an alternate direction to that of Derrida – indirectly absorbing the cultural accretions of Western metaphysics (for instance the quasi-Christian cosmology) as an acknowledged feature of Being, rather than stating that such discourse (and ultimately the discourse of Being itself) is a kind of conceptual trap which mires thought.

However, such a particular form of interpretation is quite *partial* in cultural terms, one is reminded perhaps of the limitations (which Ricoeur himself acknowledged) of the cultural categories which are the appurtenances of Being[28]. Being as a category – the topic of Heidegger – is in fact presupposed by Ricoeur as an area of argument capable of being placed in a semantic context. This indicates the limits of the application of a particular derivation of phenomenological theorization derived from a semantic improvisation upon the ideas of both Heidegger and Kant. One might feel in some sense an element of sympathy at times with Derrida, in that he would be free of such limitations, freeing thought of its mythic accretions.

Metaphor's Epistemic Status in Complexity and Local 'Truth' Claims

A recent aversion has occurred at a more distinctly conceptual level, towards metaphor in complexity theorization. More usually metaphor has been side-stepped by theorists. De Landa's (2002) conception of non-metaphoric thinking is representative of an approach critical of metaphor. In a certain sense, it may be conceived as a kind of intellectual Protestantism against the image, relative to his claim for complexity to specify "relevant and important problems," and of the idea of a "concrete universal" (De Landa, 2002, quoted in Mckenzie, 2005: 54). This latter relies on an discussion of the extrapolation of mechanism-independent topological transformations related to phase-space. De Landa relies on a series of abstractions from an initial physical state which is modeled, and expressed metaphorically in an initial conception of a physical system. De

28. See for instance the brief synopsis of Ricoeur's hermeneutics as expressed in Cutrofella (2005: 293-303). This shows the complexity of Ricoeur's position: "Ricoeur suggests that it is necessary to 'demythologize' the symbol (so as to avoid the allegorizing temptation)." It is easier said than done, however; as Ricoeur realizes – noting the tension between reflective thought and allegorization which he argues – realizes a cusp or focus in symbolism.

Landa's point is that mathematical variation is not itself necessarily metaphor dependent, but the significance of the successive abstractions relies on their initial portrayal in a metaphoric context (i.e., in this case the Lorenz attractor as a spatially conceived entity)[29]. In an important sense, the process of abstraction relies on an image, or metaphor, within the complexity sciences from whence to abstract or ideationally 'purify'. Secondly, the aspect of the metaphoric expression of the Lorenz attractor is part of its potential cognitive mobility in terms of our initial grasp of the conception at an other than highly abstract (mathematical) level. So, an inference of abstraction ('concrete universals') is that the capability for a coherent absorption of such conceptions to be part of a *theory* of complexity, may be diminished.

Of course, what De Landa argues relates to the felt connection between the (seemingly flawed or contentious) universalistic theory of complexity and metaphor. On this account, the cohesive aspect of metaphor, within complexity theorization – allowing for the sense of an emergent theoretical movement (allied to themes such as relational holism) – is undercut. In terms of its axiological status, metaphor is viewed as being less sure than a process of non-metaphoric abstraction, relative to the aspect of 'theory-building'. This places the spotlight firmly on the relevance and value of the analogical process as a topic of contention relative to the construction of a 'universalistic' theory of complexity. This may result in divergent interpretations regarding the status of metaphor in any give case (I have noted Khalil's view of the unificational metaphor relative to complexity as one case of this). It is possible to debate what does and does not consist of metaphor on the basis of a particular or partial interpretation of metaphor relative to axiological prerequisites (philosophical underpinnings of perspective).

Letiche & Uden (1998)[30] interestingly note that the contrast between metaphor (in terms of the contrast between a minimal realism) and those who 'assume' underlying structures, powers, mechanisms and tendencies exist, whether or not detected, that govern actual events, i.e., what he defines as the contrast between transcendent and local realisms. However, one can at the same time be a 'local realist' relative to metaphor and its claims (if Letiche can be so described), and yet inferentially extend theories which infer complexity theory's global epistemic status – as I have argued is the case with PCT. So, in the case of Ricoeur (despite the seeming 'grass roots' basis of his theories as to the role of metaphor in sentence deployment) he is clearly a theorist who alludes to a transcendent ground indirectly in his theories of metaphor (the dynamic tension in the metaphoric copula), and more explicitly in the theories of symbolism which allude to the sacred. This can be expressed in terms of the aspect of double-reference within Ricoeur's usage of metaphor (see Stellardi, 2000)[31]. What can then

29. See the very full discussion of De Landa in Mckenzie, (2005: 48-54).
30. Letiche & Van Uden (1998) cites Lawson, (1997: 21).
31. Stellardi notes that: "*Metaphor is now founded on a double-reference: a first degree reference to reality, and a second degree reference of a poetic nature. The metaphorical power plays in the space of these two limits, and must be understood not in the form of a rhetorical figure centred on the word, or as semantic imperti-*

happen is that a (seemingly fashionable) case for local knowledge or modesty of claim is made, which cannot be sustained as the theoretical movements so defined are everywhere imperialistic in the extensibility or mobility of their cases.

The case, made recently by Stengers (2000), tends to fit with the idea remarked above of a local or contingent realism. It is a valid point to note the limitations of complexity in order to avoid blasé extension of its theoretical points, but the further inference of the local case is a defence mechanism for theoretically ruling out a contrary case to complexity. Such a "contrary case" is implicit in too ambitious or blasé an extension of complexity theory. The issue would not have been raised if the metaphor had not been very mobile in the first place. If followed to a logical conclusion, one might ask: "how is it that a complexity theory exists at all if all knowledge is local?" Should not then complex features be remarked as the distinct features of disparate areas such as dissipative systems, chaos theory, etc? Surely the hallmark of a coherent discipline is that distinct areas have a sense of co-extensive principles of science which infer the prospect of a series of thematic commonalities. Relational holism for instance can tie a variety of aspects of complexity theorization, as well as an emphasis on the striking characteristics at the local or contingent level of particular cases – such as remarkable self-organizing phenomena.

An argument that complexity seems to have an element of common cause with poststructuralism, is made by Dillon (2000), relative to what he defines as "an anteriority of radical relationality" (Dillon, 2000: 5). By "anterior" I think we mean a kind of *arche* or presupposition of an axiological nature. Of course, as theorists such as Derrida define *différance*, its status as *arche* is resisted. Dillon identifies "implicate orderliness" (following Bohm, 1980, cited in Dillon, 2005: 4) as complexity theory's particular form of relationality. This relates to the aspect of extensibility of different levels of complex activity, which infers a common principle (holism) as generative of these forms of order. In contrast, he notes that poststructuralism is marked by relationality as 'disruptive movement' – an interminable feature of alterity is its hallmark, which one might argue is a quasi-symbolic leitmotif in the case of Derrida's difference. That latter drives apart, and never consolidates, its presence (hence its antithetical nature as a countervailing 'force' or sequence of forces to what Derrida characterized as the metaphysics of presence). In between the two distinct approaches represented by Dillon, we have a distinct attitude to the issue of ontology. His point is, of course, that the two movements have a different mode of relationality – one which forces apart and one which joins (holism). The ontologization inferred by *différance* proceeds to undercut the possibility of a transcendent ground for thought, but one might suggest it is also antithetical in nature to a holism

nence internal to the linguistic code but rather as a redescription of reality with the space of one or many possible "modes of discourse". This type of discourse requires its own structure and activates its own specific kind of truth: a tensional truth, a "metaphorical truth" implemented in the copula the is which means at the same time is not *and* is as" (Stellardi, 2000: 80).

which binds – in that *différance* constantly divides, and in that sense therefore *cuts,* or disperses.

The stress on relational holism (Dillon, 2000) points to a distinction between one of the currents which is part of the unification metaphor of complexity (Khalil, 1998), and the poststructuralist theorization (as epitomized in *différance*). This shows the kind of innate tension which lies within complexity theorization (at a global level, wherein it has been created by analogical transfer of locally evinced scientific exemplars), towards its localization. One could argue the issue regarding the salience of relational holism (Dillon's point), but most definitions of complexity tend to look for thematic similarities between disparate disciplines and levels of discourse related to complexity, whatever they may be. A purely localized series of complexity instances loses the sense of asking "what is complexity?" On this account, we could have no overarching theorization therefore that sense of unity would be replaced by incompleteness. Cilliers suggested that "the view from complexity means that we cannot have perfect knowledge of complex systems" (Cilliers, 2005: 263), but there is a difference between that comment of the empirical phenomena of complexity (incompleteness of data at a local level) and a generalization that complexity is intrinsically incomplete as completeness is not possible in a complex world. Incompleteness would suit a poststructural interpretation in that it stresses the contingency of complexity outcomes and thereby avoids the need to deal with the issue of whether complexity theory is a kind of global epistemic order (or contains the seeds of that argument incipiently).

Metaphor is, after all, not just a particular form of analogical extension, but built on the extension of imputed significance from one category to another on the basis of their difference (in some sense), as well as of similarity. Metaphor, "through analogy, establishes connections not only between different objects, but also through different semantic fields and linguistic games" (Stellardi, 2000: 48), and plays a part in the mobilization of discourse. An illustration of this process perceived as 'force', is the appeal of Derrida's conception of *difference* – exemplified in Cilliers's imaginative extrapolation of the neural network and its traces to a series of differences – whose structure is erasure and whose lack of presence implies volatility (Cilliers, 1998: 82). But, such a structure of similarity between the nature of *différance* and neural connections is, apart from an attempt to discern an analogical resemblance, (which, albeit ironically, hints at a literal 'truth') also a resemblance of source to target of metaphoric application. The 'source' being poststructuralist theory – the target for the metaphor being cognitive science (connectionism). How we interpret this metaphoric application – whether metaphor as incidental vehicle or essential feature of thought – does not avert the reality of metaphor being the glove that fits around the fit of the poststructuralist thesis to the connectionist case (neural firing in the brain). In this sense, we face the tenacity of acknowledging the value of metaphor in our thinking processes, but that does not discharge the aspect of how we 'weight' or read metaphor, or which theorizations we choose to 'read' metaphor as representing. The poststructuralist case tends to avoid the metaphoric 'freight' by engagement in the conceptual play of the ideas, but the feature of considering *différance* as self-reflective for instance – a looser version of Cilliers's point – itself engages the *différance* process with the charge of modeling human identity and, inferentially, as proving a key or *arche,* to the dilemma of doing this.

Conclusion

This resistance towards transcendence is the true reason why the correlation between the symbolic and the process of differentiation, associated with causal power in the theories of Ricoeur (for instance), is viewed as illusion – hence the view that "Derrida's deconstruction ends up by dissolving the symbolic within the imaginary" (Dupuy, 1998: 83). That is a distinct case from a suspicious viewing of Ricoeur's arguments as underscored by theological presuppositions (which I would share). I suspect also the view of complexity as 'local' lends itself to the anti-symbolic camp precisely because if we can render tenuous the issue of a causative point (or the matter of transcendent perspective from whence to decide what is or is not the case – relative to the truth of complexity), then we can avoid the entire issue of argument by a quasi-philosophical presupposition (for instance Derrida's) as to the significance of metaphor. Plainly, this will not do. All that is then occurring is that the doubts as to the status of the complexity theory as an *arche,* at an epistemic level, are being transposed as a feature of ontological ambiguity relative to poststructural perspectives.

Conversely, the view of a kind of 'implicate order', or of co-extensive ranges of principles (such as holism) which might generalize complexity towards the position of a strong naturalism, has resulted in a sense that complexity is overstretched in its claims relative to its scientific evidence (at a local level). Nevertheless, these types of ambitious theorization have been the basis of the advancement of complexity towards areas of conscious life, and mark the first significant phase of complexity theory advancement. The shift towards higher cognitive levels of absorption has also brought an encroachment of complexity theorization into areas of philosophy, for example phenomenology. This, in turn leads to an involvement in the aspect of ontology (the nature of things 'in themselves', or the concern with Being, as defined in Heidegger's reflections on the Kantian view).

A strong problem with Derridean theorization is that it resists categorization in terms of presence; it alludes to a pre-epistemic base in its emphasis on origin and indecidability (the conception of différance). Ricoeur is perhaps more at home with presence, though the device of the metaphoric copula he deploys, alludes to both non-being and Being (in that respect we see the resonance of Heidegger). It could be suggested, however, that Derrida's *différance* claims a truth status, as a form of 'arche' or origin. Derrida self-consciously uses the term 'arche-writing' to indicate a language prior to speech, but "negating the sense of an origin as primordial source" (Protevi, 2005: 33). Hence, this is an issue which Derrida is aware of. Derrida's concern is with writing, and the 'movement of difference' marks his view of language as system of 'traces'. These ideas may be said to resist classical correspondence theories of language; they play with the aspects of the nature of an 'origin', and hence the causal basis for language equivalence and definitions. It might be suggested that Ricoeur's approach is more direct, and in that sense avoids the potential charge of being a claim of truth which self-consciously undercuts the basis of its own 'truth-claim'. These two philosophies are cases whose value lies at the cusp of the expansion of complexity theorization within the terms of debate I describe. These occur in terms of metaphor being engaged as a methodological feature implicated by semantics; but in turn, attitudes to these features of discourse reflect particular philosophical postures relative to the status of ontology.

These thoughts seem to indicate the inevitability of the source analogy (complexity theory at a primary level of scientific level of analysis) becoming colored with metaphor as it is conceived, not purely as it is withdrawn from the womb of analysis. Precisely how that coloring might take flesh, the take on metaphor in complexity following primarily Ricoeur and his hermeneutical theorization (with some further consideration of Lakoff and Johnson) is one aspect of this paper. Mobility of complexity theory is tenacious and, in part, occurs because of the close affinity of analogization and holism. Recent debate, however, with the 'linguistic turn' has tended to import the idea of 'local' rather than transcendent 'truth'. This maybe fashionable, but as a mode of ideology rather than argument, its claims ought to be debated. Metaphor, itself a topic of critique of Western metaphysics in the post-structural precursor theorizations of Nietzsche (for instance), has become a basis by which complexity theory can extend itself from its areas of scientific inception, towards the social sciences. This 'mobility' of metaphor can therefore give rise to the claims of complexity theory to act as a 'universalistic' theory which cuts across disciplines. However, there is of course a distinction between complexity as a global epistemic theorization – implied by metaphoric extensions – and the alternate use of metaphor as a kind of critique of 'transcendent' truth claims. This twist, in the construction of the value attached to metaphor, has come as a result of the recognition that a global epistemic base of complexity is a close cousin to a claim for a transcendent basis of knowledge (eschewed at an ontological level by poststructuralism). It is suggested that one avenue away from this nexus of controversy, is to adopt a phenomenological view of complexity which can avoid these difficulties by virtue of its not having a direct problem with the idea of an arche, at an ontological level – hence the presentation of Ricoeur's theorization.

References

Bruns, G. (1993). "Against poetry: Heidegger, Ricoeur, and the originary scene of hermeneutics," in D. E. Klemm and W. Schwieker (eds.), *Meanings in texts and actions: Questioning Paul Ricoeur,* Charlottesville, VA: University Press of Virginia, ISBN 0813914116.

Capra, F. (1996). *The web of life: A new synthesis of mind and matter,* London, England: Harper Collins, ISBN 0002554992.

Cilliers, P. (1998). *Complexity and postmodernism: Understanding complex systems*, New York, NY: Routledge, ISBN 0415152879 (2000).
Cilliers, P. (2005). "Complexity, deconstruction, and relativism", *Theory, culture and society*, ISSN 0263-2764, 22(5): 255-267.
Coupe, L. (1997). *Myth: The new critical idiom,* London, England: Routledge, ISBN 0415134943.
Cutrofella, A. (2005). *Continental philosophy: A contemporary introduction,* New York, NY: Routledge, ISBN 0415242096.
De Landa, M. (2002). *Intensive science and virtual philosophy,* London, England: Continuum, ISBN 0826479324 (2005).
Derrida, J. (1982). "White mythology," in A. Bass (trans.), *Margins of philosophy*, Chicago, IL: University of Chicago Press, ISBN 0226143260 (1984).
Dillon, M. (2000). "Poststructuralism, complexity and poetics," *Theory, Culture and Society,* ISSN 0263-2764, 17(5): 1-26.
Dupuy, J.-P. (1998). "The autonomy of social reality: On the contribution of systems theory to the theory of society," in E. L. Khalil and K. E. Boulding (eds.), *Evolution, order and complexity,* London, England: Routledge, ISBN 0415137284 (1996).
Evans, F. (2003). "The beginnings of postmodernism: Phenomenology and Bergsonism, Derrida and Deleuze," in L. Lawlor (ed.), *Thinking through French philosophy: The being of the question,* Bloomington, IN: Indiana University Press, ISBN 0253215919.
Haskell, R. E. (1987). *Cognitive and symbolic structures: The psychology of metaphoric transformation,* Norwood, NJ: Ablex, ISBN 0893913685.
Holland, J. H. (1998). *Emergence: From chaos to order*, Oxford, England: Oxford University Press, ISBN 0192862111 (2002).
Johnson, G. (1995). *Fire in the mind: Science, faith, and the search for order,* London, England: Penguin Books, ISBN 0140174222.
Kauffman, S. A. (1992). *Origins of order: Self-organization and selection in evolution,* New York, NY: Oxford University Press, ISBN 0195079515 (1993).
Kearney, R. (1991). *Poetics of imagining: Modern to postmodern,* Edinburgh, Scotland: Edinburgh University Press, ISBN 0823218724 (1998).
Kearney, R. (2004). *Debates in continental philosophy: conversations with contemporary thinkers,* New York, NY: Fordham University Press, ISBN 0823223183.
Khalil, E. L. (1997). "Biological metaphors, socio-economic theory and reductionism," *Economic Issues*, ISSN X000-9529, 2(2): 45-57.
Khalil, E. L. (1998). "Social theory and naturalism: An introduction," in E. L. Khalil and K. E. Boulding (eds.), *Evolution, order and complexity,* London, England: Routledge, ISBN 0415137284 (1996).
Kofman, S. (1983). *Nietzsche and metaphor,* London, England: Continuum, ISBN 0485120984 (1993).
Lakoff, G and Johnson, M. (1980). *Metaphors we live by,* Chicago, IL: Chicago University Press, ISBN 0226468011.
Lakoff, G and Johnson, M. (1999). *Philosophy in the flesh: The embodied mind and its challenge to Western thought,* New York, NY: Basic Books, ISBN 0465056733.
Lawlor, L. (2002). *The basic problem of phenomenology,* Bloomington, IN: Indiana University Press, ISBN 0253215080.
Letiche, H. (2000a). "Phenomenal complexity theory as informed by Bergson," *Journal of Organizational Change Management*, ISSN 0953-4814, 13(6): 545-57.
Letiche, H. (2000b). "Self-organization, action theory and entrainment," in *Emer-*

gence: A Journal of Complexity Issues in Organizations and Management, ISSN 1521-3250, 2(2): 24-37.

Letiche, H. and Van Uden, J. (1998). "Answers to a discussion note: On the 'metaphor of the metaphor'," in *Organization Studies,* ISSN 0170-8406, 19(6): 1029-33.

Maasen, S. (2000). "Metaphors in the social sciences: Making use and making sense of them," in F. Hallyn (ed.), *Metaphor and analogy in the sciences,* Boston, MA: Kluwer Academic Publishers, ISBN 0792365607.

Mckenzie, A. (2005). "The problem of the attractor: A singular generality between sciences and social theory," *Theory, Culture and Society,* ISSN 0263-2764, 22(5): 45-65.

Medina, J. (2005). *Language: Key concepts in philosophy,* London, England: Continuum, ISBN 0826471676.

Merleau-Ponty, M. (1962). *The phenomenology of perception,* London, England: Routledge, ISBN 0415278414 (2002).

Moran, D and Mooney, T. (2002). *The phenomenology reader,* London, England: Routledge, ISBN 0415224225.

Morçöl, G. (2001). "What is complexity science? Postmodernist or Postpositivist?" in *Emergence: A Journal of Complexity Issues in Organizations and Management,* ISSN 1521-3250, 3(1): 104-119.

Olson, M. (1981). "Symbol and metaphorical truth," in *Myth, symbol and metaphorical truth,* Paris, France: University of Notre Dame Press, ISBN 0268013497 (1982)

Prigogine, I, and Stengers, I. (1984). *Order out of chaos,* London, England: Flamingo, ISBN 0394542045.

Protevi, J. (2005). *The Edinburgh dictionary of continental philosophy,* Edinburgh, Scotland: Edinburgh University Press, ISBN 0748617167.

Ricoeur, P. (1974). "The hermeneutics of symbols and philosophical reflection: I," in D. Ihde (ed.), *The conflict of interpretations: Essays in hermeneutics,* London, England: Continuum, ISBN 0485300613.

Ricoeur, P. (1979). "The metaphorical process as cognition, imagination and feeling", in S. Sacks (ed.), *On metaphor,* Chicago, IL: University of Chicago Press, ISBN 0226733343, pp. 141-159.

Sheard, S. (2001). "Slain God: A symbolist perspective on the new science and related management theorization," *Emergence: A Journal of Complexity Issues in Organizations and Management,* ISSN 1521-3250, 3(2): 78-112.

Sheard, S. (2005). "Continental philosophy and complexity - part 1: A review of Letiche's Phenomenological Complexity theory," *Emergence, Complexity and Organization,* ISSN 1521-3250, 7(3-4): 105-118.

Stellardi, G. (2000). *Heidegger and Derrida on philosophy and metaphor: Imperfect thought,* New York, NY: Humanity books, ISBN 1573928240.

Stengers, I. (2000). *The invention of modern science:Theory out of bounds,* Minneapolis, MN: University of Minnesota Press, ISBN 0816630569.

Thelen, E. and Smith, L. B. (1994). *A dynamic systems approach to the development of cognition and action,* Cambridge, MA: MIT Press, ISBN 0262200953.

Waldorp, M. M. (1992). *Complexity: The emerging science at the edge of chaos,* New York, NY: Simon and Schuster, ISBN 0140179682.

Appendix 1: Summary of Points of Distinction PCT and Ricoeur's Phenomenological Hermeneutic of Metaphor

In this appendix I summarize these features and also make further discussion of the conception of difference, explored in the first paper in this series.

If once conceives of PCT (Letiche) relative to these aspects, (see Letiche, 2000a, 2000b) the metaphor of 'emergence' would be temporalized within the type of theorization that Ricoeur presents. It would be ontic (inside time). Its 'emergence' would then relate to the aspect of transition between distinct phases of metaphoric transference, which flickered between Being and non-Being (according to Ricoeur's incorporation of Heidegger's theorizations).

Hence on this account the metaphoric process would correlate Being and time. However, of course, Letiche makes no such correlation with metaphor. Rather, Letiche eschews a temporalizing of the satellite metaphors of complexity – such as 'emergence' or 'the edge of chaos'. As far as Derrida was concerned, Heidegger reached out towards (but had not completely defined) the problem of the origins of metaphysical thought (if the problem can be so defined). Derrida's conception of difference is notable in this regard: "différance is not, does not exist, is not a present-being (on) in any form..." and that: "In a conceptuality adhering to classical strictures 'différance' would be said to designate a constitutive, productive and originary causality, the process of scission which would produce or constitute different things or differences..." (Derrida, 1982: 9). That aspect of a split (or schism) is antecedent to the conception on Being (on this account) (Derrida, 1982).

Hence, at the risk of reiteration: Letiche cannot define Being as within or without of time because he follows Derrida who indicates différance as a mode of description of a process which hovers at that point between a temporal and atemporal perspective (as indicative of an originary or motive force within perception). Hence, complexity metaphors within Letiche's PCT could not possibly be construed as either one thing or another in terms of time; they are caught in the différance process. Emergence then becomes synonymous with différance on that score. However, there is a difference, in that one is not purely dealing with a view of emergence relative to phenomenology, but also looking to impute that phenomenological basis of understanding relative to the complexity sciences, of which the metaphors are representatives.

A phenomenological perspective which is underpinned by Ricoeur's viewpoint might tend to locate emergence, and other associated complexity metaphors more clearly in an ontic or ontological realm – ontic to the extent that the metaphors are conscious creations (poetical metaphor transposed into a social realm) or ontological to the extent that they are symbolic features (cosmic symbols). That said, the problem remains as to whether we construe the symbols as features within or out of time, (it depends on the definition of Being one adopts I think). In contrast, PCT (following Derrida) must locate on the fault line between time and an atemporal state (as Letiche notes). This said, the scientific activities to which complexity metaphors refer are clearly occurring within time; it is our perception of their activities which can be portrayed in terms of *différance*. In this sense, the complexity metaphors which Letiche locks into, and equates with PCT, are epistemologically bound as activities, but in their descriptive sense (relative to his particular philosophical constructions of PCT on the influence of Derridean theory) the metaphors are locked out of the epistemological realm by virtue of their abiding (ceaselessly) on the ontico-ontological divide.

14. Complexity in a Complex Europe: Reflections on the Cultural Genesis of a New Science

Damian Popolo
Durham University, UK

***Cause and effect**: such a duality probably never exists,—in truth we are confronted by a continuum out of which we isolate a couple of pieces, just as we perceive motion only as isolated points and then infer it without ever actually seeing it. The suddenness with which many effects stand out misleads us; actually, it is sudden only for us. In this moment of suddenness there is an infinite number of processes that elude us. An intellect that could see cause and effect as a continuum and a flux and not, as we do, in terms of an arbitrary division and dismemberment—would repudiate the concept of cause and effect and deny all conditionality.*

Nietzsche, *The Gay Science*, Book III, 112.

Introduction

This article wishes to present three related arguments:

1. Complexity – understood as a set of knowledge practices which rely on a concept of time irreversibility[1] – represents one of the many paths to knowledge embedded in the European intellectual tradition. In a sense, this article wishes to introduce the idea that the approach used by Fritjof Capra in his *Tao of Physics* (1979) can be more usefully applied to Europe. In fact, not only does a particular strand of European philosophy have a lot in common with the ethos of Complexity, but some important research reveals that the precursors of Complexity theory can be found in specific intellectual traditions within European philosophical thought.
2. Given the intellectual origins of Complexity, Complexity has the potential to re-open a traditional European debate regarding the exact role

1. It would be impossible to summarize in this article what is meant exactly by this 'definition' of Complexity. Arguments supporting this interpretation of Complexity have been presented in detail elsewhere (Popolo, 2003). Suffice it to say that this article adopts a highly 'Prigoginian' understanding of Complexity as a set of knowledge practices that accept the concept of time irreversibility. Arguments explaining why this feature ought to be considered as the defining characteristic of Complexity are presented elsewhere (Popolo, 2003). In a nutshell, this 'definition' of Complexity is useful insofar substantial research in the philosophy and history of science point at temporal linearity as the defining characteristic of modern scientific epistemology, and it is precisely in this context that Complexity's opposed approach represents an important change.

of Reason after the Enlightenment, a debate that saw the Old World fighting what we could call an 'Epistemic Civil War'. This 'war' was eventually won by a particular understanding and interpretation of the Enlightenment, which can be found in classical (Newtonian) scientific epistemology. Complexity is regarded precisely as a set of knowledge practices that challenge the most important features of such epistemology. Interestingly, whereas thinkers such as Foucault (1970) identify a concept of linear temporality as the cornerstone of the modern configuration of knowledge, important exponents of Complexity Science identify the novelty of Complexity in its recognition of time irreversibility (or the 'arrow of Time') as a real state of affairs in nature (Prigogine & Stengers, 1997). A detailed analysis of this relationship has been conducted elsewhere (Popolo, 2003).

3. Differing visions on the role of Reason after the Enlightenment have come with related understandings of how polities and societies should be organized. Ultimately, the European 'Epistemic Civil War' is what characterizes differing visions on the 'European Spirit', and the consequent diverging opinions on the nature of the European project. On the other hand, Complexity is regarded as being reminiscent of arguments presented by a specific camp in this Epistemic Civil War (the 'loosing' side, i.e., Natural Philosophy). It follows that the ethos of Complexity has direct implications for our understanding of Europe's integration process, or lack of thereof. In this context, the Husserlian understanding of the relationship between approaches to knowledge and the formation socio-political organizations is particularly relevant precisely because it tackles the related issues of 'European crisis' and European integration.

At first sight, it may appear as though the seemingly extravagant claims outlined above must be beyond the scope of a single article. Perhaps a more concise way of appreciating such claims can emerge by summarizing in this introduction the main seemingly unrelated points (i.e., books) that are joined by the dotted lines represented in this article's arguments.

The first of these points is James Gleick's (1998) seminal work, *Chaos*, in which we learn that two of the most prominent pioneers of the new Science were fascinated by outdated paradigms crystallized in Goethe's Romantic theories on the nature of plants and colors. The second is Eric Hobsbawm's (2003) *Age of Revolutions*, in which we learn that differences between German Romanticism and Anglo-French rationalistic theories came with very different approaches to policy, governance and political ideology. Furthermore, in the *Age of Extremes*, Hobsbawm (2003) directly links German Natural Philosophy and Romanticism with the emergence of 'Chaos Theory'. Finally, as stated above, Foucault (1970) argues that the French philosopher Henri Bergson represents the beginning of a new configuration of knowledge by virtue of his new conceptualization of nonlinear time. Incidentally, Progogine (1997) takes Bergson as a starting point for his enquiry into the 'arrow of time' and the 'end of certainty'. This article constitutes an attempt to link these points in order to present debates regarding the role of Reason before and after the (re-)emergence of Complexity, and to understand what the consequences of such re-emergence may be for the future of European cultures and societies in the context of European integration.

On the Intellectual Genesis of Complexity

Before embarking on a study of the genesis of Complexity for the purposes of its comparison with 'Newtonian' science, we should bear in mind Michael Dillon's warning: to 'compare' complexity inevitably leads to the temptation of defining it. Dillon (2000), however, reminds us of the Nietzschean belief according to which only things which do not have a history can be defined. Complexity, as a set of loosely defined principles, does have a history, a history which enables us to capture its ethos rather than its precise (and necessarily abstract, for necessarily essentialist) definition. Since a critical assessment of Complexity is becoming more and more urgent, it is equally urgent to give careful consideration to the genesis of Complexity's underpinning ideas, which do shed light on concepts related to European identity, for example.

The urgency is due to the fact that in recent years there has been an explosion of work concerned with the implications of Complexity Science. There is no subject that has remained unscathed. Within the areas covering International Relations and European Studies, for instance, it seems that everything can be understood in novel ways thanks to the insights of Complexity. We have discovered that Complexity can provide concepts to better understand the dynamics of globalization (Urry, 2002), of war (Alberts & Czerwinski, 2003, Beyerchen, 1992) and even of regional security regimes (Clemens, 2001). Theorists have attempted to construct models of international politics on Complexity-inspired concepts of turbulence (Rosenau, 1992) and system interaction (Jervis, 1999). One way or another, Complexity has fascinated and influenced the work of scholars whose names are well known to the International Relations research community, such as Robert Cox, Janet Abu-Lughod, Immanuel Wallerstein and John Lewis Geddis. This is not to mention the enormous impact of Complexity on other branches of the social sciences, especially sociology.

Yet, with so much debate over relatively novel terms, there is a serious risk that the term 'Complexity Science', and with it the concept, will eventually be diluted to a set of rather shallow notions. At best, the term would then be considered to be useful only insofar it can provide appealing metaphors for old concepts that one may want to present in original ways. Worse, the term could be adopted to justify rather dubious concepts as having benefited from the latest insights of the new science. It is difficult not to note the inherent similarities between some recent academic publishing on Complexity and the myriad of publications underlined by ideas such as 'improve you sales thanks to the insights of Complexity'!

It is in these contexts that the most simple questions, such as 'what is Complexity Science?', tend to become the most difficult to answer. These are the questions that increase the temptation for a standard, and therefore an inadequate and confusing definition. This is evident in the latest round of the 'Science Wars', in which Sokal (1999: 134-147) attempts to deliver us from the heresy of a humanities-based interpretation of Complexity by stating that Complexity is, contrary to what it may seem, essentially Newtonian. This is of course in contradiction with the arguments of other well-established scientists who, despite disagreeing on what Complexity 'is', agree on what it is not: Newtonian. Sokal's neat definition of Complexity seems to be at odds, for example, with Prigogine and Stengers's (1997).

Moreover, an avalanche of volumes, both in 'popular science' and in academia seek to clearly present Complexity as a set of strictly defined concepts. The creation of a 'Complexity checklist' used to label research as 'Complexity research' is all too evident, for example, in Alberts and Czerwinski's (2003) work. These quests for precise criteria are not only artificial, but contribute to a general failure to critically assess the philosophical, historical and indeed epistemic impact of Complexity. The objective of producing a short genesis of Complexity is not therefore related to the task of 'defining', or 'presenting' the subject, but to capture its inherent ethos.

For the reasons enumerated above the task of contextualizing Complexity science in a broad philosophical framework is becoming increasingly important. There has been some debate recently surrounding the question of whether Complexity could be seen as a manifestation of 'post-modern' science (Morçöl, 2002). Some scholars have sought to underline the similarities between Complexity science and post-structuralist philosophy (Cilliers, 1998; De Landa, 2002; Rasch & Wolfe, 2000; Pearson, 2002). Some others have perceived Complexity as being simply the manifestation of what Lakatos would call a 'progressive paradigm'. In this view, the emergence of notions such as 'deterministic chaos' suggest that Complexity is nothing more than a more sophisticated tool for the understanding of patterns that just happen to appear as being completely random – but that remain patterns nonetheless (Byrne, 1998). Others have suggested that Complexity, in a number of fields, has changed many paradigmatic assumptions to such a point that it can now claim to be a legitimate scientific revolution in the Kuhnian sense (Gleick, 1998). As it will be appreciated by the end of this article, a far more ambitious argument could be put forward. Complexity science quite simply questions a Newtonian understanding of the role of Reason after the Enlightenment. In a nutshell it will be argued that according to this Newtonian vision, Reason was simply regarded as a means to provide certainty and security through the elaboration of metaphysical science[2], whilst a Complexity 'ethic' would regard Reason as that

2. Arguments surrounding the metaphysical nature of modern science have been explored at length. In a nutshell, this article defines the concept of 'metaphysical' as an approach to knowledge that relies on essentialism, reductionism and determinism. References are made throughout the article to works that better capture this definition of metaphysics. The differences between scientific metaphysics and

which allows human beings to accept and deal with uncertainty.

A timely reminder of Complexity's origins can be found in the works of Eric Hobsbawm (1999). Hobsbawm rightly asserts that the principles of Complexity did not 'appear' but re-appeared from the European intellectual heritage under the misleading tag of 'chaos theory'. The re-appearance of such ideas was possible thanks to the increasing calculating powers of computers. According to Hobsbawm the re-emergence of 'Complexity' has profound implications for the concept of causality. Such an approach does have the potential to undermine several 'political, economic and social' assumptions (Hobsbawm, 1999: 696-697).

Apart from thoroughly describing the effects of Complexity in the world of the natural sciences, Hobsbawm proceeds to describe how, drawing on Max Plank, 'scientific' and social, economic and political histories are necessarily intertwined, indeed, many social events are underpinned by the way in which scientific methodology portrays itself as a new path towards certainty in all fields of inquiry. Hobsbawm (1999: 673-717) masterfully demonstrates how modern science, seen as a path to certainty, and the 'Age of Catastrophes' (or the 'short twentieth century') are inherently related. Crucially, Hobsbawm identifies the genesis of Complexity in the truly European 'Epistemic Civil War', which was characterized by two contrasting interpretations of the role of reason after Enlightenment, and therefore of European modernity. The following quote (Hobsbawm, 1999: 756, my translation) is of fundamental importance to critically assess the ethos and origins of Complexity:

"*Developments within 'chaos theory' in the 1970s and 1980s are not without commonalities with the emergence, in the beginning of the 19th century, of a 'romantic scientific school', of which Germany was the primary focus (Naturalphilosophie); that reacted to the dominant 'classical' science, essentially represented by practices in France and in Great Britain. It can be interestingly observed that two eminent pioneers of the new science (Feigenbaum and Libchaber) were in fact inspired by Goethe's – radically anti-Newtonian – theory of colors, and also by Goethe's treatise* On the Transformation of Plants, *which can be considered as a precursor of anti-Darwinian and anti-evolutionist theories.*"

Hobsbawm retrieves this crucial information on Goethe's fundamental influence on the two precursors of Complexity in Gleick's (1998: 163-65) seminal book. Armine (1998) has already pointed out the significance of this passage in Gleick's work.

The genesis of Complexity is therefore to be located in a moment of dislocation between two opposing visions of 'scientific' knowledge, ultimately based upon two different visions regarding the role of reason after the Enlightenment. This article wishes to suggest that the emergence of Complexity may represent the re-introduction of German romanticism

Complexity Science have been presented elsewhere (see Popolo, 2003).

in modern science. It is tempting to interpret recent developments within the epistemology of science as the re-introduction of romanticism's ethos, should we say, 'through the back door'. As such, the future of Complexity should be seen in terms of the *'longue durée'* as one of the possible paths to knowledge opened by the European philosophical heritage.

Existing research in a number of disciplines has pointed to natural philosophy and German romanticism as precursors of Complexity science. Bronowski's (1977) *A Sense of the Future: Essays in Natural Philosophy* tackles the issue at length[3]. As Butz (1997: 5) points out, Capra's (1996) *Web of Life* is a holistic history of systems that goes from Goethe to Lovelock's Gaia hypothesis. Baake (2003: 185) notes how Morowitz stated that "Goethe may be saying that the one emerges from the many, that unity arises out of diversity – an idea found in modern complexity theory." Farnsworth (2001: 216) elucidates the relation between Goethe and Complexity theory in the context of water dynamics whilst Reahard and McCarthy (2001) go a step further and do not hesitate to introduce the notion of "Goethe's complexity theory." In this context, the contribution of Rowland's (2001) monograph *Goethe, Chaos and Complexity* is also significant.

Modes of thought and modes of social organization

The broad intellectual movement referred to as Natural Philosophy is credited as being the main influence on the precursors of Complexity, yet it is important to note that it played an important role in shaping the modern European system of thought and modern European polities and societies. Hobsbawm locates in the 'classicism – romanticism' nexus another battleground for the 'dual Revolution' in the 'Age of Revolutions'. The dual Revolution consisted of economic (industrial revolution) and political (French revolution) developments that re-configured every aspect of thought in the Old World. As such, Hobsbawm interprets the 'classicism – romanticism nexus' as an 'Epistemic Civil War' which takes place during this crucial revolutionary period of European history. All these developments are related to a dramatic re-evaluation of the values of the Enlightenment. Hobsbawm (2003: 335, emphasis added) explains that,

"*If developments in the field of the sciences parallel those elsewhere, it is not because each of them can be hooked on to a corresponding aspect of economic or political ones in any simple way. Yet the links are hard to deny. The main currents of general thought in our period have their correspondence in the specialized field of science, and this is what enables us to* establish a parallelism between sciences and arts or between both and socio-political attitudes. *Thus 'classicism' and 'romanticism' existed in the sciences, and …* each fitted in with a particular approach to human society. *The equation of classicism (or, in intellectual terms, the rationalist, mechanist Newtonian universe of the Enlightenment) with the milieu of bourgeois liberalism, and of romanticism (or, in intellectual terms the so-called 'Natural Philosophy')*

3. See chapter 13, *New Concepts in the Evolution of Complexity.*

with its opponents, is obviously an over-simplification, and breaks down altogether after 1830. Yet it represents a certain aspect of truth. Until the rise of theories like modern socialism had firmly anchored revolutionary thought in the rationalist past, such sciences as physics, chemistry and astronomy marched with Anglo-French bourgeois liberalism. For instance, the plebeian revolutionaries of the Year II were inspired by Rousseau rather than Voltaire, and suspected Lavoisier (whom they executed) and Laplace not merely because of their connections with the old regime, but for reasons similar to those which led the poet William Blake to excoriate Newton. Conversely, 'natural history' was congenial, for it represented the road to the spontaneity of true and unspoiled nature. The Jacobin dictatorship, which dissolved the French Academy, founded no less than twelve research chairs at the Jardin des Plantes. Similarly, it was in Germany, where classical liberalism was weak, that a rival scientific ideology to the classical was most popular. This was Natural Philosophy ... it was speculative and intuitive. It sought for expressions of the world spirit, or life, *of the mysterious organic union of all things with each other, and good many other things which resisted precise quantitative measurement or Cartesian clarity."*

Few doubt that in the end it was the "rationalist, mechanist Newtonian universe of the Enlightenment," and its "particular approach to human society" that prevailed. Some attribute to this triumph the worst crimes of the twentieth century – particularly with regards to the metaphysical essence of this approach. Indeed, Hardt and Negri (2001) see a constant cleavage between metaphysics and philosophies of immanence throughout European history, a cleavage that is also embedded in the 'classical – romantic science' nexus. Dupré (1993: 219-244) offers some interesting thoughts about the links between modern metaphysical science and contemporary developments in politics, economics and societies. On the other hand there can be little doubt that the scientific method has been largely determined by broader cultural contexts, which, for example, inform what research questions can or cannot be considered. For instance, it would have been quite impossible for any 'scientific method' to emerge in any society that did not possess a particular concept (and word to express the idea of) causality. These are research questions that characterize the discipline of sociology of science (see in particular the very important contributions of Latour, 1988).

It is impossible not to consider the intrinsically German nature of these events. Heidegger related the triumph of mechanistic, metaphysical epistemology to his contemporary political situation. Heidegger (Inwood, 2000: 6) always linked political events to 'the Forgetfulness of Being' and to technology. Indeed, "The willful construction of a world empire to last four millennia shows, he argued, a preference for quantity over quality that is alien to genuine creators like the Greeks. *Empire-building stems not primarily from dictators and authoritarian states, but from the metaphysical essence of modernity.*" This 'verdict' on Nazism was "delivered in the summer of 1941, when Hitler's power was at its height."

Heidegger was not the only German thinker who attempted to conceptualize the links between modes of knowledge (in particular metaphysics) and his contemporary political context (in Heidegger's case, this attempt would lead to dramatic consequences). Edmund Husserl's insights on the nature of European approaches to knowledge and the related issue of European crisis are fundamental if we wish to understand how knowledge practices and modes of social organization interact, and how Complexity – understood as a knowledge practice – could influence the future evolution of European polities.

As Europe was descending into the abyss of the Holocaust, Husserl (1965) delivered a lecture in Vienna in 1935 on 'Philosophy and the Crisis of European Man'. Husserl started the lecture by reflecting on the nature of the humanistic sciences (*Geisteswissenschaften*). Husserl described such sciences as being concerned with life as "purposeful living, manifesting spiritual creativity – in the broadest sense, creating culture within historical continuity" (Husserl, 1965: 150). Yet, despite the evidence of sickness afflicting purposeful living in all European nations, Husserl lamented that the "so luxuriantly developed humanistic sciences here fail to perform the service that in their own sphere the natural sciences perform so competently" (Husserl, 1965: 151). Much to Husserl's concern, the humanistic sciences had not developed a "medical science concerned with nations and with international communities" (Husserl, 1965: 150).

The methodological difference between the humanistic and natural sciences, which is responsible for this lack of humanistic medical practice, consists in the latter's possibility to abstract and on the impossibility of the former to do so. According to Husserl (1965: 152) mathematical exact sciences can abstract and operate metaphysically,

"*but only nature can be handled as a self-contained world; only natural science can with complete consistency abstract from all that is spirit and consider nature purely as nature. On the other side such a consistent abstraction from nature does not, for the practitioner of humanistic science who is interested purely in the spiritual, lead to a self-contained 'world', a world whose interrelationships are purely spiritual, that could be the theme of a pure and universal humanistic science, parallel to pure natural science.*"

Husserl maintains that it is therefore "understandable that the practitioner of humanistic science, interested solely on the spiritual as such, gets no further than the descriptive, than a historical record of spirit, and this remains tied to intuitive finitudes." This has been a recurring argument in philosophy of science before Husserl and ever since. For example, Bergson (Gutting, 2001: 50), Foucault (1970), Kuhn (1970: 24), and more recently Dupré (1993); have all pointed to the metaphysical nature of natural science (the 'self-contained' world, or 'paradigm') as a requisite for its progression. However, Husserl (1965: 153) admitted that this way of thinking may rest on "fatal prejudices":

"What is to be said, then, if the whole mode of thought that reveals itself in this presentation rests on fatal prejudices and is in its results partly responsible for Europe's sickness? I am convinced that this is the case, and in this way I hope to make understandable that herein lies an essential source for the conviction which the modern scientist has that the possibility of grounding a purely self-contained and universal science of spirit is not even worth mentioning, with the result that he flatly rejects it."

The fatal prejudice is the belief according to which only 'self-contained' and universal sciences are capable of grounding knowledge, thus the modern scientist flatly rejects the idea that a "universal science of the spirit can be created." On the contrary, Husserl sees in the creation of philosophy understood as a "science of the whole" the original manifestation of what he calls "Spiritual Europe." The parallelism with Complexity Science is striking: in a sense Complexity's drive towards anti-reductionism is itself an attack on 'self-contained' sciences. In Complexity's inherent multidisciplinarity one can detect the embryos of what Husserl calls "philosophy," that is, a science of the whole. And here lies the original phenomenon of 'Spiritual Europe'. In other words, Europe's sickness is a consequence of a 'fatal prejudice' which regards the construction of 'self-contained' worlds as a legitimate approach to knowledge.

Husserl's thinking and subsequent attempt of delineating a pure science of the spirit touches on some of the most contemporary issues regarding European policy and identity as well as on recent developments in scientific epistemology embedded in Complexity. Indeed, the effort of creating a spiritual medical practice is not only meant to prevent Europe from facing the abyss once again, but also to create a truly European union, beyond the realm of national interests. Such a practice, however, requires a radical transformation in the way we perceive – and attain – knowledge, in the way we think of 'disciplines' or even 'interdisciplinarity'. This is about re-inventing knowledge itself, or, as Foucault would put it, transforming the modern episteme. Indeed, according to Husserl's (1965: 159) vision,

"[Philosophy], correctly translated, in its original sense, this bespeaks nothing but universal science, science of the world as a whole, of the universal unity of all being ... in the emergence of philosophy in this sense, a sense, that is, which includes all sciences, I see – no matter how paradoxical this may seem – the original phenomenon of spiritual Europe."

According to such vision, only such a spiritual form of Europe can be responsible for the emergence of a "supernationality of a completely new kind," a transformation of Europe, which is "no longer a number of different nations bordering on each other, influencing each other only by commercial competition and war. Rather a new spirit stemming from philosophy and the sciences based on it, a spirit of free criticism providing norms for infinite tasks, dominates man, creating new, infinite ideals" (Husserl, 1965: 177). This Husserlian vision calls for a humanistic understanding of knowledge as a whole (universal science, emergency of philosophy as a science of the whole, etc.). But how does this Husserlian approach relate to Complexity? Recent research aims to show how a humanities-based understanding of Complexity reveals the extent to which Complexity represents a particular tradition of thought and understanding of societies (Jackson, 2003). Moreover it can be argued that the issues which Husserl identifies as problematic (reductionism, essentialism or naturalism, and so on) are precisely the notions that Complexity seek to overcome. More importantly, the arrow of time in Complexity is what allows an understanding of indeterminism in nature, which is precisely what allows us to accept novelty in a critical way. It is in the acceptance of this indeterminacy that Prigogine & Stengers (1997) envisage a new concept of 'freedom' and the production of novelty in nature. This is *precisely* what Husserl calls the "production of new, infinite ideals." Such ideals are produced as we revert to the original phenomenon of Spiritual Europe (i.e., philosophy as a science of the world as a whole) and abandon the concept of 'self-contained worlds'.

Therefore according to Husserl, the production of new, infinite ideals (i.e., novelty) is a medicine. This has to be prescribed when the major danger facing modern European man appears to the surface: intellectual weariness. It is not, however, the weariness of knowledge, but rather the laziness created by a quest for certain knowledge that fails to consider the creative novelty of nature, and man's role in it. It is the laziness arising from the technicalization of science, which, according to Husserl, is the product of an excessive 'mathematization' of knowledge. Such technicalization of science leads to naturalism, which is the *causa prima* of Europe's sickness: "The crisis of European existence can end in only one of two ways: in the ruin of a Europe alienated from its rational sense of life, fallen into a barbarian hatred of spirit; or in the rebirth of Europe from the spirit of philosophy, through a heroism of reason that will definitely overcome naturalism. Europe's greatest danger is weariness" (Husserl, 1965: 192).

The Future of Knowledge and the Future of Europe

Husserl's remarks on the intimacy between modes of knowledge and European crises did not emerge in a vacuum. In this context, Husserl's predecessors' main motivation for understanding the nature of modern science was precisely related to a desire to understand their contemporary socio-political situation. Antimo Negri (1994) focuses on this Husserlian concept of European crisis as he discusses the important relation between Nietzsche's philosophy and modern science. Negri (1994: 27) elucidates how Nietzsche's concept of 'living dangerously' influenced the Husserlian fear of intellectual weariness:

"The meaning of the Nietzschean concept of 'living dangerously' is evident: it is not possible to live 'resolutely', 'dangerously', if one does not fight with a 'vigorous spiritual disposition' the extreme danger embodied in the weariness resulting from an exceedingly mathematized science, which gives the illusion

of providing firm foundations to knowledge – those which the Newtonians call principia mathematica in philosophia naturalis – robust and definitive. The true danger is to believe or to lead other people to believe that such foundations exist. The 'resolute' and 'dangerous' task that Nietzsche undertakes consists in destroying such belief."

It should be clear that we are dealing with the recurring theme of western metaphysics. The technicalization of science – a necessary component in the quest of developing certain knowledge – is a product of western metaphysics, the metaphysics that Heidegger and Husserl, largely inspired by Nietzsche, sought to overcome. It should be clear now that the main feature of the metaphysical approach consists in the creation of 'self-contained' worlds, or what Kuhn would call paradigmatic science.

Nietzsche's philosophy and its implication for the concept of 'Europe' has been explored in some length by Stefan Elbe (2003). Yet, Elbe's effort ignores the fundamental role that considerations over the nature of modern science played in Nietzsche's philosophy in general and in Nietzsche's treatment of the 'good European' in particular. Reviewing the book, Strong (2004) writes that Elbe "notes both Nietzsche's and Max Weber's argument (in *Science as a Vocation*) that whatever science can provide, it will not provide meaning to its or any other enterprise."

This is not helpful: the real tragedy, according to Nietzsche, is that despite his conviction that modern metaphysical science cannot replace God, men are bound to force the belief in the infallibility of reason – embedded in mathematical logic and modern science – into God's empty throne. That is why Nietzsche exhorts men of science to build their foundations on the Vesuvius in a letter to Jakob Burckhardt (Negri, 1994: 3). The Vesuvius is of course a volcano, and the idea is that men of science should be ready to accept that the foundations of their knowledge ought to be disintegrated from time to time. It is why, in the same gist, Heidegger (Negri, 1994: 26) maintains that "under the power of the pretences advanced by the principle of reason we find the fundamental traits of actual human existence, which works ad infinitum to obtain certainty and security." It is thus the forced triumph of rationalistic scientific reason after the death of God that is responsible for Europe's sickness – this is Nietzsche as an obvious precursor of Adorno's and Horckheimer's *Dialectic of Enlightenment*. Nietzsche, in this light, should also be seen as the epicenter of later developments including Foucault and Deleuze. That great lover of Goethe's Natural Philosophy, Nietzsche, is the kernel that unites French 'Philosophy of Freedom'[4] and the intrinsically German critical experience of modern rationalistic thought.

We have plenty of evidence indicating that Nietzsche perceived

4. The influence of Nietzsche on Foucault, Derrida and Deleuze is undeniable. See Gutting (2001). Tiles (2005: 173-4) equally reveals how the French tradition of philosophy of science, particularly the writings of Bachelard, anticipates in striking manner many of the intellectual dilemmas presented by Complexity. In this context it is worth remembering that Foucault's own work on intellectual history was heavily influences by Canguilhem's and Bachelard's (Gutting, 2001).

this role of modern science as one of the main causes of the hecatombs to come. One of the most pertinent passages in *The Gay Science* laments the deification of nature through science. Paragraph 108 (book III) illustrates the idea that God's shadows are still around[5], while paragraph 109 reveals where such shadows are[6]. It is however difficult to put is as clearly as Keith Ansell Pearson (1994: 36) does: "One of the consequences of humanity's faith in morality is the cultivation of truthfulness, of a *will* to truth (think of the Christian confession, for example). Over time this will to truth in Christianity is transmuted into the intellectual conscience which underlies modern scientific inquiry."

Science is the new metaphysics after the death of God: such is Nietzsche's diagnosis. To be sure, Nietzsche believes that European philosophy (and society?) since Plato has been infected by metaphysics – however, its recent evolution into 'science' is particularly cunning. On such basis Michel Foucault proceeded to analyze modern modes of metaphysical knowledge: an analysis of the importance of such research for the philosophy of Complexity has been conducted elsewhere (see Popolo, 2003). Any genealogy of modernity which inquired into the Genesis of Complexity would establish that Complexity's ethos, since its origins, has been profoundly anti-metaphysical. It is not a coincidence if we cannot imagine the birth of science proper outside of Christianity (see, for example, 2`1, 2003 on the spiritual drives behind Newton) or if Complexity is intimately linked with 'philosophies of immanence' (De Landa, 2002; Deleuze, 2001; Pearson, 2002). Finally, it should be noted that Dillon's (1996) research into the origins of the security paradigm in political thought reveals an intimate connection between metaphysical approaches to knowledge and the provision of physical and intellectual certainty. As he puts it (Dillon, 1996: 13, emphasis added),

"*Metaphysics first allows security to impress itself upon political thought as a self-evident condition for the very existence of life... [S]ecurity thereby became the value which modern understandings of the political and modern practices of politics have come to put beyond question, precisely because* they

5. "God is dead; but given the way people are, there may still for millennia be caves in which they show his shadow. – And we – we must still defeat his shadow as well!"

6. "Let us beware of saying that there are laws in nature. There are only necessities: there is no one who commands, no one who obeys, no one who transgresses. Once you know that there are no purposes, you also know that there are no accidents; for only against a world of purposes does the word 'accident' have any meaning ... when will all these shadows of God no longer darken us? When will we complete our de-deification of nature? This quote is particularly revealing as Nietzsche insinuates that the scientific effort to attribute human concepts to nature (i.e., law) are in reality reminiscent of a Christian heritage which was precisely informed by a 'will to truth'. In other words, Nietzsche is here suggesting that science has replaced religion as the provider of certainty to which modern Europeans had become accustomed, and in this sense one can understand the Nietzschean remark according to which scientists had become the new priests.

derived its very requirement from the requirements of metaphysical truth itself."

These are philosophical preoccupations for the social-political implications of transcendental philosophies embedded in modern science, preoccupations that find Goethe, Natural Philosophy and German Romanticism (which heavily influenced some important precursors of Complexity) as their starting point. They represent a fear that humans may come to be governed by fear, a fear and an intellectual laziness founded on a naïve trust of the certainty provided by metaphysical knowledge – of any kind. This is a concern that should be revived, urgently, today. For example, apart from Negri's contribution and a study related to quantum physics (Planck, 2002), there does not seem to be a coordinated effort to research Nietzsche's reflections on the nature of knowledge practices in general, of modern science in particular and of the socio-political implications of these.

Europe must accept Husserl's challenge of producing "new, infinite ideals" to counter the comfort provided by metaphysical knowledge. Complexity, viewed not merely as a 'scientific approach', or paradigm; but as an *ethos*, should be understood as an intellectual background capable of producing such ideals. The established links between Complexity and Continental philosophies of immanence (see De Landa, 2002; Pearson, 2002), which, for the most part, find their roots in Nietzsche; could and should reinvent the concept of 'freedom'[7]. Despite considerable progress in addressing questions related to identity formation and other humanities-related research in European Union (EU) programmes, there is still room for improvement.

For example, the EU's programme on Science and Society could be of fundamental importance for the diffusion of Complexity's ethos in the European public domain. As a programme interested by the interaction between science and society, this is a great opportunity to readdress the simplistic vision of 'science' that is still being adopted in all too many sectors of public life. Science could be presented precisely as the production of infinite ideals, as an endeavour that seeks to understand an evolving process of becoming, rather than a static reality of being. Thus the message will reinforce the idea that all our concepts are built on volcanoes, that men and women of knowledge understand that a pursuit of true understanding must entail 'living dangerously'. Such an understanding of the world would inform a new concept of freedom in the terms of how Prigogine envisaged it: the absence of certainty is in itself a tremendous opportunity for humankind to come to terms with the consequences of its actions and not an excuse to fall into the bottom-less pit of nihilism.

It is only by ensuring that the next generation of Europeans generate and deploy knowledge through a Complexity-informed ethos (holism, anti-reductionism, immanence, and so on) that the Husserlian vision of Spiritual Europe will be realized. One needs to look just how many recent developments related to European integration have been undermined by

7. For Prigogine and Stengers's concept of freedom after Epicuru's Dilemma see *The End of Certainty* (1997).

the absolute failure to deal with uncertainty – the vision of French voters refusing a Constitution partially due to fear of Polish plumbers proves the point. Everywhere it seems that fear of change motivates European attitudes towards further integration. Amongst the arguments that were presented against the Constitution we find this idea that by halting the integration process, the credentials of Turkey as a new Member State would have been debilitated. Turkey as the unfamiliar, (therefore uncertain) 'other', fellow Europeans see as a threat for job security; these are issues that gain prominence on the basis that certainty ought to be the basis for all things. Unfortunately, this may well be due to the assumption that certainty is what characterizes the natural world 'out there'. This is far from the best approach to life in general given the rapidly changing environment in which we find ourselves. To re-inverse a famous Massimo D'Azeglio quote ("now that we have created Italy we must create Italians") one could say that now that we have created a complex world we need to create Complex Europeans. If the Husserlian vision according to which the "original phenomenon of spiritual Europe" lies on its understanding of knowledge as knowledge of the whole, it follows that what characterizes most Europeans is the way in which they approach knowledge. The defining periods of history in which approaches to knowledge were established were truly European in the sense that these knew of no national boundaries. If Europe is to be forged through common understandings of knowledge, the creation of a common European Research Council, for example, is probably more significant than most aspects of European integration.

This is akin to Nietzsche's attempt to overcome metaphysics without falling into the despair that must necessarily follow the death of God; pessimism and nihilism must be overcome with the creation of art and culture understood as modes of knowledge that accept the tragic reality of this world: uncertainty. Europe must disseminate its 'post-modern' ethos in order to combat apathy and a simplistic approach to life – more importantly, to combat a predisposition to seek security, in all its forms; at any cost. Complexity, as an immanent but anti-reductionist ethos for knowledge would thus correspond to Husserl's "universal science, science of the world as a whole." Complexity's ethos is indeed at the heart of a re-invention of naturalism as that which produces novelty and uncertainty: Complexity's naturalism is anti-mechanistic. Under complexity, nature can no longer handled in a 'self contained world'.

Conclusion

European modernity has been characterized by a fragmentation of knowledge which produced a vast array of different 'sciences'. Perhaps this is due to the death of God, to the death of a belief in a unitary explanation for the world. However, thinkers such as Nietzsche, Husserl, Heidegger and others have warned about the fragmentation and technicalization of knowledge, foreseeing the consequences of such developments for Europe. At the heart of the problem, according to Nietzsche, lies a predisposition for 'truthfulness' or security created by nearly two millennia of

Christianity, and the replacement of God with a new type of metaphysics embedded in modern scientific methodology.

Complexity has its roots, on the other hand, on the epistemic civil war that shocked Europe in the 'age of Revolutions'. Arguably, the precursors of Complexity were in the losing camp, defeated by emerging visions of rationalistic and mechanistic thought. Nonetheless, the political dimension of Complexity and precursors of philosophies of immanence have had important insights about European identity. For example, Husserl's vision of a spiritual Europe emerging from the concepts of "science of the world as a whole," as a destruction of the belief in a "self contained natural world" and the consequent mechanistic view of nature is a crucial one. Not too long ago the former President of the European Commission convened a research group on the 'Spiritual and Cultural Dimension of Europe'. Debates surrounding the future identity of the Union cannot afford to ignore important developments embedded in the ethos of Complexity, as these find their roots in a common European intellectual heritage and now revolutionize, also through concrete results and scientific discovery, the way we look at the world. For these reasons, European institutions must ensure that this understanding of Complexity's ethos is shared across all strata of European societies, and that a 'Complexity approach' to knowledge is inculcated in young generations of Europeans form the earliest stages of their curricula.

References

Alberts, D. and Czerwinski, T. (eds.) (2003). *Complexity, Global Politics and National Security,* Washington, DC: National Defense University Press, ISBN 1579060463.

Armine, F. (1998). "The metamorphosis of the scientist," in D. Seamon and A. Zajonc, *Goethe's Way of Science: A Phenomenology of Nature*, Albany, NY: SUNY Press, ISBN 0585090629.

Baake, K. (2003). *Metaphor and Knowledge: The Challenge of Writing Science,* Albany, NY: SUNY Press, ISBN 0791457435.

Beyerchen, A. (1992). "Clausewitz, nonlinearity and the unpredictability of war," *International Security*, ISSN 0162-2889, Online ISSN 1531-4804, 17(3): 59-90.

Bronowski, J. (1977). *A Sense of the Future: Essays in Natural Philosophy,* Cambridge, MA: MIT Press, ISBN 0262021285.

Byrne, D. S. (1998). *Complexity Theory and the Social Sciences: An Introduction,* London, England: Routledge, ISBN 0415162955.

Butz, M. R. (1997). *Chaos and Complexity*, Taylor & Francis Publishing, ISBN 1560324198.

Capra, F. (1979). *The Tao of Physics*, London, England: Fontana Publications, ISBN 0006357075.

Capra, F. (1996). *The Web of Life*, New York, NY: Anchor Books, ISBN 0385476752.

Cilliers, P. (1998). *Complexity and Postmodernism: Understanding Complex Systems,* New York, NY: Routledge, ISBN 0415152860.

Clemens, W. (ed.) (2001). *The Baltic Transformed: Complexity Theory and European Security,* Lanham: Rowman and Littlefield Publishers, ISBN 0847698580.

De Landa, M. (2002). *Intensive Science and Virtual Philosophy,* London, England:

Continuum, ISBN 0826456227.
Deleuze, G. (2001). *Pure Immanence: Essays on a Life,* Cambridge, MA: MIT Press, ISBN 1890951242.
Dillon, M. (1996). *The Politics of Security,* London, England: Routledge, ISBN 0415129605.
Dillon, M. (2000). "Poststructuralism, complexity and poetics," *Theory, Culture & Society,* ISSN 0263-2764, 17(5).
Dupré, J. (1993). *The Disorder of Things: Metaphysical Foundations of the Disunity of Science,* Cambridge, MA: Harvard University Press, ISBN 0674212606.
Elbe, S. (2003). *Europe: A Nietzschean Perspective,* London, England: Routledge, ISBN 0415369754.
Farnsworth, R. (2001). *Mediating Order and Chaos,* Rodopi Publishing, ISBN 904201427X.
Foucault, M. (1970). *The Order of Things: An Archaeology of the Human Sciences,* London, England: Tavistock Publications, ISBN 0422718904.
Gleick, J. (1998). *Chaos,* London, England: Vintage, ISBN 0749386061.
Gleick, J. (2003). *Isaac Newton,* New York, NY: Pantheon Books, ISBN 0375422331.
Gutting, G. (2001). *French Philosophy in the Twentieth Century,* Cambridge, England: Cambridge University Press, ISBN 0521665590.
Hardt, M. and Negri, A. (2001). *Empire,* Cambridge, MA: Harvard University Press, ISBN 0674006712.
Hobsbawm, E. (1999). *L'age des extremes,* Bruxelles: Editions Complexes, ISBN 2870279922.
Hobsbawm, E. (2003). *The Age of Revolutions,* London, England: Abacus, ISBN 0349104840.
Husserl, E. (1965). "Philosophy and the crisis of European man", in *Phenomenology and the Crisis of Philosophy,* Q. Lauer (trans.), New York, NY: Harper Torchbooks, ISBN 0061311707.
Inwood, M. (2000). *Heidegger: A Very Short Introduction,* Oxford: Oxford University Press, ISBN: 0192854100
Latour, B. (1988). *Science in Action,* Cambridge, MA: Harvard University Press, ISBN 0674792912.
Jackson, W. (2003). *Heaven's Fractal Net: Retrieving Lost Visions in the Humanities,* Bloomington, IN: Indiana University Press, ISBN 0253216206.
Jervis, R. (1999). *System Effects: Complexity in political and Social Life,* Princeton, NJ: Princeton University Press, ISBN 0691005303.
Kuhn, T. (1970). *The Structure of Scientific Revolutions,* Chicago, IL: University of Chicago Press, ISBN 0226458083 (1996).
Morçöl, G. (2002). "What is complexity science? Postmodernist or postpositivist?," *Emergence,* ISSN 1521-3250, 3(1): 104-119.
Negri, A. (1994). *Nietzsche: La Scienza sul Vesuvio,* Bari: Laterza, ISBN 8842044431.
Pearson, K. A. (1994). *An Introduction to Nietzsche as a Political Thinker,* Cambridge, England: Cambridge University Press, ISBN 0521417228.
Pearson, K. A. (2002). *Philosophy and the Adventure of the Virtual: Bergson and the Time of Life.* London, England: Routledge, ISBN 0415237270.
Planck, W. (2002). *The Quantum Nietzsche: The Will to Power and the Nature of Dissipative Systems,* Writers Club Press, ISBN 0595209521.
Popolo, D. (2003). "French philosophy, complexity, and scientific epistemology: Moving beyond the modern 'episteme'," Emergence, ISSN 1521-3250, 5(1): 77-98.

Prigogine, I. and Stengers, I. (1997). *The End of Certainty: Time, Chaos, and the New Laws of Nature*, New York, NY: Free Press, ISBN 0684837056.

Rasch, W. and Wolfe C. (eds.) (2000). *Observing Complexity: Systems Theory and the Postmodern Episteme,* Minneapolis, MN: University of Minnesota Press, ISBN 0816632987.

Reahard, J. and McCarthy, A (eds.) (2001). *On Chaos and Order in the Enlightenment*, Amsterdam, Holland: Rodopi Publishing, ISBN 9042005505.

Rowland, H. (2001). *Goethe, Chaos and Complexity*, Amsterdam, Holland: Rodopi Publishing, ISBN 9042015667.

Rosenau, J. (1992). *Turbulence in World Politics, a Theory of Change and Continuity,* Princeton, NJ: Princeton University Press, ISBN 0691023085.

Sokal, A (1999). *Fashionable Nonsense: Postmodern Intellectuals' Abuse of Science,* New York, NY: Picador, ISBN 0312204078.

Strong, T. (2004). "Europe: A Nietzschean perspective," *Notre Dame Philosophical Reviews*, http://ndpr.nd.edu/review.cfm?id=1398.

Tiles, M. (2005). "Technology, science and inexact knowledge: Bachelard's non-Cartesian epistemology," in G. Gutting (ed.), *Continental Philosophy of Science*, Oxford, England: Blackwell Publishing, ISBN 0631236104.

Tomasello, M. (2003). *The Cultural Origins of Human Cognition*, Cambridge, England: Harvard University Press, ISBN 0674005821.

Urry, J. (2002). *Global Complexity,* Cambridge, England: Polity Press, ISBN 0745628184.

15. Human Knowing and Perceived Complexity: Implications for Systems Practice

Sandro Luis Schlindwein[1] & Ray Ison[2]
1 Research Group on Environmental Monitoring and Appraisal, Federal University of Santa Catarina, Brazil
2 Open Systems Research Group, The Open University, UK

"*Even when the individual trees are highly interesting and picturesque, it has use to see what the forest looks like in the large*" (Rescher, 1998: xvii).

Introduction

One of the strongest claims of the scientific revolution is that science provides an objective and better description of the natural world than other ways of knowing. However, the 'real-world' of human affairs seems to us to be different than the world simplified by science - we experience it as complex, or more complex than the world and the issues that are usually addressed by 'normal' science and its methods. For a long time complexity has been ignored by classical science, in which scientists described an objective world following deterministic laws. But we live embedded in 'situations of complexity'. As a result we increasingly become aware of manifestations occurring in natural and social phenomena for which traditional epistemologies based on mechanistic explanations are no longer valid. Instead of considering complexity as a temporary shortcoming arising from our limited or partial understanding of reality, or as something that has to be eliminated in order for scientific progress to proceed (Stengers, 2004), complexity has now been recognized as an emergence in the world we help to co-create and in which we live. Very often we hear the claim that we live in an era of 'expanding complexity'. However, for Morin (1999a) the problematique of complexity is still marginal in scientific, epistemological and philosophical thinking, although for him "complexity is not only in the foam of real phenomena. It is in its own principle." This is indeed very different from the perspective adopted by Hegel for whom "only in the surface reigns the game of irrational chance," and which has been the predominant and guiding perspective in science since its beginning.

To accept complexity means to turn upside down the view of classical science and a belief in the lawful structure of the world underlying the phenomenal surface. It means also to break an epistemic border. Therefore, the notion that all complex processes could be understood in terms of underlying simple universal laws is indeed an article of faith (Rosen, 1977). The growing acceptance of complexity, also as a worthwhile scientific issue, brought with it the emergence of complexity science. For Dent (1999: 5) "complexity science is an approach to research, study, and perspective that makes the philosophical assumptions of the emerging worldview." This growing 'movement' towards complexity has brought with it some challenging questions which must be addressed, like: what is complexity? Do we identify or do we experience complexity? How do we decide that something is complex, or how do we know (or learn) that something is complex? Are there objective criteria to decide that something is complex? How complex is complexity, or how complex is the discourse of complexity? What implications does a particular explanation of complexity have for our doing in the world?

It is not the aim of this paper to explore possible answers to these questions, but to discuss how they are dependent on the epistemological choices made by those answering them. It means that the explanations made about complexity might be built on distinct epistemologies, on different ways of knowing about the world. Furthermore, the (re)introduction of complexity into scientific discourse has raised issues, the significance of which goes far beyond its scientific interest, since they might determine the way we construct and engage with 'real-world' problems. All these issues will be discussed in this paper as constitutive parts of the epistemological problem of complexity. We also explore the different explanations of complexity which are current by pointing to what they reveal and conceal.

Current Understandings of Complexity

The word 'complexity' derives from the Latin word *complexus* and literally means 'to weave together'. However, since complexity has been (re) introduced into scientific discourse[1], it has been understood in different ways. According to Rosen (1977) our views regarding the concept of complexity have tended to be as richly varied as complexity itself. For Ottino (2004), for example, complex systems can be identified by what they do, and also by how they may or may not be analyzed. But there is no agreement upon a definition of complexity. Therefore, instead of proposing or assuming a particular definition of complexity, we will adopt here two general categories into which the existing explanations about complexity might be included.

Broadly we can say that in some explanations complexity has been understood as an intrinsic property of a certain kind of system, or as occurring in a certain kind of natural and social phenomena[2]. The kind of complexity emerging from this understanding can be called 'descriptive complexity'. All the attempts that have been made to quantify it or to achieve a quantitative measure of complexity might be included in this category. A very common example is the observation that forms the basis for most

1. For Le Moigne (1999) *complexity* appeared officially in the scientific literature after an historic paper by Warren Weaver published in *American Scientist* in 1948 (pp. 536-544) and entitled "Science and complexity" - this article is reprinted in this issue of *E:CO*.

2. As an aside it is worth noting that among the plethora of recent authors on complexity the notion of 'system' is often un- or under-theorized when linked to complexity, as in 'complex system' or 'complex adaptive system'. This of course has practical implications.

characterizations of the complexity of an object: the complexity is directly proportional to the length of the shortest possible description of that object, as is reported by Casti (1995). This seems to be also the perspective of Cilliers (1998) for whom "complex systems do have characteristics that are not merely determined by the point of view of the observer." From an epistemological perspective, 'descriptive complexity' is based on the assumption of the existence of an objective reality, external and independent of us, and to which we can have privileged access, resulting in the assumption that complexity can be objectively measured.

In other explanations complexity has been understood as the result of a distinction or resulting from a particular perception of a situation (of complexity) made by an observer, what can be denominated as 'perceived complexity'. Casti (1995) recognized the role of the observer in the acknowledgment of complexity. For him:

"*[T]he complexity is an inherently subjective concept; what is complex depends upon how you look. When we speak of something being complex, what we're doing is making use of everyday language to express a feeling or impression that we dignify with the label complex. But the meaning of something depends not only on the language in which it is expressed (i.e., the code), the medium of transmission and the message, but also on the context. In short, meaning is bound up with the whole process of communication and doesn't reside in just one or another aspect of it. As a result, the complexity of a political structure, a national economy or an immune system cannot be regarded as simply a property of that system taken in isolation*" (pp. 269-270).

This kind of reasoning in which the observer is part of the description was firstly developed by Heinz von Förster, and as he was asked to give an explanation of what complexity is all about, he said that it resides more in the eye of the beholder than in the observed thing (according to an interview reported by Passis-Pasternak, 1993). However, Casti (1995) definitely admits that "whatever complexity a given system might have it is a joint property of the system and its interaction with another system, most often an observer and/or controller... So just like truth, beauty, good and evil, complexity resides as much in the eye of the beholder as it does in the structure and behavior of a system itself." Le Moigne (1990) in his book about the theory of a general system also claims that complexity results from the perception of an observer. Therefore, for us, and based on what we have already said, 'perceived complexity' does not result from cognitive incompetence, as suggested by Rescher (1998).

In contrast to 'descriptive complexity', the epistemological assumptions of 'perceived complexity' are related to the epistemologies based on the assumption that reality results from the distinctions made by an observer. According to this epistemology the explanations we make about the world are not independent of us as explainers. However, this kind of epistemology has been associated with subjectivism, which according to the mainstream scientific thinking is opposed to objectivism, and therefore very often has been considered non-scientific, since objectivism is one of the core assumptions of classical scientific thought. It must be remembered that as complexity has been expurgated from the dominant scientific discourse, which is inspired by the intelligibility criteria of classical dynamics, until very recently also human subjectivity has been expurgated from the same discourse, where realism still is the predominant epistemology. To some extent, it seems to be compelling to deny subjectivism in the same way as complexity has been denied for a long time, since the shift to a scientific discourse which not only acknowledges situations of complexity but also accepts the uncertainty linked to it, is not straightforward. But we also reject the simplistic dichotomy (dualism) objectivity-subjectivity, and in this paper subjectivity will be employed only to remind us that it is not possible to separate the observer from the world s/he describes. Furthermore, it must be remembered that the adoption of one or another of these epistemologies does not necessarily follow rational criteria, but seems to be a matter of preference or emotion.

In his attempt to define modes of complexity, Rescher (1998) distinguished three 'modes', namely *epistemic*, *ontological* and *functional* complexity. Among these modes of complexity, the epistemic embraces three categories: *descriptive*, *generative* and *computational* complexity. Although he made an effort to define in different terms these three types of epistemic complexity, in our view they are very similar and closely related to each other. For him, however, descriptive complexity is the most fundamental form. It is remarkable that in the distinctions made by Rescher (1998) there is no place for what is discussed here as 'perceived complexity'. This may be interpreted as strong evidence for the dominant objectivistic epistemology held, whether implicitly or explicitly, by the majority of those who are searching for an explanation for complexity.

These different understandings and explanations about complexity are nurturing the emergence of a science of the complex or a science of complexity. For Casti (1995) the emergence of a scientific theory of complexity "would be a major step toward the development of a framework within which we can begin to understand how to control and manage what our [cognitive] maps tell us are complex processes." Despite Casti's intention of controlling complexity, revealing the extent to which he still is a hostage of classical epistemologies, we will discuss later one possible framework for managing complexity in the world, what we understand as 'systems practice'. Furthermore, it is also necessary to develop what could be called an 'epistemology of complexity', as a further step to lay down a common ground over which to develop 'complexity thinking' or 'complexity practice'.

An Epistemology of Complexity

For the purpose of this paper the two general categories of complexity we have defined, as well as their epistemological and ontological implications, will be understood as complementary parts of the epistemological problem of complexity. One of the first and more important authors to address this issue and to make claims for an epistemology of complexity was the French philosopher Edgar Morin (Morin, 1999a, 1999b, 2002). For him (Morin, 1999a) complexity is marginalized among the classic scholars of Anglo-Saxon epistemology (Popper, Kuhn, Lakatos, Feyerabend, among others), and from the perspective of epistemology there is only one exception: Gaston Bachelard, who considered complexity a fundamental issue, and that there is nothing simple in nature, only the simplified.

The epistemological problem of complexity raises some fundamental cognitive issues of how human beings know about the complexity of the world they live in. For instance, from the perspective of 'descriptive complexity', and considering the strong influence of objectivity in science, the verification of whether the behavior of a '(natural) system' is 'linear' or 'nonlinear' frequently has been used as a validation criteria to decide about its complexity. However, we are not keen to provide an 'externalist' explanation of complexity. Instead, we assume here that none of us share a common experiential world. All we have at our disposal is our ability to communicate about our worlds of experience and, sometimes, a history of living in a common culture, including language, over a period of time. The sharing of a common culture allows us to appreciate the apparent paradox between our individual and unique cognitive histories and our experience that collectively we do not experience the world in relative or subjective (meaning here the lack of regularities) ways. And it is this unique cognitive history we each have as human beings that is denied when only an objective explanation of complexity is pursued and accepted as scientifically valid.

Therefore, and following the explanations coming from the biology of cognition, we assume that the way human beings know about the complexity of the world is a biological phenomena. Human beings make distinctions about their world according to their biological cognitive structure, and not according to the structure of the world around them. As argued by Martinez (2001) "biology creates thought and thought creates biology. Just as mind and body cannot be separated, to attempt a separation of mind and world would create an artificial split between observer and observation that assumes we can 'step out' of the world we are attempting to observe." As a result, we claim that complexity is not, as commonly accepted, a specific attribute of physical, biological and social phenomena, but instead a transdisciplinary epistemological assumption, and a way of knowing about the world.

A different cognitive approach can be found in Rescher (1998). In his book *Complexity: A Philosophical Overview* he dedicates considerable space to discuss the cognitive aspects of complexity - or how human beings can know about the complexity of their world - particularly in Chapter 3. For him, in the development of knowledge, progress is always a matter of complexification, since nature is ontologically complex. According to him (p. 58) "to claim the ontological simplicity of the real is somewhere between the hyperbolic and absurd," and the commitment of scientific method to simplicity is nothing else than a matter of the procedural principle of least effort. Now, to assume that the commitment to simplicity results from the procedural principle of least effort [sic] is not only too simplistic, but also seems to deny any epistemology behind this approach. Furthermore, this kind of thinking reveals an epistemological commitment to objectivity, which is the basis of 'descriptive complexity'. This is his typical stance, as he says that "substantial extensions in our cognitive access to new sectors of nature's parametric space is made possible through the enhancement of investigative technology." In other words, our cognitive limitations regarding complexity might be overcome by technology. This seems to miss the point.

To admit human experience as key to our understanding of complexity does not mean that everything said or done is valid or even that we will find as many distinctions of complexity as living human beings. Although the number of possible distinctions of complexity may appear to be infinite, the diversity of these distinctions will be constrained through human community life, the sharing of culture and history, and through collective interests and preferences. The contrary is nothing more than a fallacy frequently remembered by those who adopt an objectivist epistemology to make explanations about the world, and deny any other approach as plausible. The claim being made here about the understanding of complexity is, therefore, that its recognition is a cognitive process prescribed by the biological structure of human beings, rather than as already existing in the objects of the world and which can be identified and measured. It is therefore of great interest to investigate the interrelationship between biology and epistemology, and how to address the epistemological problem of complexity from the perspective of the biology of cognition, also because we must be aware that the extent to which we as human beings can know about the complexity of the world is constrained by our cognitive limitations.

'Perceived complexity' is therefore necessarily related to complexity thinking. Complexity thinking is based on the assumption that subject and object, although not being the same, also are not radically separate. A complete (total) separation would make knowledge impossible (Ciurana, 2004). As claimed by Casti (1995), complexity resides as much in the eye of the beholder as it does in the structure and behavior of a system itself. And for Morin (1983) the kind of complex thinking he is suggesting requires the reintegration of the observer in his observation. For him, whereas the traditional epistemologies are characterized by disjunction and fragmentation, complexity thinking is a kind of thinking capable of rejoining, or contextualizing knowledge (Morin, 1999b).

To some extent the epistemological problem of complexity can be considered a particular case of the more general epistemological problem of how we human beings know about the world, and what constitutes evidence about that world. Therefore, most of the issues highlighted in this paper are undecidable, i.e., we cannot give definitive explanations about them, and our claim is restricted only to the reintroduction of the role of the observer into the explanations about complexity.

Implications for Systems Practice

Our concern in this paper is not only with the way human beings know about complexity (what is addressed here as the epistemological problem of complexity), or to discuss only the philosophical-theoretical aspects of this issue, but with its practical significance especially as it relates to building capacity for systems practice, understood as a way for managing in situations of complexity. We are adopting here a different perspective than that of Rescher (1998) for whom the complexity of real problem-situations outruns our capacity for their management. Our motivation in this field is to build a praxiology for 'systems practice' as a generic form of managing complexity in the world. Some attempts at developing methodological devices combining insights coming from systems thinking and complexity can be found in the literature, as in the systemic intervention methodology of Midgley (2003).

One of the most intuitive insights regarding complexity is that 'complex systems' [and complex processes] or situations of complexity are irreducible, i.e., "you cannot start slicing up systems of this type into subsystems without suffering an irretrievable loss of the very information that makes these systems a system" (Casti, 1995). A remarkable aspect is that the emergence of some complex features can only be recognized if we look at the whole system and not its parts. The adoption of this perspective originates from systems thinking, a kind of thinking which deals with wholes, boundaries and emergent properties instead of parts. We claim therefore that systems thinking needs to be situated at the origins of the (re)introduction of complexity into scientific discourse, and should also be considered the theoretical basis of 'complexity thinking' and 'complexity practice'. As Checkland (1999) has pointed out, in the process of coping with the complexity of the world, the word 'system' is no longer applied to the world, it is instead applied to the process of our dealing with the world. Systems thinking and complexity thinking share therefore the same claim according to which a mechanistic worldview gives way to an organic worldview, despite the fact that some authors, such as those cited by Midgley (2003), contrast complexity and systems thinking.

For Morin (1999a) 'complexity' appeared as a marginal stream between engineering and science, in cybernetics and systems theory. This may explain why complexity was marginal among the classic names of Anglo-Saxon epistemology, since systems thinking itself can be considered only marginal in relation to the main streams of scientific thinking. But systems thinking in its many traditions has evolved into an approach for making sense of complexity, for managing complexity, although we must be aware, as advised by Senge (1990: 281), that "systems thinking shows that there simply is 'no right answer' when dealing with complexity." Therefore, we also avoid the terms 'manage' and 'managed', with their deterministic overtones, in favor of 'managing' which is an active process associated with our daily living.

Ison (2004) describes the fundamental epistemological choice to be made as to see system or complexity either:

1. As something that exists as a property of some thing or situation; and that, therefore, can be discovered, measured and possibly modeled, manipulated, maintained or predicted, or;
2. as something we construct, design, or experience in relationship to some thing, event, situation, or issue because of the distinctions - or theories - we embody.

These options can be clearly associated with the two modes of complexity - descriptive and perceived complexity - discussed above. The choice of any one of those options has profound implications for our being and becoming in the world - and thus for our daily living as we do what we do. It means that the adoption of an understanding of complexity will determine the nature of the engagement of a systems practitioner with a 'real-world' situation. For Cilliers (1998: ix) engaging with complexity entails engaging with specific complex systems. For us, however, and from the epistemological perspective we adopt, engaging with complexity entails engaging in situations of complexity and using systems or complexity thinking to learn our way towards purposeful action that is situation improving.

Some implications in making an epistemological choice are suggested by Ison (2004). According to him, by choosing the first option we seduce ourselves into the belief that we can say a lot about the nature of the thing, but this choice says almost nothing about the practitioner concerned with the thing. Choosing the latter position, on the other hand, tells us something about the practitioner and about what they know and are able to do, as well as about their relationship to the thing or event they experience - how the 'thing' itself arises in our act of distinction. Ison (ibid) emphasizes that, "making a choice of one epistemological position or another in a given context is not an act of discarding or deciding against the other position - it is an act of being aware of the choice being made" and taking responsibility for it. Being epistemologically aware opens up more choices for action. Following Maturana (1988) responsibility replaces objectivity as the ethical basis of praxis.

Acknowledgements

Part of this paper was written when the first author was a Visiting Research Fellow at the Systems Department, Centre for Complexity and Change, The Open University, Milton Keynes (UK). The hospitality and support of staff of the Department as well the support of CAPES (Ministry of Education, Brazil) are highly appreciated.

References

Casti, J. L. (1995). *Complexification: Explaining a Paradoxical World Through the Science of Surprise*, New York, NY: Harper Collins Publishers.

Checkland, P. (1999). *Systems Thinking, Systems Practice*, Chichester, UK: John Wiley.

Cilliers, P. (1998). *Complexity & Postmodernism: Understanding Complex Systems*, London, UK: Routledge.

Ciurana, E. R. (2004). "Complexidade: Elementos para uma definição," in *Ensaios de Complexidade 2*, Carvalho,E. d. Assis; Mendonça,T. (org.), Porto Alegre: Sulina, pp. 48-63.

Dent, E. B. (1999). "Complexity science: A worldview shift," *Emergence*, 1: 5-9.

Ison, R. (2004). "Part 3: Understanding systems approaches to managing complexity", in *T 306 Managing Complexity: A Systems Approach. Block 1. Juggling with Complexity: Searching for System*, 2nd edition, Milton Keynes, UK: The Open University, pp. 59-139.

Le Moigne, J. -L. (1990). *A teoria do sistema geral: Teoria da modelização*, Lisboa: Instituto Piaget.

Le Moigne, J. -L. (1999). "Inteligência da complexidade," in *O Pensar Complexo, Edgar Morin e a Crise da Modernidade*, Pena-Vega, Alfredo and do Nascimento, Elimar P. (org.), Rio de Janeiro: Garamond, pp. 47-88.

Martinez, M. E. (2001). "The process of knowing: A biocognitive epistemology," *The Journal of Mind and Behavior*, 22(4): 407-426.

Maturana, H. (1988). "Reality: The search for objectivity or the quest for a compelling argument," *The Irish Journal of Psychology*, 9(1): 25-82.

Midgley, G. (2003). "Science as systemic intervention: Some implications for systems thinking and complexity for the philosophy of science," *Systemic Practice and Action Research*, 16(2): 77-97.

Morin, E. (1983). *O problema epistemológico da complexidade*, Publicações Europa-América.

Morin, E. (1999a). *Ciência com consciência*, Rio de Janeiro: Bertrand Brasil.

Morin, E. (1999b). "Por uma reforma do pensamento", in O *Pensar Complexo. Edgar Morin e a Crise da Modernidade, Pena-Veja, Alfredo and do Nascimento*, Elimar P. (org.), Rio de Janeiro: Garamond, pp. 21-46.

Morin, E. (2002). "Complexidade e Ética da Solidariedade", in *Ensaios de Complexidade, de Castro, Gustavo; Carvalho, Edgar de A. and de Almeida*, Maria da C. (org.), Porto Alegre: Sulina, pp. 11-20.

Ottino, J. M. (2004). "Engineering complex systems," *Nature*, 427: 399.

Pessis-Pasternak, G. (1993). "Heinz von Förster, Pioneiro da Cibernética," in *Do Caos à Inteligência Artificial: Quando os Cientistas se Interrogam*, SãoPaulo: Editora da UNESP, pp. 197-206.

Rescher, N. (1998). *Complexity: A Philosophical Overview*, New Brunswick: Transaction publishers.

Rosen, R. (1977). "Complexity as a systems property," *Int. J. General Systems*, 3: 227-232.

Senge, P. (1990). *The Fifth Discipline: The Art & Practice of the Learning Organization*, London, UK: Random House.

Stengers, I. (2004). "The challenge of complexity: Unfolding the ethics of science. In Memorium Ilya Prigogine," *Emergence: Complexity and Organization*, 6(1-2): 92-99.

16. On the Importance of a Certain Slowness

Paul Cilliers
University of Stellenbosch, ZAF

"In philosophy the winner of the race is the one who can run most slowly. Or: the one who gets there last."

Wittgenstein (*Culture and Value*)

Introduction

As a result of a whole range of what one could call "pathologies" in contemporary culture, the idea of "slowing down" has of late been mooted in a number of contexts[1]. A few can be named briefly. The "Slow Food" movement, which started in Italy but has a worldwide following, extols the virtues of decent food made from decent ingredients without compromise. The resistance shown to "junk food" is not only based on aesthetic considerations, but also on ethical (and nutritional) ones. The movement promoting "Slow Cities," also of Italian origin, fosters an understanding of cities that is more humane. Such cities should encourage walking rather than driving, have small shops with local products rather than shopping malls, and, in general, provide opportunities for the community to interact, not to live in isolation. "Slow schooling" is a movement that questions educational processes in a world geared for instant results. It emphasizes the contextual nature of knowledge and reminds us that education is a process not a function. On a more personal level, "slow sex" involves attitudes that try to prevent that the values of the marketplace ruling in our intimate relationships. We need to recognize that the journey is more important than the destination, and that takes time. An immediate or perpetual orgasm is really no orgasm at all.

There are a number of very important issues at stake in these examples. In what follows, however, the focus will not be on these social movements as such, but on some of the underlying principles that make the debate on slowness an important one. Through an analysis of the temporal nature of complex systems, it will be shown that the cult of speed, and especially the understanding that speed is related to efficiency, is a destructive one. A slower approach is necessary, not only for the survival of certain important values or because of romantic ideals, but also because it allows us to cope with the demands of a complex world in a *better* way.

The argument will be made initially by briefly analyzing current distortions in our understanding of time. These distortions result, on the one hand, from the rational and instrumental theories we have about a modern world, and, on the other, from the effects of certain technologies,

1. See Honoré (2004) for a discussion of the emergence of several movements that challenge the "cult of speed."

especially communication and computer technologies. In order to show why these are "distortions," or at least to show why these distortions are problematic, the temporal nature of complex systems will be discussed. The relationship between memory and anticipation will be central to this discussion, but attention will also be paid to the importance of delay and iteration. These characteristics of complex systems have important implications for our understanding of the formation of identity, both individual identity as well as the identity of groups[2]. In closing, a number of general cultural issues involving the fast and the slow will be looked at.

It is important to realize that the argument for slowness is *not* a conservative one; at least not in the political sense of the word. It is not merely backward looking nor a glorification of what has been. Although it emphasizes the historical nature of knowledge and memory, the argument for slowness is forward looking: it is about an engagement with the future as much as with the past. Slowness is in itself a temporal notion, and in many ways the opposite of the notion "static." In point of fact, it is actually an unreflective fastness that always returns you to the same place.

It should also be stated up front that there is no argument against an *appropriate* fastness. A stew should simmer slowly, but a good steak should be grilled intensely and briefly. The argument is against unreflective speed, speed at all cost, or, more precisely, against speed as a virtue in itself: against the alignment of "speed" with notions like efficiency, success, quality, and importance. The point is that a system that has carefully accumulated the relevant memories and experiences over time will be in a better position to react quickly than one that is perpetually jumping from one state to the other.

Perhaps "slow" and "fast" are not exactly the correct terms to use. Terms like "reflective" and "unreflective," or "mediated" and "unmediated" may be more accurate. Nevertheless, the debate taking place uses "slow" and "fast," and the terms do have a certain rhetorical significance. If we stay with their use, it is done in a metonymical way. The whole point of this paper is to give them a richer meaning.

Living in the Present

In *Time: The Modern and Postmodern Experience,* Helga Nowotny (1994) argues for a certain shift in our experience of time. In short, in my paraphrase, and incorporating insights from Bauman (e.g., Bauman, 1992), the argument is the following: One of the main aims of the instrumental rationality flowing from the Enlightenment was to create conditions in which we are not controlled by contingency. To achieve these conditions, it is necessary to understand, and preferably *control,* the future. This demands coordinated and goal-oriented action in the present. Modernism becomes a

2. The fact that complex systems demand plural descriptions is in itself something that takes time. Although the link between plurality and delay has to my knowledge not been made explicit, it is implied where there are arguments for multiple descriptions and when "process" is emphasized. See for example Richardson, 2005.

project that demands our total commitment against the forces of irrationality and chaos.

The modernist project has two important effects on our understanding of time. In the first place, our actions need to be coordinated. This can only happen if time is universalized in such a way that we all live in the "same" time. This was achieved mainly through technology – that is, the construction of accurate clocks – and by regulating time globally. Instead of each person or local community living in their own time, it was necessary to synchronize time in such a way that activities in, say, New York and Paris could be correlated. The effects of this, however, go much further than merely synchronizing time in different parts of the globe. It also means that private time and public time are synchronized. We have to live our lives according to a generalized and controlled understanding of time. A subjective, or should one say phenomenological, experience of time has to be sacrificed in order to generate a universal temporal framework in which we can operate efficiently.

The second effect of instrumental rationality on our understanding of time is a result of the desire to control the future; for the future to be made knowable. This would only be possible if the future, in some essential way, resembles the present. We cannot anticipate what we do not know, and therefore we should do everything in our power to create a future that does not disrupt the steady progress we are making toward a better world. This modernist strategy is perhaps exemplified best in Hegel's dialectic of history, which is supposed to converge toward an ultimate solution. The actual result of this ideology is to extend the present into the future, causing us to live in a perpetual "present." This collapse of the diachronic into the synchronic allows instantaneous interaction between everybody; it creates a world that is fast and efficient. The sacrifice made in order to achieve this, however, is nothing short of sacrificing the very notion of temporality. Nowotny (1994: 16) calls it "the illusion of simultaneity."

The way in which contemporary society lives in an eternal present, or what Eriksen (2001) calls the "tyranny of the moment," is made possible, and augmented, by the surge in technology, especially computer and telecommunication technology. We are instantaneously in contact with everybody, everywhere. Not only has the distinction between home and the workplace collapsed, but also the distinction between work time and private or leisure time. It is expected of many of us to be available, always and everywhere[3]. This state of affairs may have been less detrimental if it did not also demand instant response. The very reason for mobile phones and email lies in the fact that immediate response is possible. It is in this "immediate" that the main problem lies. There is less and less time for reflection. Reflection involves delay, and in a cult of speed, delay is unacceptable. This move away from reflection to immediate response has profound implications for our understanding of what it is to be human (see Parkins, 2004: 376-379), to which we shall return.

3. When arrangements are made for more flexible working hours to facilitate parenting, for example, this is already a form of resistance to speed and efficiency.

The "illusion of simultaneity," the idea that if we live quickly and efficiently in the present we are somehow closer to reality, is nevertheless exactly that: an illusion. We cannot escape our temporal nature, and our persistence in trying to do so can only lead to pathology. The necessity of delay and reflection needs to be re-evaluated. This can be done from a number of perspectives. A Freudian analysis would show that instant gratification is actually a destruction of pleasure. More sublime pleasure can be found only if desire is delayed, anticipated as a memory of something still to come, yet something that should also in principle be able to surprise us. Derrida calls the illusion of living in the present, of thinking that we have access to an objective understanding of reality if we live "in" it, the "metaphysics of presence" (Derrida 1976: 49). He introduces the notion of *différance* specifically to undermine the metaphysics of presence (62). *Différance* is a notion that intertwines difference (as a spatial notion, one could say) and delay (to defer, a temporal notion) as the engines of meaning (Derrida, 1982). The present consists only as a combination of memory (of what has been) and anticipation (of what is to come).

In his novel *Slowness,* Milan Kundera (1996) uses the metaphor of somebody riding on a motorcycle as being constantly in the present. Speed and the demands of the machine reduce his horizon to something immediate. Someone walking, however, is moving at a pace that allows for a much wider horizon. The stroll unfolds in time in a way that opens up reflection about where we are coming from and where we are going to as we walk. This theme of both the past and the future being present in a meaningful experience of the present could be pursued in much more detail from both a Freudian and Derridean perspective – and several others too – but the argument for a meaningful temporality – that is, something slower – will be made here from the perspective of the dynamics of complex systems.

Complex Systems, Temporality and Memory

An important aspect of complex systems, one that certainly complicates our understanding and modeling of such systems, is their temporal nature. Complex systems unfold in time, they have a history that co-determines present behavior and they anticipate the future. Moreover, as we know at least since the work of Prigogine, the behavior of complex systems is not symmetrical in time. They have a past and a future that are not interchangeable. This being "situated in time" does not always receive adequate attention in our analysis of complexity.

The central notion at stake when we talk of time and complexity is that of "memory." Memory is the persistence of certain states of the system, of carrying something from the past over into the future. It is not merely the remembering of something in the past as if belonging to that past, it is the past being active in the present. We should therefore not think of memory in abstract terms, but of memory as something *embodied* in the system. In many respects the system *is* its memory. If one accepts an understanding of complexity that emphasizes the relational nature of the system, it is useful to think of systems as networks where the connections between the nodes are more important than the nodes themselves. The nature of these connections is a result of which states of the network are "retained," thus the structure of the system is a result of the sedimented history of the system[4].

It is important to remember that memory is not merely a cumulative process. The structure in the network of relationships can only develop if certain states of the network are *not* maintained. Memory is a result of a process of selection. The states that are significant are repeated more often and therefore form more permanent links in the network. Less significant states will fade away over time. Memory is only possible if the system can also forget[5]. What is important to note at this stage is that memory is not an instantaneous thing, it takes time to develop, it is slow.

If one characterizes memory as the past being carried over into the future, it follows that the future can only be anticipated in terms of the memory of the system. Anticipation is *not*, or at least should not be, simply an extrapolation of the present. It is a complex, nonlinear process that tries to find some trajectory, some way of "vaulting" from that which has already been experienced to that which has to be coped with. The quality of the anticipation is a function of the quality of the memory. A more varied, richer, deeper, and better-integrated memory will open up more sophisticated anticipatory capabilities.

The obvious question now would be to ask *how* such a rich memory is formed. This is a complex issue, but for the sake of the argument at stake here, one can say the following: Memory is information from the environment that has been filtered, it is that which has been interpreted – by the memory already sedimented in the system – as significant. The identity of the system is, in some sense, its collection of dynamic memories. The implication is that the system cannot reflect, or act on, everything that is going on in the environment at a given moment. If that were the case, the system would always be merely a reflection of its environment and would have no identity of its own. In order for it to be a system at all, a system that has its own identity, that can react to the environment and not merely mirror it, a certain hysteresis is required. The system must be slower than its environment[6].

4. This argument can also be made using the example of the brain, and links with many Freudian arguments in an interesting way. See Cilliers (1998: 45–47, 92, 108) for further discussion.

5.This process is known as the "use principle" or Hebb's rule. For more detail see Cilliers (1998: 17-18, 93-94).

6. Perhaps this claim needs to be qualified somewhat. Strictly speaking, the system must change at a different rate than its environment. There are many activities within as well as outside the system (without introducing the problem of boundaries here; see Cilliers, 2001), thus it is difficult to talk of the "rate of change" as a single measure. Some processes in the system may be faster than the environment, but they are possible only within the context of more general structures that have been formed through iterative processes that involve recycling and delay. One could perhaps say that the system should be slower than its environment "on average," but this could lead to other confusions since "average" is a problematic notion

The notion of hysteresis is an important one[7]. It links to the notions of delay and *différance* discussed above. An event in the environment of the system does not have inherent and immediate significance for the system. Its significance is established in time as it is re-enacted in the system and carried over into the future. In a way, the significance of an event has always already been established (in terms of the memory of the system), but never completely or finally, since the significance is always also to be determined by what is still to come. The system has to hang on to some aspects with a certain tenacity: not let go of them too quickly. There is risk involved in this, of course. The system has to invest resources in this process. It cannot maintain everything; it has to select. If too many of the wrong things are carried over it will impair the system's performance. However, if not enough is carried over, it will also fail.

To put it in slightly different terms: The system has to find a way to discriminate between information and noise[8]. If it follows every trend in its environment, it will also be following noise. If it reacts too slowly it will only follow the low-frequency trends, which may also be just noise. The system must be stable enough not to be buffeted around by every fluctuation, and it must be flexible enough to be able to adapt when necessary. Where this optimal point lies is not a question that can be answered from an objective viewpoint. The balance between stability and change is a contingent thing that plays itself out in time. What one can say, though, is that merely to be fast will destroy the system.

The argument for a certain slowness should start to take shape now. A viable system has to be able to *resist* some of the dynamics in its environment. There should be a temporal space in which the past is allowed to play itself out in interaction with the present. There must be time for reflection and interpretation. The faster the system becomes, the shallower its resources will be. Ultimately quick behavior will be no more interesting than Brownian motion.

It must be stressed again that the argument for a certain slowness is not a conservative argument. A certain amount of conservation is a prerequisite for a system to maintain itself, of course. The important point, to which we shall return, is that a "slow" strategy is not a backward-looking one. If a somewhat slower tempo allows a system to develop a richer and more reflective memory, it will allow the system to deal with surprises in its environment in a better way. The argument of slowness is actually an argument for appropriate speed. There is no objective or immediate rule for what that speed is. If anything, it is a matter of experience, and experience

---

in this context. At the very least one could say that processes involving delay are crucial in identity formation.

7. Hysteresis is the "lagging of effect when cause varies" (*Oxford Concise English Dictionary*).

8. The distinction between information and noise is, for the system, a strategic choice. There is no pre-determined criterion by which they can be separated. This, however, does not imply that all noise is information just waiting to be framed in the appropriate way. See the discussion of Taylor below.

(as Aristotle urged) has to be gained, it cannot be "given" in an immediate way. It is experience that determines which piece of meat should be fried quickly and which should simmer slowly in the stew. She who fries everything quickly will only have something nice to eat now and then, and then purely by chance.

Integrity, Identity and Reflection

In his novel *The Discovery of Slowness,* Sten Nadolny (2003) gives us a fictionalized account of the life of John Franklin. Franklin, a 19th century explorer primarily obsessed with finding the Northwest passage, is slow. His advance (in the Royal Navy) is also slow, mainly because being slow is confused with being stupid. Since he is not stupid he is gradually awarded command, and those working with and under him discover the advantages of being slow. Franklin is persistent, dependable, and trustworthy. Even in war, thorough reflection pays dividends that are not always immediately apparent. His political career, as governor of Van Diemen's Land (now Tasmania), ends badly only because he is disgraced by those out for quick and selfish results. His principles are not negotiable.

Franklin is a worthwhile human being because he has integrity. There is a substance to his personality that may seem opaque at first, but eventually shows itself as solidity. The nature of his integrity is directly coupled to his slowness. He assimilates, integrates, and reflects before he acts. This is sometimes a ponderous process, and he pays a price for it. Under normal circumstances it is easy not to notice someone like this, or to pass him by, but when there is a crisis, it is him people turn to. He can be trusted, he will come up with something. This is most significant. It is exactly when one would think that being fast is what is required that slowness proves its worth.

The link between slowness and integrity is also an issue in J. M. Coetzee's (2005) novel *Slow Man.* Here we have a character who resists change, despite the cruel demands being made on him. He clings to a set of values that are important to him, and this gives his personality substance. However, he is too stubborn, and eventually he cannot adapt to new circumstances. One has tremendous sympathy for him, but he turns out to be too slow, and pays the price for it. Even so, it is clear that when there is a choice between the loneliness of the slow and the superficial companionship of the quick, the author sides with the slow. Integrity is more important than a certain kind of success.

Despite Coetzee's darker view, there is no reason why slowness should be solitary and sad. Quite the contrary is true. In his novel *Slowness,* Milan Kundera (1996) shows with great conviction how a certain slowness is a prerequisite for being fully human. What is at stake in this novel is not moral integrity, or a kind of Calvinist dependability, but the sensuality of human interaction, the beauty of a relationship that unfolds in time, the ecstasy of a love that has a history and a future. Being human implies having a body, something with its own rhythms and demands. If we reduce all of this to something merely instrumental, to transactions written in legal terms (not in lyrical prose), if we demand results *now*, then we will stop being human. Language cannot be reduced to a code; it plays itself out in a certain context. What is more, even if we immerse ourselves in the context we have to wait beyond the last sounds. When all is said, the meaning has not finally arrived yet. It is the anticipation of what it could yet mean that draws us forward. *Einmal ist keinmal.*

Many may feel that the novel is an outdated art form, something to be replaced with the fast and immediate communication of the digital code. In his book *The Moment of Complexity,* Mark Taylor (2003) seems to lean in this direction. For him, something of a paradigm shift has occurred in the last few decades. We live in a new world with new forms of communication and new forms of learning and human interaction – something he seems quite willing to sell. He resonates with a fast world, something new and exciting.

Taylor's emphasis on the new goes hand in hand with a nearly interchangeable use of the notions "noise" and "complex." This problematic conflation is the result of an understanding of complexity primarily informed by chaos theory and of information as entropy. Such an understanding, inspired by the theories of Shannon and Chaitin, will attribute the highest information content to a purely random sequence[9]. Although these notions are important in the context of computation, they are less useful when talking about complex systems in general. Living systems, including the social systems that Taylor explores, are neither random nor chaotic. Despite the fact that they are constituted through nonlinear interaction and that they are capable of novel and surprising behavior, they are well structured and robust. They persist through time and maintain themselves. When we encounter behavior that we do not understand, or cannot decode, it often looks like noise, but once it is understood we can see the patterns. These patterns are not merely or only an order imposed by the observer, but also characteristics of the system itself. Complexity may look like noise, but all noise is not something complex waiting to be decoded. Sometimes noise is just noise[10]].

Taylor's argument is seductive, but, to my mind, wrong if not harmful. In his fervor to embrace the posthuman, he looks at the history of being human with a certain disdain. It seems as if he thinks that complexity is a recent discovery and forgets that being human has always been complex. He embraces the present and wants to deal with it quickly and efficiently. We can be educated instantaneously by electronic means and thus we should make a radical break with old methods. In his excitement he forgets that complex systems, even those in our postmodern world, are constituted historically, that they develop and change, and that one of their primary functions is to distinguish between information and noise. This cannot be done at the press of a button.

9. See Hayles (1999) for a discussion of these issues. Primary sources are Shannon (1949) and Chaitin (1987).
10. A similar, and more detailed argument is made in Cilliers (2005).

The ideas of the posthuman and the cyborg are of undeniable importance[11], but in our enthusiasm to embrace new modes of being we should be careful not to effect a transformation into something inhuman[12]. Machines are fast, but they are machines. The present argument is not for an *a priori* rejection of the possibility of machines with human capabilities, or one that denies the intimate relationship that humans have always had with technology. Our cultural existence presupposes the use of tools. The difference between using a quill and a word-processor may have huge implications on a practical level, but they also share some essential features[13].

The notion "posthuman" is thus an ambiguous one. If it signifies a tight coupling between the body and technology, we have always been posthuman. If it signifies the obsolescence of the body, perhaps in the sense that a "person" could be downloaded instantaneously as software and run on a machine, it becomes a problematic notion at odds with the idea that human identity is also the result of a certain temporal embodiment. The general argument presented here maintains that any complex system, organic or not, would have to incorporate a certain slowness.

The need for slowness, and a warning against the embracing of the fast, can perhaps be motivated best from the perspective of philosophy. Philosophy, in its most general form, is essentially the art of reflection. Wendy Parkins (2004) analyses contemporary culture as one moving away from reflection, and argues that what we need is an "ethics of time." She does not elaborate much on what such an ethics should look like, but it is something that needs careful attention, not only from a moral perspective but also from the purely pragmatic perspective of how to live and survive in a fast world. Such an "ethics" will be complex in itself. It will have to unfold in time and be conscious of its own temporal nature.

For now, instead of rushing around like the Red Queen in a world where change is virtuous merely because it is change, we can start by taking some time out to reflect. At this point the argument for slowness becomes a political one: We should put up some resistance to a culture in which being fast is a virtue in itself. We should say "no" with a little more regularity.

11. See Badmington (2000) for a collection of philosophical essays on the posthuman.

12. See Hayles (1999) for a detailed discussion of cybernetics, the development of the posthuman, and the importance of embodiment. See Braidotti (2005) for an affirmative discussion of the posthuman that is neither a euphoric, uncritical acceptance of advanced technology, nor a nostalgic lament for the decline of classical humanism.

13. There would definitely be a lot less drivel to wade through if it were not possible to write so quickly.

References

Badmington, N. (ed.) (2000). *Posthumanism*, London, England: Palgrave, ISBN 0333765389.

Bauman, Z. (1992). *Intimations of Postmodernity*, London, England: Routledge, ISBN 0415067502 (1991).

Braidotti, R. "Cyberfeminsim with a difference," http://www.let.uu.nl/womens_studies/rosi/cyberfem.htm.

Chaitin, G. J. (1987). *Algorithmic Information Theory*, Cambridge, England: Cambridge University press, ISBN 0521616042 (2004).

Cilliers, P. (1998). *Complexity and Postmodernism: Understanding Complex Systems*, London, England: Routledge, ISBN 0415152879.

Cilliers, P. (2001). "Boundaries, hierarchies and networks in complex systems," *International Journal of Innovation Management*, ISSN 1363-9196, 5(2): 135-147.

Cilliers, P. (2005). "Complexity, deconstruction and relativism," *Theory Culture & Society*, ISSN 0263-2764, 22(5): 255–267.

Coetzee, J. M. (2005). *Slow Man*, London, England: Seeker and Warburg, ISBN 0670034592.

Derrida, J. (1976). *Of Grammatology*, Baltimore, MD: John Hopkins University Press, ISBN 0801858305.

Derrida, J. (1982). "Différance," in J. Derrida, *Margins of Philosophy*, Chicago, IL: The Harvester Press, ISBN 0226143260 (1985), pp. 1-27.

Eriksen, T. H. (2001). *Tyranny of the Moment: Fast and Slow Time in the Information Age*, London, England: Pluto Press, ISBN 074531774X.

Hayles, N. K. (1999). *How We Became Posthuman: Virtual Bodies in Cybernetics, Literature, and Informatics*, Chicago, IL: The University of Chicago Press, ISBN 0226321460.

Honoré, C. (2004). *In Praise of Slowness: How a Worldwide Movement is Challenging the Cult of Speed*, London, England: Orion, ISBN 0060750510 (2005).

Kundera, M. (1996). *Slowness*, London, England: Faber and Faber, ISBN 0060928417 (1997).

Nadolny, S. (2003). *The Discovery of Slowness*, Edinburgh, Scotland: Canongate, ISBN 1589880242 (2005).

Nowotny, H. (1994). *Time: The Modern and the Postmodern Experience,* Oxford, England: Polity Press, ISBN 0745608922.

Parkins, W. (2004). "Out of time: Fast subjects and slow living," *Time and Society*, ISSN 0961-463X,13(2): 363-382.

Richardson, K. A. (2005). "The hegemony of the physical sciences: An exploration in complexity thinking," *Futures,* ISSN 0016-3287, 37: 615-653.

Shannon, C. E. (1949). "Communication in the presence of noise," *Proc. IRE.*, 37: 10-21.

Taylor, M. C. (2003). *The Moment of Complexity: Emerging Network Culture*, Chicago, IL: The University of Chicago Press, ISBN 0226791173 (2002).

Detailed Contents

D

E

F

G

H

I

J

K

L

M

N

O

P

R

S

T

www.ingramcontent.com/pod-product-compliance
Ingram Content Group UK Ltd.
Pitfield, Milton Keynes, MK11 3LW, UK
UKHW041639031026
15587UKWH00001B/389

ANALOG AND MIXED-SIGNAL BOUNDARY-SCAN

A GUIDE TO THE IEEE 1149.4 TEST STANDARD

Edited by

ADAM OSSEIRAN
Engineering Institute of Geneva

KLUWER ACADEMIC PUBLISHERS
BOSTON / DORDRECHT / LONDON

A C.I.P. Catalogue record for this book is available from the Library of Congress.

ISBN 0-7923-8686-8

Published by Kluwer Academic Publishers,
P.O. Box 17, 3300 AA Dordrecht, The Netherlands.

Sold and distributed in North, Central and South America
by Kluwer Academic Publishers,
101 Philip Drive, Norwell, MA 02061, U.S.A.

In all other countries, sold and distributed
by Kluwer Academic Publishers,
P.O. Box 322, 3300 AH Dordrecht, The Netherlands.

Printed on acid-free paper

Printed in the Netherlands.

DISCLAIMER

The material in this book, is not intended as a complete description of the IEEE 1149.4 standard, and does not provide all the information contained in that standard. If you wish to produce designs compliant with the standard, you will need to purchase a copy from the IEEE, as well as a copy of IEEE 1149.1, the standard that IEEE 1149.4 augments. You should not assume that by following the standard, and the suggestions in this book, that your design will be free of trouble. Analog design and analog testing are difficult arts, and no single book will tell you how to do them. Hopefully, this book will increase your understanding of the IEEE 1149.4 standard, and of mixed signal test, so that you will be less likely to make mistakes.

TABLE OF CONTENTS

LIST OF FIGURES

LIST OF CONTRIBUTORS

Brian R. Wilkins *(Chapter 1)*
Department of Electronics and Computer Science, University of Southampton, Southampton, SO17 1BJ, England
brw@ecs.soton.ac.uk

Colin M. Maunder *(Chapter 2)*
British Telecom Advanced Communications Technology Centre, MLB 2/1, Adastral Park, Martlesham Heath, Ipswich, IP5 3RE, England
colin.maunder@bt.com

Rodham E. Tulloss *(Chapter 2)*
PO Box 57, Roosevelt, New Jersey 08555-0057, USA
ret@njcc.com

Steve Sunter *(Chapter 3)*
Logic Vision, 1525 Carling Avenue, Ottawa, K1Z 8R9, Canada
Sunter@lvision.com

Mani Soma *(Chapter 4)*
Departement of Electrical Engineering, University of Washington, Seattle, WA 98195, USA
Soma@ee.washington.edu

Keith Lofstrom *(Chapter 5)*
Keith Lofstrom Integrated Circuits, 5290 SW Elm, Beaverton, OR 97005, USA
keithl@klic.rain.com

John McDermid *(Chapter 6)*
Hewlett-Packard, 815 SW 14th Street, R&D, Loveland, CO 80537, USA
mcd@lvld.hp.com

PREFACE

This book contains more than the IEEE Standard 1149.4. It also contains the thoughts of those who developed the standard. Adam Osseiran has edited the original writings of Brian Wilkins, Colin Maunder, Rod Tulloss, Steve Sunter, Mani Soma, Keith Lofstrom and John McDermid, all of whom have personally contributed to this standard. To preserve the original spirit, only minor changes were made, and the reader will sense a chapter-to-chapter variation in the style of expression. This may appear awkward to some, although I found the lack of monotonicity refreshing.

A system consists of a specific organization of parts. The function of the system cannot be performed by an individual part or even a disorganized collection of the same parts. Testing has a system-like characteristic. Testing of a system does not follow directly from the testing of its parts, and a system built with testable parts can sometimes be impossible to test. Therefore, testability of the system must be organized.

Some years ago, the IEEE published the boundary-scan Standard 1149.1. That standard provided an architecture for digital VLSI chips. The chips designed with the 1149.1 architecture can be integrated into a testable system. However, many systems today contain both analog and digital chips. Even if all digital chips are compliant with the standard, the testability of a mixed-signal system cannot be guaranteed. The new Standard 1149.4, described in this book, extends the previous architecture to mixed-signal systems.

The book has six main chapters. *Chapter 1* cites examples of testing of interconnects in a mixed-signal printed circuit board, and shows how the new architecture facilitates testing. *Chapter 2* reviews the previously published boundary-scan Standard 1149.1 that applies to digital systems. The knowledge of 1149.1 is essential for using the new standard. *Chapter 3* gives the complete architecture and instruction set for 1149.4. *Chapter 4* discusses test methodologies for systems designed with the new standard. *Chapter 5* describes the design of the peripheral cell that is an essential component needed for implementing the 1149.4 test architecture. *Chapter 6* takes a slight deviation from earlier chapters. Here, a structural test methodology using the new mixed-signal test architecture is described. Components are tested in structural testing as opposed to the testing of input-output behavior in the more prevalent functional testing. Structural testing, as supported by 1149.4, allows efficient diagnosis of faulty components.

The material in this book is not intended as a complete description of the standard. It is recommended that the designer should obtain the complete documents on the 1149.1 and 1149.4 standards from the IEEE.

IEEE 1149.4 builds on the solid foundation provided by IEEE 1149.1 and has been driven by intense efforts of many capable participants from industry and academia all over the world. The credit for the technical accomplishments described in this book belongs to the members of the IEEE 1149.4 Working Group, and to other members of the test community who have directly or indirectly assisted in the standard development.

Vishwani D. Agrawal

Consulting Editor

Frontiers in Electronic Testing Book Series

Chapter 1

Introduction to the IEEE 1149.4

Brian R. Wilkins
University of Southampton, brw@ecs.soton.ac.uk

BACKGROUND

Figure 1.1 represents an electrical circuit constructed as a printed circuit assembly (PCA) consisting of a substrate carrying a pattern of conductors (the interconnect) on which separately manufactured components are mounted so that the component pins make electrical contact with the interconnect. In normal functional operation, the PCA connects to other parts of the system by way of a set of contacts such as the edge-connector shown in Figure 1.1.

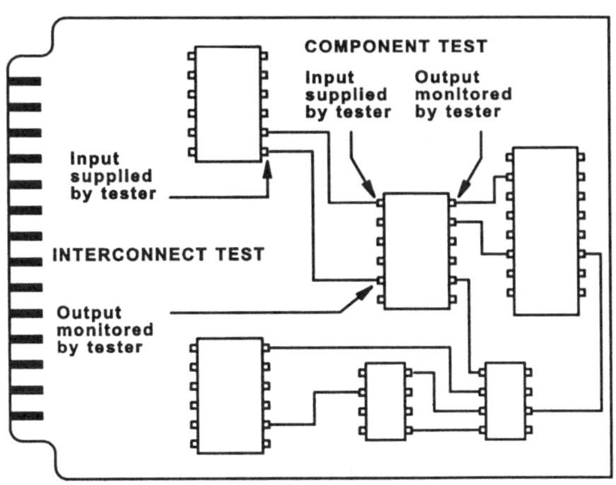

Figure 1.1. **Printed circuit assembly and in-circuit testing**

The principle of constructing electronic circuits in this way has been a standard practice for several decades. For much of this time, the testing of such PCAs has relied on the use of an in-circuit tester [1], which applies mechanical probes to the individual component pins. This allows each component to be tested effectively in isolation, by applying test signals to the component inputs and monitoring the test responses at the component outputs as shown in Figure 1.1. Also indicated in Figure 1.1 is the way of testing the interconnect by applying test signals to the output pins of driving components and monitoring responses at the input pins of driven components.

This procedure has been made increasingly difficult as advances in technology have resulted in components that are more complex, while being contained in smaller packages. The density of components over the board has increased dramatically, while at the same time each individual component carries increased numbers of smaller (and in many cases inaccessible) pins. Mechanical probing is becoming more difficult, and in some cases (such as with ball grid array packages) impossible unless additional probe pads are incorporated. This is the problem that was directly addressed by the boundary-scan standard.

Although the original intention of the boundary-scan standard was to make provision for chips and boards of all kinds, the difficulties posed by some particular circuit features (notably, but not exclusively, the problem of analog circuits) led to the task being sub-divided. IEEE Std 1149.1 (Standard Test Access Port and Boundary-Scan Architecture), hereafter referred to as "dot1", which had previously been brought to an advanced stage by JTAG (the Joint Test Action Group) was taken forward to cope with purely digital circuits; analog and mixed-signal circuits were deferred for separate consideration. The development of the IEEE Std 1149.4 (Mixed-Signal Test Bus), hereafter referred to as "dot4", began with a preliminary meeting of interested individuals in the summer of 1991, at which it was agreed that there was a need for a test structure that could be incorporated into mixed-signal integrated circuits. From the beginning, it was assumed that the new standard would be constructed as an extension of dot1, and would operate in harmony with dot1, while making provision for the testing needs of a wider range of circuits.

Because a dot4 chip is basically a dot1 chip with additions, and because the dot1 features and the overall control protocols are subject to the rules of dot1, it is essential, before starting to implement dot4, to become familiar with the provisions and workings of dot1. The following sections give a very brief review of the underlying principles of dot1, mainly in order to draw attention to the limitations of dot1 when applied to practical mixed-signal circuit boards.

1.1 Dot1 principles

1.1.1 Electronic probing

The basis of dot1 is that a boundary-scan cell is associated with every pin on a chip. Each boundary-scan cell contains at least one flip-flop, and these flip-flops can be reconfigured into a single shift register (the boundary-scan register), the ends of which are brought out to dedicated pins on the chip (TDI and TDO) as illustrated in Figure 1.2. Two other dedicated pins (TMS and TCK) complete the Test Access Port (TAP), which allows test equipment to gain access to every pin of every component on a board (assuming that every chip complies with dot1), and so to load and inspect data at the pins without needing to pass test signals through the chip [2]. This is electronic probing, achieving test access without mechanical contact.

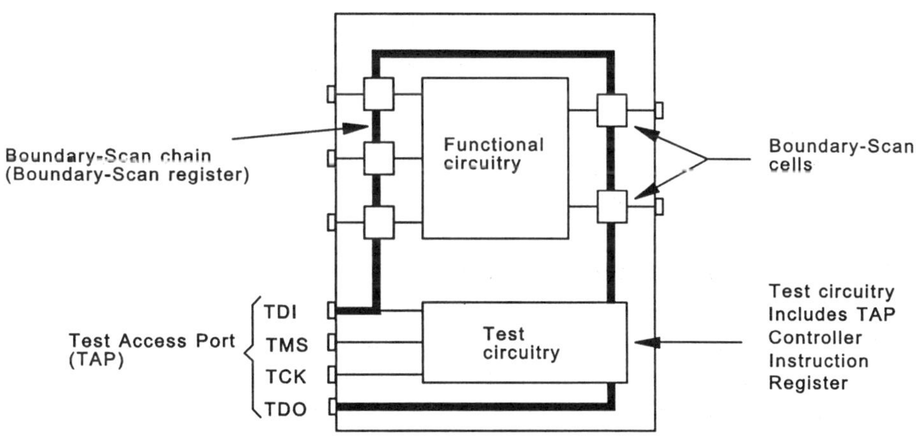

Figure 1.2. **An IEEE 1149.1 digital chip**

1.1.2 Register structure

In addition to the boundary-scan register, there are two other registers that are required in a dot1 chip: the bypass register (a single-stage register that allows a chip to be effectively removed from the board-level scan chain) and an instruction register. The manufacturer may optionally fit any number of additional registers to provide extra user-defined test capabilities.

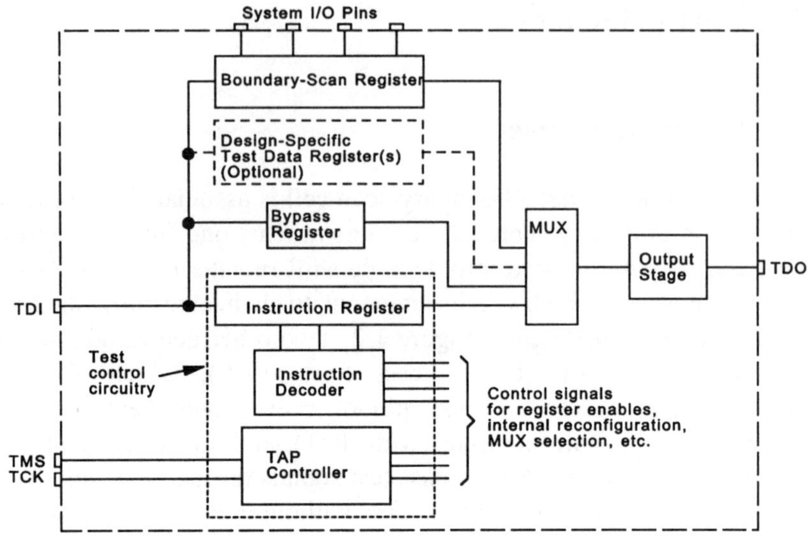

Figure 1.3. **Test register structure**

The overall register structure of a dot1 chip is shown in Figure 1.3. Test equipment is able to load any of these registers (serially) through TDI, and to inspect the contents of any register through TDO.

1.1.3 Control of the test structure

Control signals are required both for the transfer of data and for the choice of which of the alternative parallel paths is active: these signals are supplied, as shown in Figure 1.3, partly by the TAP controller and partly by the instruction decoder, interpreting the particular instruction loaded into the instruction register.

The TAP controller is a synchronous finite state machine whose state transition diagram is shown in Figure 1.4. State transitions, selected by the value present on TMS, are initiated by a rising edge on TCK.

An instruction is a bit-pattern that is loaded into the instruction register serially through TDI, and is then decoded by the instruction decoder. It defines the set of test data registers that operate while the instruction is selected, and it defines the test data register path used to shift data between TDI and TDO. There are some mandatory instructions that must be supported by a dot1 chip, and some defined instructions that are optional. In addition, any number of user-defined instructions may be included at the discretion of the manufacturer. One such instruction, for example, might be to permit a test of the core circuitry, using a conventional internal scan chain.

The scan chain itself, in this case, would appear as one of the optional design-specific test data registers.

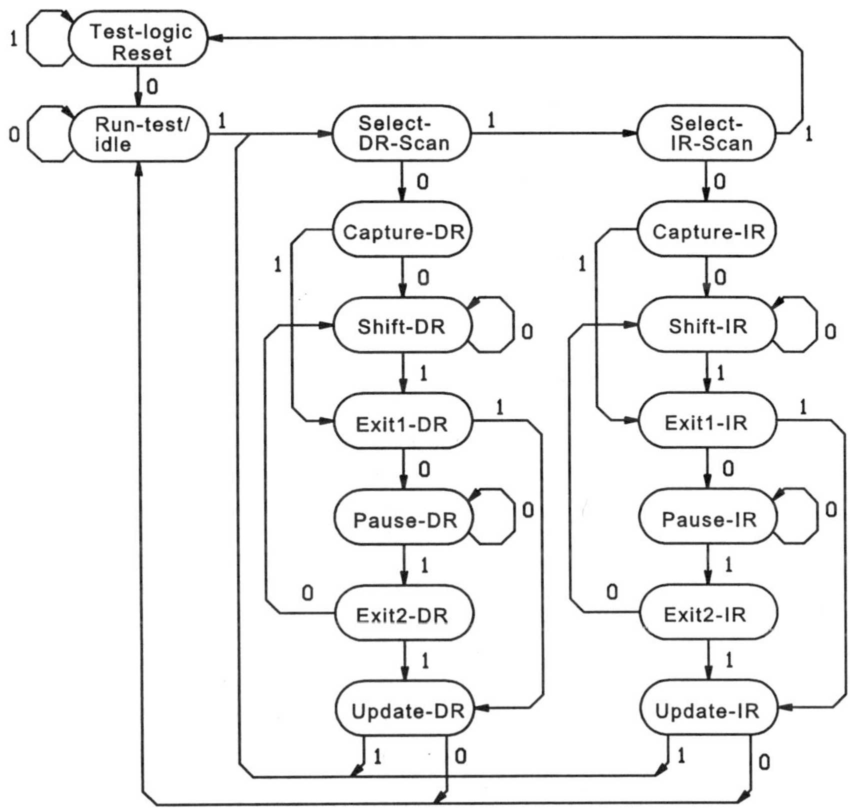

Figure 1. 4. **TAP controller state transition diagram**
[Redrawn from IEEE Std 1149.1]

1.1.4 Analog and differential pins

Although dot1 is primarily concerned with digital boards loaded with digital chips, it needed to take some account of the possibility that some analog pins might be incorporated into a predominantly digital chip. In the absence of a structure tailored to the needs of analog pins, the decision taken by dot1 was to exclude them from core circuitry.

Figure 1.5 shows how mixed-signal chips are dealt with in dot1: there are no boundary-scan cells on the analog pins, but internal boundary-scan cells are introduced at the digital/analog interface. This gives the best possible access to the digital part of the core, and allows most of the interconnect to be tested, but leaves some gaps in the coverage when applied to a mixed-signal board.

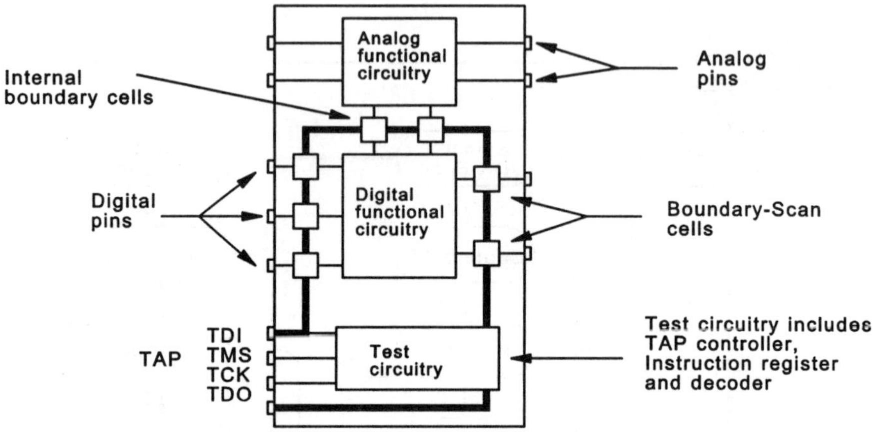

Figure 1.5. **A mixed-signal dot1 chip**

Figure 1.6, in which analog and digital pins are represented by "A" and "D" respectively, shows some of the defects that might occur on a practical mixed-signal board. The gaps in the connecting lines represent open circuits, and the blobs represent short-circuit (bridging) faults. Notice particularly that short circuits can occur between analog and analog, digital and digital, and analog and digital lines, as well as across discrete components in interconnection networks. The fundamental requirement for interconnect testing is to detect open circuits in the interconnections between components, and to detect and diagnose bridging faults anywhere in the interconnect regardless of whether the lines involved normally carry digital or analog signals. Without access to the analog pins, the ability to detect and diagnose these faults is limited.

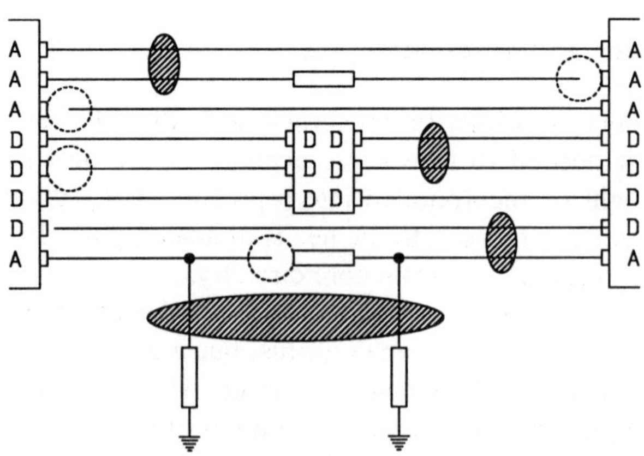

Figure 1.6. **Defects on a mixed-signal PCA**

Dot1 implicitly hopes that the board being tested has only a small minority of analog connections, so that the limitation of fault coverage will probably be acceptable, but when dealing with a fully mixed-signal environment the results could be more serious.

Another feature of analog circuits, also included in Figure 1.6, is that groups of one or more discrete components are often interposed between chips, performing functions such as level shifting, passive filtering, and ac coupling. The inclusion of discrete components in the interconnection pathway gives rise to the concept of "extended" (as opposed to "simple") interconnect, as illustrated in Figure 1.7. Simple interconnect can be tested, provided we have access to the pins at each end, by applying digital signals and collecting digital responses: both the required connections and the possible faults are digital in nature. The testing of components within an extended interconnect requires analog measurements, allowing, for example, impedances of discrete components to be computed. This is a facility that is provided in dot4.

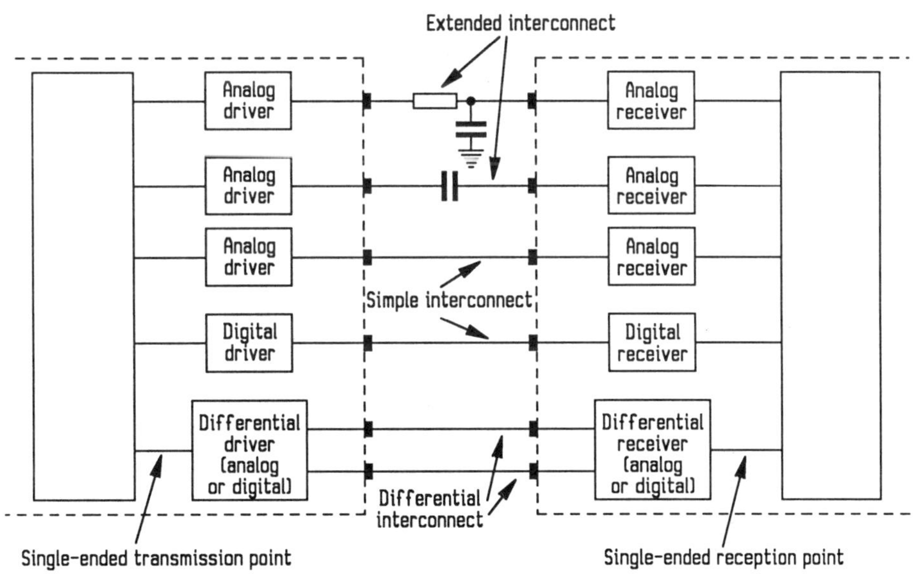

Figure 1.7. **Simple, extended and differential interconnects**
[Redrawn from IEEE Std 1149.4]

Also illustrated in Figure 1.7 is an example of differential interconnect, which is a pair of pathways carrying signals whose information content is defined by the pair of signals rather than by either one signal individually. Dot1 does not address the problems of differential interconnect: it defines all differential pins as analog, whether attached to analog or digital differential circuitry, which then requires it to be dealt with as shown in Figure 1.8.

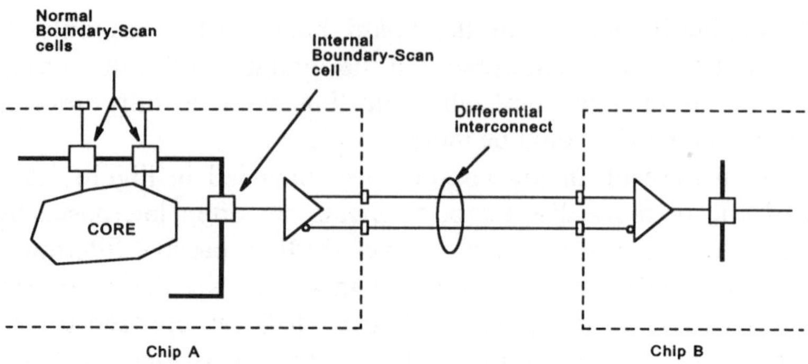

Figure 1.8. **Differential interconnect in dot1**

In the case of digital differential interconnect, these pathways would normally be excited by a single signal, and the activity would be converted back to a single signal after reception. In dot1, with the internal boundary cells shown in Figure 1.8, the functional connection between single-ended transmission point and single-ended reception point can be tested, but this would not necessarily detect either open circuit faults in the differential interconnect or bridging faults involving other connections.

1.2 Test needs of mixed-signal boards

In developing dot4, it was considered important to recognize that the components mounted on a practical mixed-signal PCA would not consist exclusively of those equipped with either dot1 or dot4.

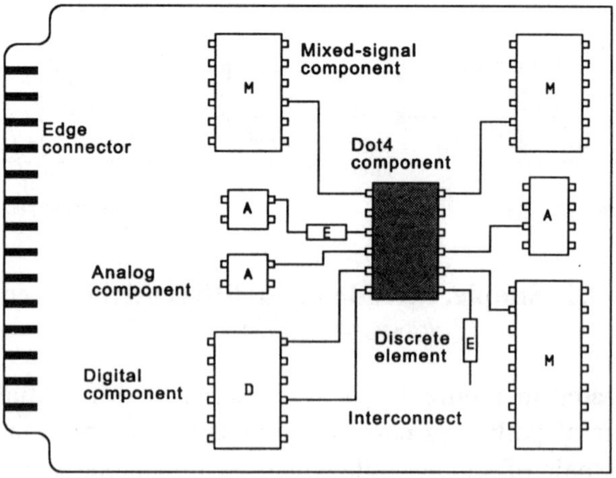

Figure 1.9. **Mixed-signal printed circuit assembly**

Figure 1.9 shows a dot4 component mounted on a typical mixed-signal PCA. Its pins may be connected to

- other mixed-signal components (labeled M), which may or may not conform to dot4;
- digital components (labeled D), which may or may not conform to dot1;
- analog components (labeled A), analogous to digital glue logic, and which could be anything from a passive analog signal processing circuit to an operational amplifier or active filter, or even a single transistor. All such elements would be unlikely to contain any associated testability features;
- passive circuit elements (labeled E), such as coupling capacitors or pull-up resistors, which will not have any associated testability features.

1.3 Dot4 boundary modules

1.3.1 Function

In dot4, every pin is required to have a boundary module associated with it. In the case of the digital pins, the boundary modules (DBMs) are identical to the boundary-scan cells defined in dot1, except that there is an option to extend the facilities using a form of analog boundary module (ABM) if enhanced (analog) measurement capabilities are required.

A standard ABM consists of a switching network connected between the pin and the core circuitry, allowing the pin to be put into the core disconnect (CD) state, in which it is isolated from the core, and to be connected to a dc voltage or an internal analog test bus line.

Interconnect testing has always been seen as the most important test capability of dot1. With boundary modules on all pins, giving the ability to apply logic values and to monitor logic responses at all inputs and outputs, all simple interconnect on a dot4 mixed-signal board can be tested using the conventional dot1 protocols.

Extended interconnect can be tested by making analog (parametric) measurements using the analog test buses to apply test stimuli and to monitor analog test responses.

1.3.2 Control

Each DBM contains a memory element that can store logic test stimuli or responses. These memory elements appear as part of the boundary-scan register, and are accessed by external test equipment through TDI and TDO under the control of the TAP controller.

The ABMs are also controlled digitally, the operation of the switches in each ABM being determined by the contents of its individual control register. The control register can also be used to capture digital test results, representing either logic values or digitized analog responses. All the ABM control registers are included in the boundary-scan register, which can therefore be loaded and unloaded through the TAP as with the DBMs.

1.4 Analog test bus

1.4.1 ATAP

To support the analog stimulus and response capabilities of dot4, two two-wire test buses are involved: a board-level bus carrying signals between the test equipment and the chip, and an internal one that distributes the signals round the interior of the chip. Access to the internal bus is through a dedicated pair of pins that constitute the analog test access port (ATAP).

1.4.2 TBIC

The test bus interface circuit (TBIC), illustrated in Figure 1.10, takes the form of another switching structure, allowing each of the internal test bus lines to be connected to either or both ATAP pins or to be isolated from both of them. Isolation of the internal test bus when not in use is necessary to avoid unwanted interference and cross coupling.

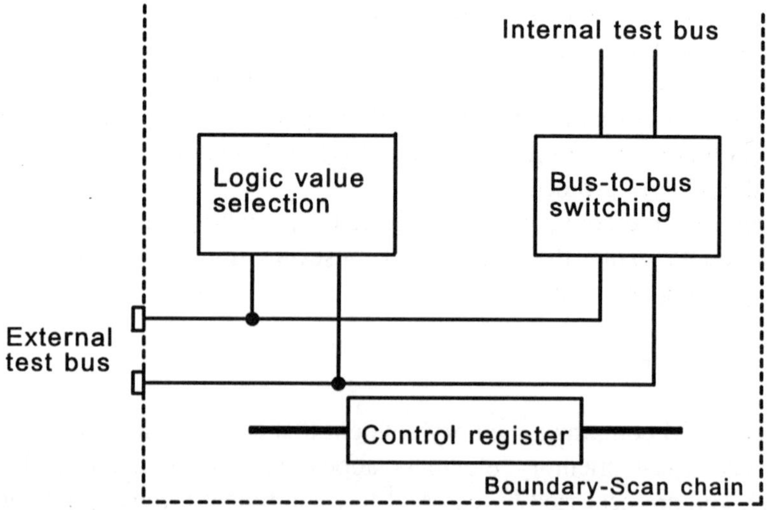

Figure 1.10. **The TBIC (Test Bus Interface Circuit)**

At board level, external test bus lines will be distributed to the ATAP pins of every chip. Testing the integrity of these bus connections is therefore an important part of the board interconnect test: the TBIC therefore allows each ATAP pin to be connected to one of two dc voltages, which act as logic values for interconnect testing.

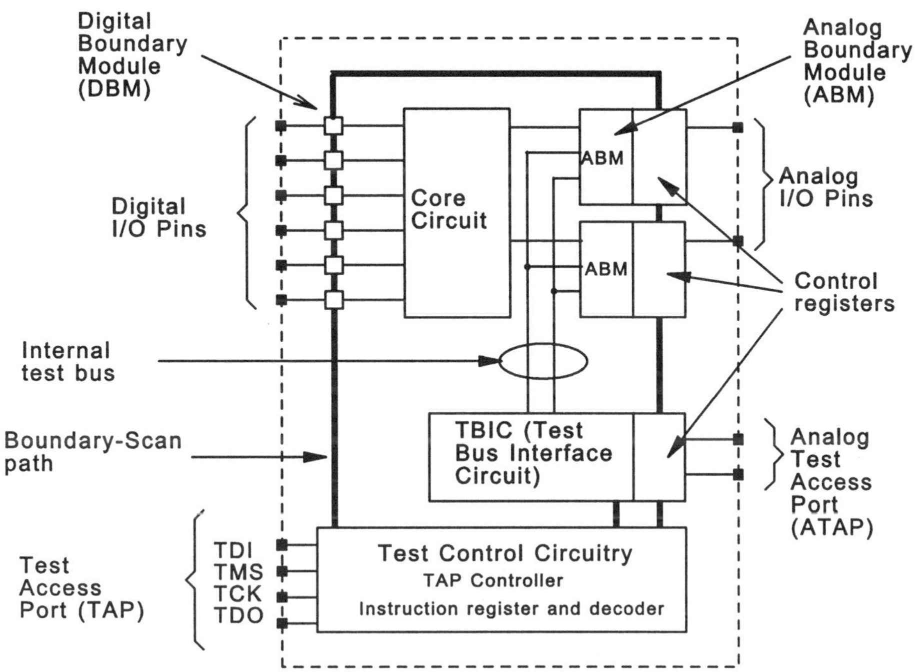

Figure 1.11. **Structure of a basic dot4 chip**

If a particular chip has one or more pairs of differential input or output pins, the manufacturer may choose to include additional test structures to provide for differential testing, or may choose to omit the additional structure and accept a reduced level of test function. The optional additional structure consists of two additional pins in the ATAP, and two corresponding additional wires in the internal bus; the control of this additional structure is exercised by an additional TBIC, identical to that shown in Figure 1.10.

There is a further optional provision in the design of the TBIC. In the simple structure shown in Figure 1.11, the internal test bus is distributed to every analog pin in the chip. There are some circuits (particularly those with multiple power supply voltages) for which it would be difficult to implement this structure without performance problems caused by cross talk, leakage, feedback, or loading. In this situation, it may be helpful to provide several separate two-wire internal test buses, all served by a single two-pin ATAP.

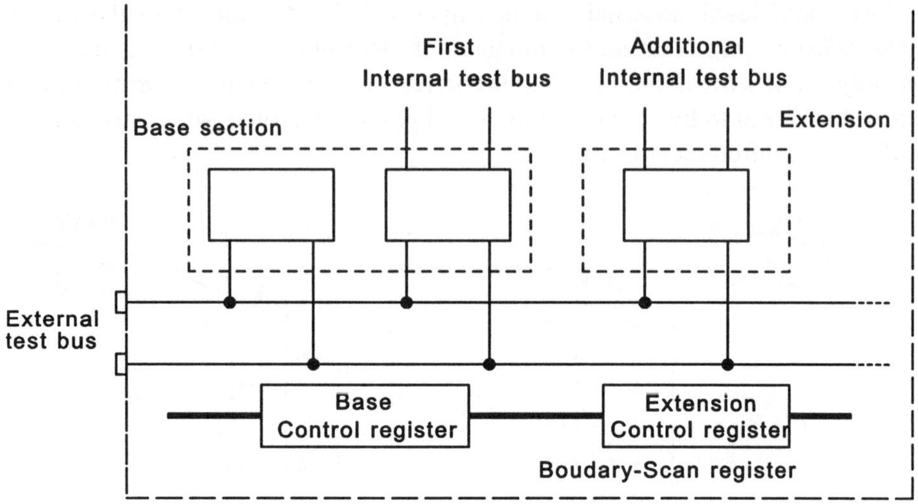

Figure 1.12. **Partitioned TBIC**

The resulting TBIC switching structure, shown in Figure 1.12, consists of a base section and any number of extensions. The base section is identical with the TBIC of Figure 1.10, consisting of two sub-sections, the first supporting board-level interconnect testing on the external test bus, and the second providing for the connections between the ATAP and one internal test bus. This second sub-section is replicated as many times as necessary to provide for the connections between the ATAP and each of the other internal test buses.

The partitioned TBIC is controlled by an extended control register, which also consists of a base section (identical in function to the control register of an unpartitioned TBIC), together with an extension associated with each additional partition. The control register and all its extensions are included in the boundary-scan register.

It should be noted that neither the differential extension nor the partitioned bus option has any effect on the ABM, which in all cases will be concerned with just two of the bus wires.

1.5 Capabilities

A PCA is tested, both in production (to verify correct manufacture) and in field service (to detect and locate faults), using automatic test equipment (ATE) to supply test signals to, and to collect test responses from, some or all of the pins of the individual chips. The aim of the test structures described in dot4 is to provide test access to individual chip pins from the edge-connector, so as to reduce or eliminate the need for the ATE to make direct physical contact via mechanical probes. The extent to which this aim can be satisfied will depend on the make-up of the PCA: the pins of those chips that conform either to dot4 or to dot1 will be accessible from the edge-connector, and this will provide at least partial access to the pins of non-conformant components. If this is inadequate for testing purposes, additional probe access would need to be employed, but even in the worst case the amount of probe access required would be reduced compared with a PCA with no boundary-scan chips.

Three classes of tests on a mixed-signal PCA are provided for on a dot4 chip.

a) Interconnect Test
b) Parametric Test
c) Internal Test

1.5.1 Interconnect test

Dot1 has been found to provide an effective mechanism for interconnect testing of a PCA populated with digital components. Dot4 provides a similar capability for an assembly such as that illustrated in Figure 1.9, populated with analog, digital, and mixed-signal components, together with networks of discrete elements.

The fundamental requirement for interconnect testing is to detect open circuits in the interconnections between integrated circuits, and to detect and diagnose bridging faults anywhere in the interconnect regardless of whether they normally carry digital or analog signals. The structure provided by dot4 allows testing in full compatibility with dot1, permitting mixed-signal interconnect testing to take place at the same time, and using the same procedures, as digital interconnect testing. The dot4 structures also provide for differential interconnects testing, whether digital or analog.

1.5.2 Parametric test

The second class of test supported by dot4, parametric test, permits analog measurements using analog stimulus and response, allowing, for example, impedances of discrete components to be computed.

By modifying the boundary modules attached to digital pins so that they include elements of the dot4 structure, components such as pull-ups associated with digital pins can also be measured.

1.5.3 Internal test

The third capability of dot4, internal test, relates to the ability to test the internal circuitry of the chip, either in isolation or while mounted on a substrate. Comprehensive internal testing may require the incorporation of internal test structures whose impact on the cost and performance of the circuit may be deemed unacceptable. The provision of facilities for internal test, as well as the incorporation of any amount of ad hoc testability features within the internal core structure, is, therefore, optional; the defined structure, however, will support the inclusion of any number of designer-defined test functions.

REFERENCES

[1] *J. Baston*, “In-Circuit Testing”, Van Nostrand Reinhold, New York, 1985

[2] *K. P. Parker*, “The Boundary-Scan Handbook, 2nd Edition, Analog & Digital”, Kluwer Academic Publishers, Boston, 1998

Chapter 2

The Boundary-Scan standard

ANSI/IEEE Std 1149.1[1]

Colin M. Maunder
British Telecom Advanced Communications, colin.maunder@bt.com

Rodham E. Tulloss
PO Box 57, Roosevelt, NJ, ret @njcc.com

BACKGROUND

ANSI/IEEE Std 1149.1 defines a standard implementation of boundary-scan now built into many catalog and application-specific integrated circuits. The standard was developed as a solution to two continuing trends that are having a significant, adverse, impact on the task of testing loaded printed wiring boards: increasing chip complexity and greater miniaturization. The former increases the difficulty of test generation, while the latter impedes access for the bed-of-nails and hand-held probes on which many established test techniques depend.

This chapter provides a guide to the principal features defined by the standard and to their operation. It is intended as a prelude to the standard itself, not as a substitute for it. In particular, it is recommended that readers who intend to implement integrated circuits, design tools, or test systems that support the standard read the standard document before doing so. For a view of the wide range of applications of the standard, the reader is directed to [2].

[1] Some material in this chapter appeared in an article in the Journal of Electronic Testing: Theory and Applications, volume 2, 1991, pages 27-42.

2.1 Motivation

Two continuing trends are having a significant, adverse, impact on the cost of testing loaded Printed-Wiring Boards (PWBs):

1. *Increasing complexity*: As integrated circuits (ICs) become more complex, the difficulty of generating a test for loaded boards increases. For in-circuit testing, the generation of a test module for a new IC design could take several man-months. For functional testing, test generation times are significantly longer, due to the need to propagate test data through the complex ICs while applying tests to other chips on the board. Test lengths also increase as complexity rises, pushing up the cost of applying the finished test. In in-circuit testing, the maximum test length is constrained by a time limit imposed to ensure that components are not damaged by backdriving, so the result of increasing complexity is often that less thorough testing is possible.
2. *Greater miniaturization*: Test techniques that became widely used during the 1980s – in-circuit, functional, cluster, and emulation testing – depend significantly on the ability to make contact with connections internal to the loaded board. Bed-of-nails or hand-held probes must be connected to such connections during test application and for fault diagnosis. The use of surface-mount assembly techniques, particularly when coupled with double-sided component mounting and the use of buried vias to connect layers of interconnect on the PWB, reduces board geometries-making the finished product smaller and, unfortunately, more difficult to probe.

ANSI/IEEE Std 1149.1 [1] - the *Standard Test Access Port and Boundary-scan Architecture* - is the basis of solutions to these problems. The key is the elimination of the need for physical probing of the loaded board, which is achieved by building an electronic test-access mechanism (the boundary-scan path) into the integrated circuits themselves.

2.2 A Chip-Level View

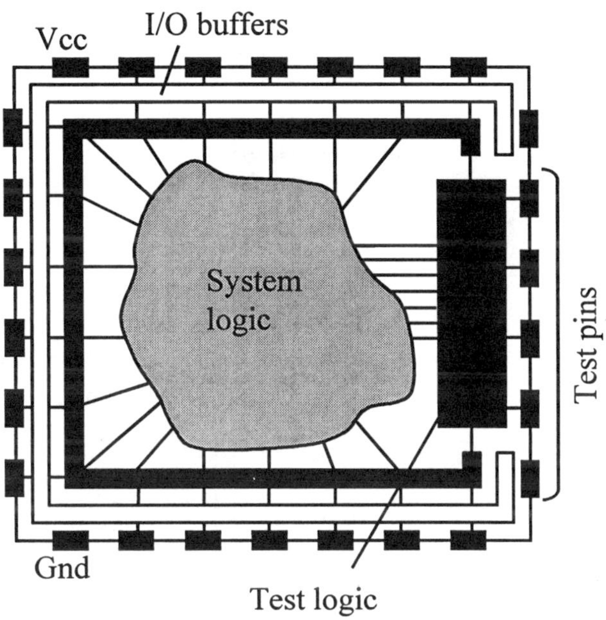

Figure 2.1. **Chip-level view**

Figure 2.1 gives an overall view of an IC design that conforms to ANSI/IEEE Std 1149.1. The circuitry can be broken down into two parts:

1. *The system logic*: This is the circuitry that performs the – normal – function for which the chip was designed. For example, it could be the logic necessary to build a microprocessor, a communications interface, or a counter.
2. *The test logic*: This is the circuitry defined by the standard and includes all the blocks in Figure 2.1 other than the system logic. The test logic is used during testing of the IC and of the board onto which the IC is assembled. It does not contribute to the normal operation of the system logic.

Some circuitry may be shared between the system and test logic. For example, a register in the system logic may have test modes of operation-perhaps it might be a BILBO or similar device [3]. Under these circumstances, the circuitry is regarded as system logic while it contributes to the normal chip function and as test logic while it participates in test operations.

2.3 The Test Logic Architecture

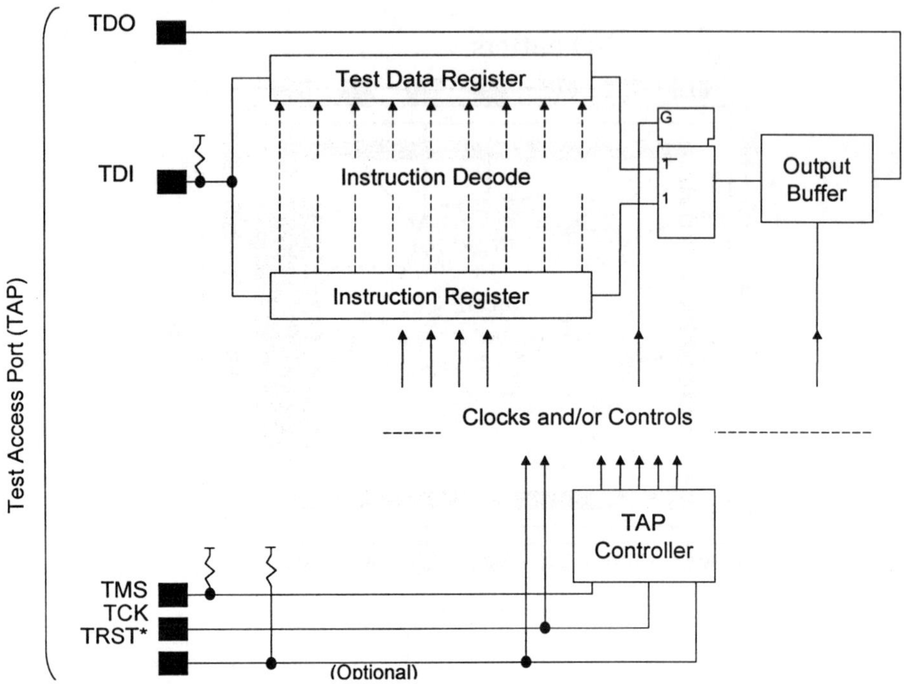

Figure 2.2. **IEEE Std 1149.1 test logic**

The top-level schematic of the test logic defined by IEEE Std 1149.1 includes three key circuit blocks (Figure 2.2):

> *The TAP controller*: This responds to control sequences supplied through the test access port (TAP – see below) and generates the clocks and control signals required for correct operation of the other circuit blocks.
>
> *The instruction register*: This shift-register-based circuit is serially loaded with the instruction that selects a test to be performed.
>
> *The test data registers*: This is a bank of shift-register based circuits (see Section 2.7.). The stimuli or conditioning values required by a test are serially loaded into the test data register selected by the current instruction. Following execution of the test, results in test data registers can be shifted out for examination.

These circuit blocks are connected to a TAP which includes the four or, optionally, five signals used to control the operation of tests and to allow serial loading and unloading of instructions and test data. The TAP on an IC is directly analogous to the "diagnostic" socket provided on many automobiles - it allows an external test processor to control and to communicate with the various test features built into the product. When ICs that implement the standard are combined on a PWB, they can be arranged in a single daisy chain with:

- the TDI input of the first IC in the chain connected to the board edge or to the output of an on-board master device;
- subsequent ICs each having their TDI connected to their predecessor's TDO; and
- the last IC's TDO being fed to the board edge or to the input of the on-board master device.

This arrangement supports the serial transfer of test data and instructions.

2.4 The TAP (Test Access Port)

The TAP contains four or, optionally, five pins. These are:

1. *The test clock input (TCK)*: This is independent of the system clock(s) for the chip so that test operations can be synchronized between the various chips on a PWB.
2. *The test mode select input (TMS):* The operation of the test logic is controlled by the sequence of Is and Os applied at this input. The sequence on TMS directs the TAP controller in its generation of the clock and control signals required by the other test logic blocks.
3. *The test data input (TDI):* Data applied at this serial input are fed either into the instruction register or into a test data register, depending on the sequence previously applied at TMS.
4. *The test data output (TDO):* This serial output from the instruction register or from a test data register depending on the sequence previously applied at TMS. During shifting, data applied at TDI will appear at TDO after a number of cycles of TCK determined by the length of the register included in the serial path. When data are not being shifted through the chip, TDO is set to an inactive drive state (e.g., high-impedance).
5. *The optional test reset input (TRST*)*: The test logic is designed so that it can be reset synchronously under control of TCK and TMS. TRST* provides a supplementary reset mechanism. The test logic is reset when a 0 is applied at TRST*.

The TDI, TMS, and TRST* inputs are either equipped with a pull-up resistor or otherwise designed such that, when they are not driven from an external source, the test logic perceives a logic 1.

2.5 The TAP Controller

A key goal during the development of IEEE Std 1149.1 was to keep the number of pins in the TAP to a minimum, based on the knowledge that many ICs are pin- (rather than silicon-) limited. As test engineers are only too aware, designers are always reluctant to allocate pins for test purposes.

The TAP controller allows this goal to be met. It is a 16-state finite state machine that operates according to the state diagram shown in Figure 2.3. Note that in the states whose names end -DR, the test data registers operate, while in those whose names end -IR, the instruction register operates. A move along a state transition arc occurs on every rising edge of TCK. The 0s and Is shown adjacent to the state transition arcs show the value that must be present on TMS at the time of the next rising edge of TCK for the particular transition to occur.

Eight of the 16 controller states determine operation of the test logic, allowing the following test functions to be performed:

- *Test-Logic-Reset:* All test logic is reset, allowing normal operation of the chip to occur without interference. Regardless of the starting state of the TAP controller, the Test-Logic-Reset controller state is reached by holding the TMS input at 1 and applying five rising edges at TCK, Alternatively, where TRST* is provided, it can be used to force the controller asynchronously into the Test-Logic-Reset controller state at any desired point during circuit operation.
- *Run-Test/Idle*: The operation of the test logic in this controller state depends on the instruction held in the instruction register. When the instruction is, for example, one that activates a self-test, then the self-test will be run in this state. In another case, if the instruction in the instruction register is one that selects a data register for scanning, then the test logic will be idle.
- *Capture-DR*: Each instruction must identify one or more test data registers that are enabled to operate in test mode when the instruction is selected. In this controller state, data are loaded from the parallel inputs of the selected test data registers into their shift-register paths.
- *Shift-DR*: Each instruction must identify a single test data register that is to be used to shift data between TDI and TDO in the Shift-DR controller state. Shifting allows the previously captured data to be examined and new test input data to be entered.
- *Update-DR*: This controller state marks the completion of the shifting process. Some test data registers may be provided with a latched parallel

output to prevent signals applied to the system logic, or through the chip's system pins, from rippling while new data are shifted into the register. Where such test data registers are selected by the current instruction, the new data is transferred to their parallel outputs in this controller state.

- *Capture-IR, Shift-IR, and Update-IR*: These controller states are analogous to Capture-DR, Shift-DR, and Update-DR, respectively, but cause operation of the instruction register. By entering these states, a new instruction can be entered and applied to the test data registers and/or other specialized circuitry. The new instruction becomes "current" in the Update-IR controller state.

Note that in the Update-DR and Update-IR controller states, the described action takes place on the falling edge of TCK. In all other states, the described action takes place on the rising edge of TCK just before the controller leaves the state (see Figure 2.4). Also note that TDO is active only in the Shift-DR and Shift-IR controller states. In the remaining eight controller states, no operation of the test logic occurs-that is, the test logic is "idle" The "pause" states (Pause-DR and Pause-IR) are provided to allow the shifting process to be temporarily halted. This might occur while an ATE or other equipment controlling the test logic fetches more test data from backup memory (e.g., disc).

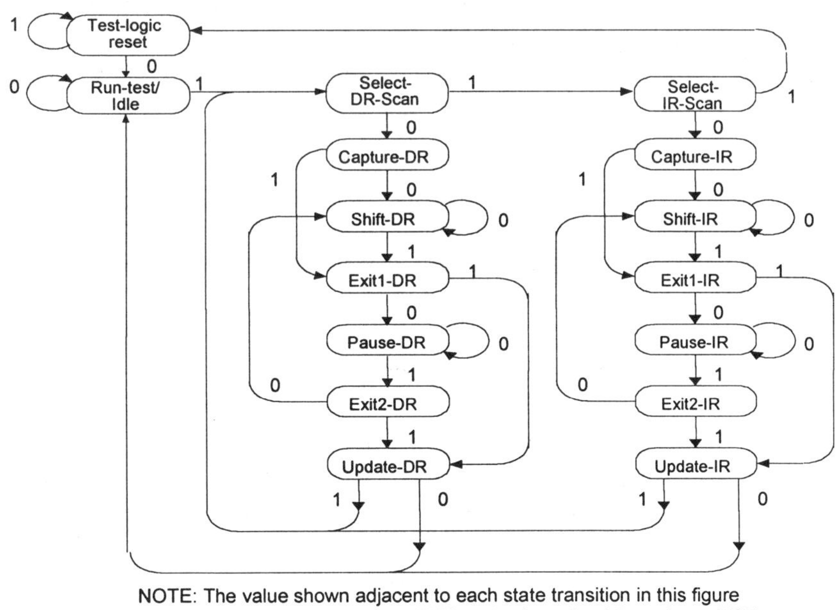

NOTE: The value shown adjacent to each state transition in this figure represents the signal present at TMS at the time of a rising edge at TCK.

Figure 2.3. **State diagram for the TAP controller**

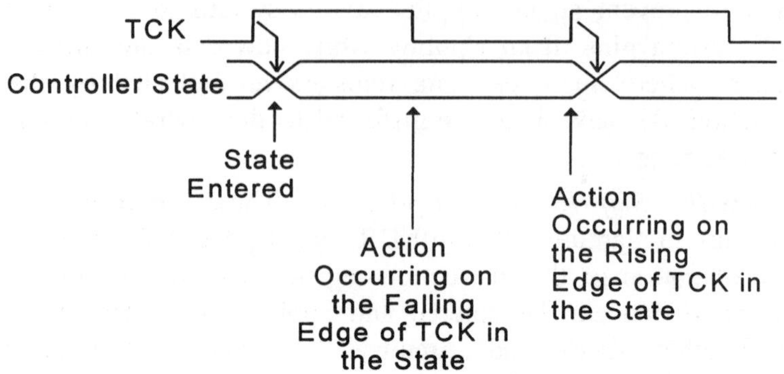

Figure 2.4. **The timing of events within a controller state**

The final six controller states (Select-DR-Scan, Select-IR-Scan, Exit1-DR, Exit1-IR, Exit2-DR, and Exit2-IR) are decision points that allow choices to be made as to the route to be followed around the controller's state diagram. For example, in the Exit-DR controller state a choice is made between entry into the Pause-DR state or entry into the Update-DR state.

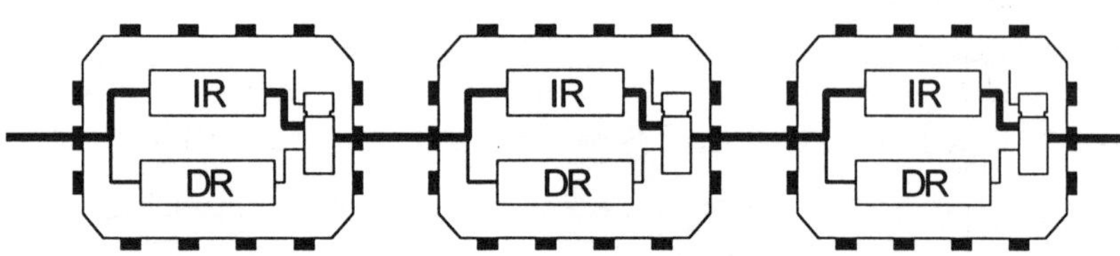

Figure 2.5. **Daisy-chain connection of instruction registers**

2.6 The Instruction Register

The instruction register provides one of the alternate serial paths between TD1 and TDO. It operates when the instruction scanning portion of the controller state diagram is entered (i.e., the portion where state names end -IR).

The instruction register allows test instructions to be entered into each IC along the board-level path. At the board level, the instruction registers are daisy chained together in the Shift-IR controller state (Figure 2.5), so a different instruction can be loaded into each chip on the path if required.

The instruction register is a parallel-in, parallel-out shift-register. The parallel output is latched so that a new instruction can be shifted in without altering the instruction applied to the remainder of the test logic. The latched output is updated from the shift-register path in the Update-IR controller state; at this time, the new instruction becomes "current." In the Test-Logic-Reset controller state, the latched output is reset-to either the IDCODE or the BYPASS instruction depending on the set of test data registers built into the particular IC (see sections 2.8 and 2.9).

The instruction register must contain at least two stages (shown cross-hatched in Figure 2.6). No maximum length is defined, since this will be determined by the number of test instructions provided by the particular IC. Stages 11 and 10 (i.e., the stages located nearest to the serial output) must be set to 0 and 1 respectively in the Capture-IR controller state. These fixed values assist in detecting and locating faults, in the serial path through chips on a board [2]. Instruction register stages numbered 12 or greater are optional and can have a parallel input from which data (typically, status information) are loaded.

2.7 The Test Data Registers

The test logic design provides for a bank of test data registers as shown in Figure 2.7.

IEEE Std 1149.1 specifies the design of three test data registers, two of which must be included in the design. The mandatory test data registers are the bypass and boundary-scan registers. The provision of a device identification register is optional and further design-specific test data registers can be added as appropriate to a given design. All test data registers are shift register based and operate according to the same principles:

- Operation of the various test data registers is controlled according to the current instruction. An instruction can place several test data registers into their test mode of operation, but it can select only one register for connection as the serial path between TDI and TDO in the Shift-DR controller state. On the other hand, it is important to note that one or more physical registers can be configured as one (virtual) test data register by a given instruction.
- Registers that are not enabled for test operation by the current instruction are configured so that they do not interfere with operation of the on-chip system logic (in the case of a register that can operate in either system or test mode, the system mode will be selected).
- Registers enabled for test operation by the current instruction will load data from their parallel inputs (if any) in the Capture-DR controller state and will make any new data available at their latched parallel outputs

(if any) in the Update-DR controller state. In other words, the results of a test are sampled in the Capture-DR controller state and the new test stimulus is available, at the latest, in the Update-DR controller state.

- Where test execution is required between the Update-DR and Capture-DR controller states (e.g., execution of a self-test), this occurs in the Run-Test/Idle state.
- The register selected by the instruction to be the serial path between TDI and TDO will shift data from TDI towards TDO in the Shift-DR controller state. All other test data registers enabled for test operation will hold their state while shifting occurs.

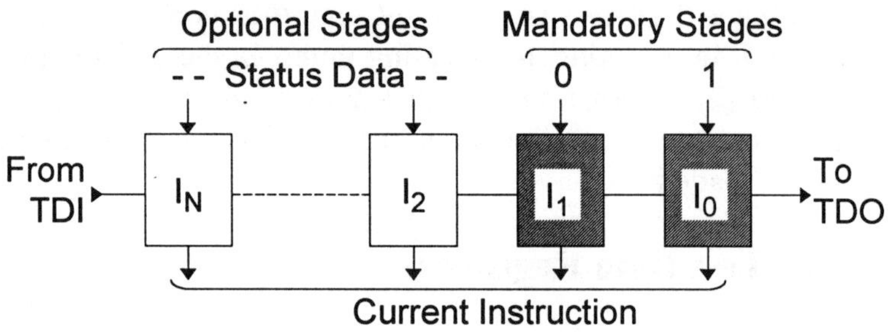

Figure 2.6. **The instruction register**

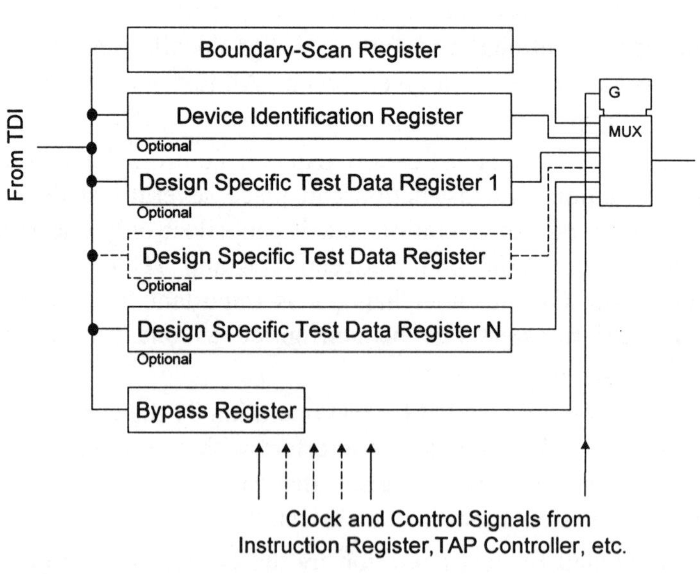

Figure 2.7. **Test data registers**

2.8 The BYPASS Instruction

Every IC that conforms to the standard must support the BYPASS instruction. A value (but not necessarily the only value) for the BYPASS instruction must be "all-1s" (i.e., a logic 1 entered into each stage of the instruction register). In ICs that do not include the optional device identification register, the BYPASS instruction is forced onto the instruction register's output in the Test-Logic-Reset state and thus becomes the current instruction whenever the test logic is reset.

The BYPASS instruction selects the bypass register as the serial path between TDI and TDO during the Shift-DR controller state. This register consists of a single parallel-in, serial-out shift-register stage that loads a constant logic 0 in the Capture-DR controller state when the BYPASS instruction is selected. The bypass register does not have a parallel data output so there is no significance to the data present in the register when shifting is completed. The operation of the register cannot interfere with that of the on-chip system logic.

As an example of an occasion when the bypass register might be used, consider a board containing 100 ICs, all with boundary-scan and connected into a single serial chain, a small part of which is shown in Figure 2.8. Assume that a need arises to access a test data register located in IC57, but that it is desired not to interfere with the operation of the remaining 99 ICs. (An example of such a situation might be when the target chip includes a "shadow" test data register that permits the state of its key internal registers to be read.) In this case, the required instruction would be loaded into IC57, with the BYPASS instruction being loaded into the other ICs. The serial bit stream shifted into TDI during the instruction scanning cycle would be:

111 .... 111CCC...CCC1111 111

where CCC ... CCC is the instruction to be loaded into IC57. As a result of use of the "all-1s" value for the BYPASS instruction, the complexity of the bit stream input to the serial path is considerably reduced. This is an important consideration, since it reduces the data storage requirement for the automatic test equipment (ATE) or bus master chip that controls the operation of the board during test.

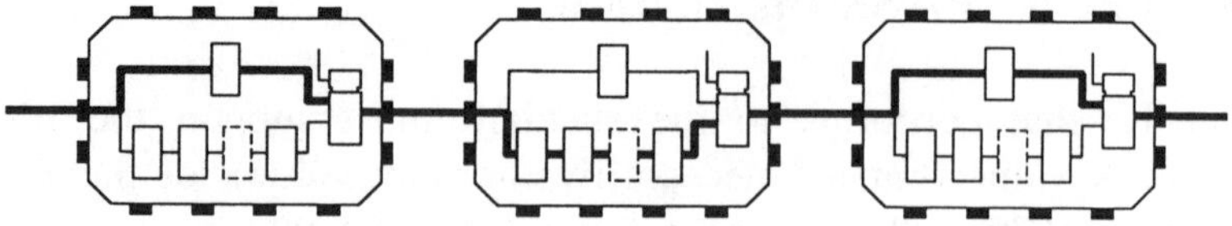

Figure 2.8. **Use of the bypass register**

Once the instructions are loaded, a minimum-length serial path to and from the target chip is set up that allows access to the chip of interest in the minimum possible time, increasing test throughput.

2.9 The IDCODE and USERCODE Instructions

The IDCODE and USERCODE instructions select use of the optional device-identification register. In every IC that includes the device-identification register, the IDCODE instruction is forced onto the instruction register's output in the Test-Logic-Reset state and thus becomes the current instruction whenever the test logic is reset.

The device-identification register allows a binary data pattern to be read from the chip that identifies the manufacturer, the part number, the variant, and (where appropriate) the programmed state. This information might be used to:

- adjust test program execution, depending on the source and/or variant of each chip present on the board;
- verify that the correct IC (or correctly programmed IC) has been mounted in each board location; or
- establish which member of a plug-compatible family of boards is being tested.

The register contains 32 parallel-in, serial-out shift register stages. Like the bypass register, the device identification register does not have a parallel output and, in consequence, there is no significance to the data contained in the register when shifting terminates. Also like the bypass register, operation of the device identification register can occur without interfering with normal chip operation.

Where an IC is programmed off-line (e.g., by blowing fuses or through some other non-reversible process), each stage must have a pair of alternative data inputs so that two different 32-bit codes can be loaded-one to identify the device and one to identify its programming. The former is loaded when the IDCODE instruction is selected, while the latter is loaded when the USERCODE instruction is selected. In all other types of IC, only one data input is required and the USERCODE instruction is not provided.

The structure of the data loaded into the device identification register in response to the IDCODE instruction is shown in Figure 2.9. There are four separate fields:

1. *The header:* ID0 loads a constant logic 1.
2. *The manufacturer code:* ID11-ID1 load an 11-bit manufacturer code. This code is derived from a scheme managed by the Joint Electron Device Engineering Council (JEDEC) [1, 4]. The code can uniquely identify up to 2032 manufacturers (since 16 codes are not used). If more than 2032 manufacturer codes are issued by JEDEC, then the scheme will result in reuse of some code values within the manufacturer code field. However, the chance that a component from an incorrect manufacturer will have the same code *and* the same test functionality is acceptably low.
3. *The part-number code:* ID27-ID12 provide a 16-bit part number, chosen by the manufacturer to distinguish a chip from the others that the company sells. In cases where more than 216 chip types are offered by a manufacturer, part-number codes might have to be reused. The objective is to minimize the chance that an incorrect chip in a given position on a board will have the same part number as the correct chip type-not to provide absolute identification of the IC.
4. *The version number code:* For chips that are manufactured in several different versions through their lives, bits ID31-ID28, can be used to distinguish up to 16 variants.

The data loaded by the USERCODE instruction may be organized as the part's programmer sees fit. It must be programmable at the same time (and in the same way) that the function of the chip is programmed.

2.10 Boundary-scan Register Instructions

The boundary-scan register is a shift-register-based structure comprising a variety of different cell designs matched onto the requirements of the particular component. Different cell designs are used according to the type of system pin concerned (input, output, 3-state, and bi-directional) and according to the set of boundary-scan instructions supported.

A simplified view of a boundary-scan register is shown in Figure 2.10. An example implementation for a cell that could be used in each of the locations shown in Figure 2.10 is given in Figure 2.11. The connections labelled PI, PO, SI, and SO in Figure 2.11 are connected to adjacent cells, the on-chip system logic, and the system pins as shown in Figure 2.10.

The cell in Figure 2.11 is designed to respond to the clock and control signals generated by the TAP controller. The mode input is controlled according to the type of pin connected to the cell (input, output, etc.) and the specific instruction selected.

Use of this sample cell design, with appropriate signals supplied to the Mode input of each cell, will result in a component that supports the SAMPLE/ PRELOAD, EXTEST, and INTEST instructions described below. Other cell designs are possible that meet the requirements of this standard for different sets of instructions. For example, if the INTEST instruction were not supported in a given design, R2 and M2 would be omitted from cells that feed data from a system pin to the on-chip system logic. The input, PI, would then be connected directly through the cell to the output, PO.

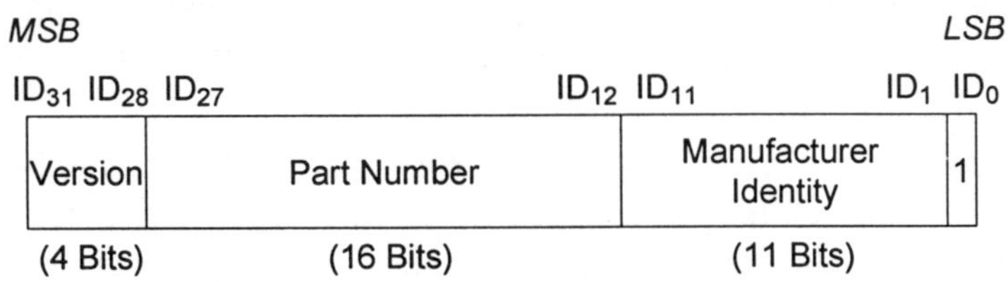

Figure 2.9. **Structure of the device identity code**

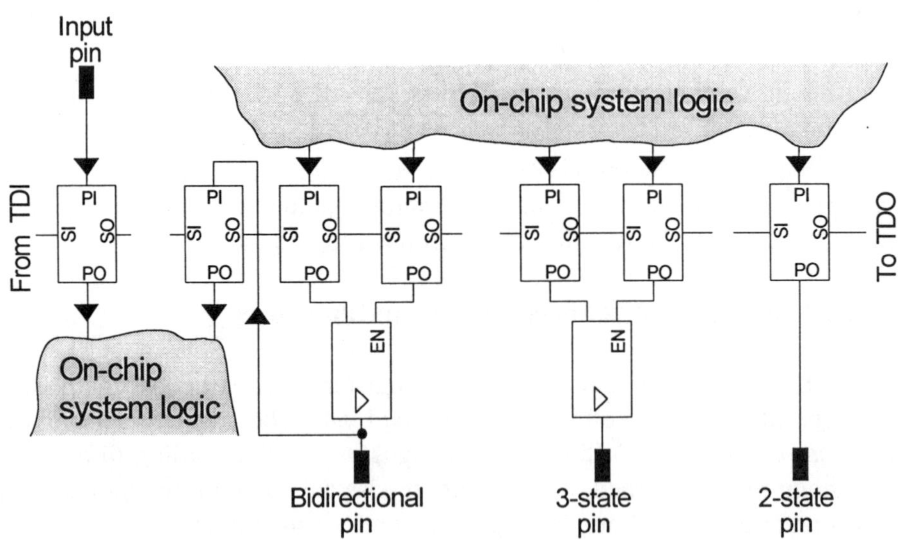

Figure 2.10. **A simplified view of the Boundary-Scan register**

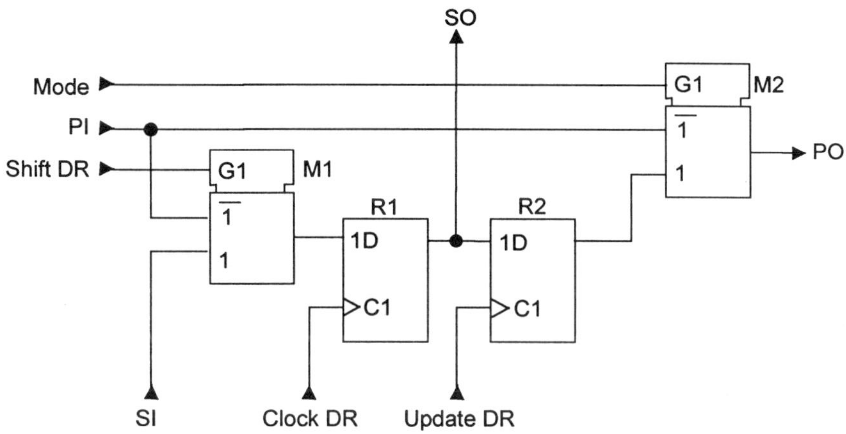

***Figure 2.11.* An example of a Boundary-Scan cell design**

As the Boundary-Scan register instructions are discussed, we note for each:

a) whether the instruction is mandatory or optional;
b) which test data registers can be connected in the serial path between TDI and TDO;
c) restrictions (if any) on the choice of binary codes for the instruction; and
d) flow of data between the component's system pins, the boundary-scan register cells, and the on-chip system logic.

2.10.1 SAMPLE/PRELOAD

This mandatory instruction allows a snapshot of the normal operation of the component to be taken and examined. The taking of this snapshot is arranged so as to have no effect on the system operation that it monitors. The instruction also allows data values to be loaded onto the latched parallel outputs of the boundary-scan shift register prior to selection of other boundary-scan test instructions. This instruction must select only the boundary-scan register to be connected between TDI and TDO in the *Shift-DR* controller state. There is no required binary value that must be decoded as the SAMPLE/PRELOAD instruction. When SAMPLE/ PRELOAD is the current instruction, test logic operation is not permitted to have any effect on operation of system logic or on the flow of signals between the IC's system pins and on-chip system logic (i.e., the Mode signals for all cells should be set to 0).

The sampling mechanism of the SAMPLE/PRELOAD instruction has the effect of loading the states of all signals flowing through system pins into their corresponding boundary-scan register cells on the rising edge of the clock feeding TCK during the *Capture-DR* controller state. This capability allows the state of the interconnect network of a machine to be captured at a desired moment (e.g., upon a failed parity check) by having the requisite edge of the signal feeding TCK triggered by a specified system event.

The preloading mechanism allows an initial data pattern to be placed at the latched parallel outputs of boundary-scan register cells (e.g., in cells connected to pins driving off-chip) prior to selection of some other boundary-scan test operation. For example, the EXTEST instruction (to be discussed next) is used for testing the interconnection of chips. It is very desirable that predetermined and non-damaging signals be driven out of chips on a board while the first test pattern to be applied to the interconnect is being scanned into the boundary-scan register. This aim can be achieved by preloading the same boundary-scan register. As soon as the EXTEST instruction has been transferred to the parallel output of the instruction register, preloaded data will be driven through the system output pins of each IC; this situation will remain in effect while scanning of external test patterns is going on.

Shifting of data for sampling and preloading can occur concurrently when required - while sampled data is shifted out, preloading of new data can occur.

(Note that it is planned to separate SAMPLE/PRELOAD into two instructions – SAMPLE and PRELOAD – in a future update to the Standard. This will not change the functionality described here.)

2.10.2 EXTEST

One of the principal motivations for the development of the Standard was the need for a non-contact method of testing board (and system) interconnect [5, 6, 7]. The EXTEST instruction is the key to the Standard's response to that need. The EXTEST instruction is mandatory. There may be more than one binary code that is interpreted as EXTEST. In the current version of the Standard, one of these codes must be the "all 0's" instruction code, although this restriction is expected to be removed at the next revision.

When EXTEST is the current instruction, the boundary-scan register is the one and only register that is to be connected between TDI and TDO for data scanning purposes. While the EXTEST instruction is selected in a chip,

a) system logic of the chip must be controlled such that it cannot be damaged as a result of signals received at the system input or system clock input pins;

b) the state of all signals driven from system output pins is completely defined by the data shifted into the boundary-scan register and changes only on the falling edge of TCK in the *Update-DR* controller state; and
c) the state of all signals received at system input pins is loaded into the boundary-scan register on the rising edge of TCK in the *Capture-DR* controller state.

The EXTEST instruction allows circuitry external to the component package to be tested. Typically such circuitry would be the board interconnect. Clusters of components that lack boundary-scan registers can be statistically tested using the same functionality although complex clusters would require more sophisticated diagnostic systems than would interconnect alone. In use of the EXTEST instruction, boundary-scan register cells at output pins are used to apply test stimuli, while those at input pins capture test results. Captured results are scanned out of the serially linked boundary-scan registers of a board while the next set of test input values is scanned in.

As was suggested above, the first test stimulus to be applied using the EXTEST instruction should be shifted into the boundary-scan register using the SAMPLE/PRELOAD instruction. This is the most judicious approach because, when the change to the EXTEST instruction takes place in the next occurring *Update-IR* controller state, known data will be driven immediately from the component onto its external connections. Where a total of N tests are to be applied using the EXTEST instruction, stimuli for tests 2 to N will be shifted in while the results from tests 1 to $N - 1$ are shifted out. Note that, while the results from the final test (test N) are shifted out, a determinate set of data should be shifted in that will leave the board in a consistent state at the end of the shifting process. This can be achieved by shifting the stimuli for test N (or indeed any other test) into the boundary-scan register a second time.

Note that the boundary-scan register cells located at input pins may optionally be designed to allow signals to be driven into the on-chip system logic when the EXTEST instruction is selected. This allows user-defined values to be established at the system logic inputs, preventing misoperation in response to noise signals arriving from the board-level interconnect. The values driven may either be constant for the duration that EXTEST is selected (e.g., by including a blocking gate at the input to the system logic) or they may be loaded serially through the boundary-scan register.

The EXTEST instruction can be entered by holding TDI at a constant low value and completing an instruction-scan cycle of sufficient duration to fill each instruction register on the board-level serial data path. As in the case of the BYPASS instruction, this may simplify demands on the ATE or bus master device that controls a test.

The data loaded into boundary-scan register cells located at system output pins (2-state, 3-state, or bi-directional) in the *Capture-DR* controller state when the EXTEST instruction is selected should be independent of the operation of the on-chip system logic. Where followed, this recommendation ensures that data shifted out of the component in response to the EXTEST instruction is not altered by the presence of faults in a chip's system logic. This simplifies diagnosis since any errors in the output bit stream can only be caused by faults in off-chip circuitry, in board-level interconnections, or in boundary-scan registers used to apply the test.

2.10.3 INTEST

INTEST is one of two optional instructions defined by the Standard that allow testing of on-chip system logic after a component is assembled on a board. The binary value(s) that are decoded as the INTEST instruction may be selected by the component designer's specification. Using the INTEST instruction, test stimuli are shifted in one at a time and applied to the on-chip system logic. To achieve this, on-chip system logic must be capable of single-step operation while INTEST is the current instruction. Internal test results are captured in the IC's boundary-scan register and are examined by subsequent shifting. Initial data setup for such a test can be achieved using the SAMPLE/PRELOAD instruction.

The INTEST instruction must select the boundary-scan register to be the one and only register connected between TDI and TDO for shifting access. When INTEST is the current instruction:

a) the state of all signals driven from system output pins must be completely defined by data previously shifted into the boundary-scan register and is allowed to change only on the falling edge of TCK in the *Update-DR* controller state;

b) the state of all non-clock signals driven into the system logic from the boundary-scan register must be completely defined by data previously shifted into the register;

c) the state of all signals output from the system logic to the boundary-scan register must be loaded into the register on the rising edge of TCK in the *Capture-DR* controller state.

The INTEST instruction allows slow-speed (static) testing of on-chip system logic. Each test pattern and response must be shifted through the boundary-scan register. While an INTEST-based test is proceeding, the logic values at the component output pins are defined from the boundary-scan register. This requirement ensures that surrounding components on an assembled board are supplied known signal levels while on-chip system logic testing is in progress. A consistent, "safe" set of data values would be shifted into the appropriate stages of the boundary-scan register using the SAMPLE/PRELOAD instruction prior to selection of the INTEST instruction. This "safe" data pattern is then reloaded into boundary-scan control cells and the cells associated with chip outputs each time a new INTEST test pattern is shifted into the boundary-scan register.

As noted above, the approach taken with the INTEST instruction requires that on-chip system logic can be operated in a single-step mode-where the circuitry moves one step forward in its operation each time shifting of the boundary-scan register is completed. Note that, for each such test step, the latched parallel output of the boundary-scan cell at the system output pin is updated from data shifted in before the contents of the shift-register is overwritten with a new test response. While the INTEST instruction is selected, an IC's boundary-scan register assumes the role of the fixture and pin electronics of an ATE used for stand-alone component testing. Cells at non-clock system input pins are used to apply test stimuli, while those at system output pins capture responses. Stimuli and responses are moved into and out of the circuit by shifting the boundary-scan register.

To achieve single-step operation, on-chip system logic can be expected to receive a sequence of clock events between application of the stimulus and capture of the response. A designer's specification of boundary-scan cells for system clock input pins would allow clocks for on-chip system logic to be provided in several ways while INTEST is the current instruction. Here are some examples:

a) The signals received at system clock pins can be fed directly to the on-chip system logic, as they would be during non-test operation. When this approach is taken, off-chip clock sources should be handled in such a way that, during internal testing via INTEST, clock signals received by the component change state only in the *Run-Test/Idle* controller state. In this way, on-chip system logic operation can be inhibited while test data is shifted through the boundary-scan register. Figure 2.12 illustrates how a system clock applied to a component should be controlled during INTEST-based testing of on-chip system logic.

b) On-chip system logic can be supplied with clock signals derived from the input signal to TCK in the *Run-Test/Idle* controller state. In all other controller states, such clocks should not change state (Figure 2.13).

c) Circuitry may be built into the component, which allows on-chip system logic to complete one step of operation upon entry into the *Run-Test/Idle* controller state. For example, if the component were a microprocessor, it would be permitted to complete a single processing cycle; this might be achieved by internal generation of a pulse on the processor's hold signal.

In such a case, the clock(s) applied at system clock pin(s) during the test could be free-running.

d) Clock signals can be loaded serially via the boundary-scan path in the same manner in which non-clock signals for the on-chip system logic are supplied. This would require the boundary-scan register to be loaded for each distinct clock signal state twice for a single-phase clock. This may be a hazard-prone operation for certain circuit designs.

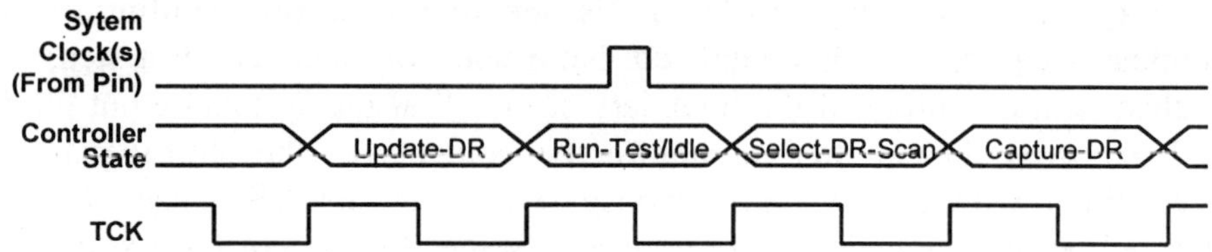

Figure 2.12. **Control of applied system clock during INTEST**

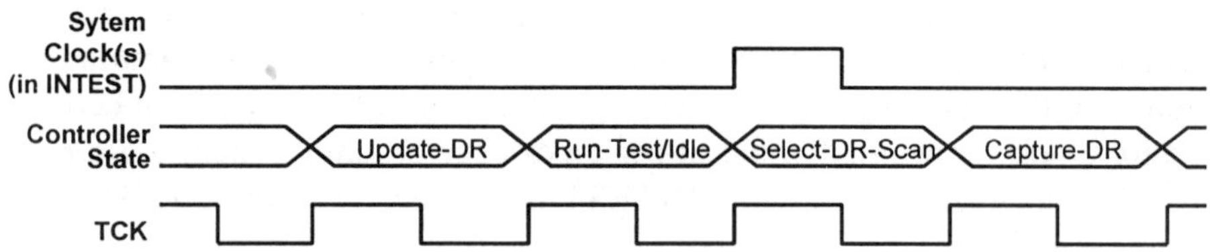

Figure 2.13. **TCK as clock for on-chip system logic during INTEST**

The Standard recommends that for boundary-scan register cells located at system input pins (clock or non-clock) or at bi-directional pins configured as inputs, the data loaded in the *Capture-DR* controller state when the INTEST or RUNBIST (see below) instruction is selected should be independent of the operation of off-chip circuitry or board-level interconnections. Where followed, this approach ensures that data shifted out of a component in response to the INTEST instruction are not altered by the presence of faults in off-chip system logic, board-level interconnections, etc. This will simplify diagnosis; any errors in the output bit-stream can be caused only by faults in on-chip system logic or the boundary-scan register.

2.10.4 RUNBIST

In many cases, it may be desirable to put more of the burden of test into the product itself, rather than on an external tester. Contemporary Built-In

Self-Test (BIST) technology is beginning to make this possible with relatively little circuit overhead or performance penalty [8]. The optional RUNBIST instruction causes execution of a self-contained self-test of the component. It's binary encoding is left to the discretion of the designer. Use of the RUNBIST instruction allows a component user to determine the health of a component without the need to load complex data patterns and without the need for single-step operation (as required for the INTEST instruction). With this capability in place, it becomes possible for all components on a board that offer the RUNBIST instruction to execute their self tests concurrently or in groups limited by power consumption or heat dissipation requirements; thus, a rapid health check for assembled boards can be provided.

RUNBIST is as an important instance of the flexibility that the Standard offers in providing a gateway (the TAP) through which powerful instructions can be passed to an IC when it has become a component of an assembled system.

The Standard allows the development of public and private instructions that could cause execution of BIST in various subsections of an IC. (Such capabilities could contribute to support of failure mode analysis). The RUNBIST instruction is intended to serve as a model for other BIST instructions and, when it is feasible to do so, to be a vehicle by means of which designers could link all self-testing circuitry within an IC into a single BIST process for the entire entity. The Standard recommends that RUNBIST be implemented wherever possible.

Self-test operation accessed via the RUNBIST instruction must execute only in the *Run-Test/Idle* controller state; and when RUNBIST is the current instruction, the test data register into which the results of the self-test(s) will be loaded must be connected for serial access between TDI and TDO in the *Shift-DR* controller state. (This register can be the boundary-scan register.) The result of the self-test(s) executed in response to the RUNBIST instruction must be loaded into the designated test data register no later than the rising edge of signal input to TCK in the *Capture-DR* controller state.

The developers of the Standard intended the RUNBIST instruction to be an encouragement for designers to create as simple an interface as possible for self testing ICs embedded in systems. Therefore, they required that where a test data register (other than the boundary-scan register) must be initialized prior to execution of the self-test, this must occur at the start of the self-test without any requirement to shift data into the component-there cannot be any requirement to enter seed values into any test data register other than the boundary-scan register. Moreover, where a component includes multiple self-test functions, these may be executed in response to the RUNBIST instruction either concurrently or in a sequence determined by the component manufacturer. In the latter case, all sequencing must be taken care of within the component without requiring alteration of instruction register contents or other external direction.

A number of the requirements in the Standard are there for purposes of isolation-to assure a repeatable self-test result for fault-free circuits and to protect off-chip circuitry during self-test. The design of the component must ensure that results of self-tests executed in response to the RUNBIST instruction are not affected by signals received at non-clock system input pins; and data shifted out of a component following completion of execution of a self-test accessed using the RUNBIST instruction must be independent of the operation of off-chip circuitry or board-level interconnections. (The latter clause ensures that BIST result data shifted out of the chip are not altered by the presence of faults in off-chip system logic, board-level interconnections, etc. As discussed under INTEST, earlier, this simplifies diagnosis.)

When RUNBIST is the current instruction, the state of all signals driven from system output pins must be completely defined by data previously shifted into the boundary-scan register (for example, by use of the SAMPLE/PRELOAD instruction as described under INTEST, above); and states of parallel output registers or latches in boundary-scan register cells located at system output pins (2-state, 3-state, or bi-directional) are not permitted to change while the RUNBIST instruction is selected. In contrast to the INTEST instruction, the data values driven through the system output pins are held while the RUNBIST instruction is selected. For a boundary-scan register cell located at a system output pin (see Figure 2.11), the *Update-DR* signal should be held at 0 while the RUNBIST instruction is selected and the Mode input should be held at 1. While the RUNBIST instruction is selected, boundary-scan register cells associated with non-clock system inputs of a chip may simply be loaded with constant, safe values. Alternatively, the boundary-scan register may act as a pattern generator and/or signature compactor in the *Run-Test/Idle* controller state-provided the states of parallel output registers or latches are unchanging as required above.

The specification of boundary-scan cells for system clock input pins allows the clocks for the on-chip system logic to be obtained in one of two ways while the RUNBIST instruction is selected.

a) The signals received at system clock pins can be fed directly to the on-chip system logic, as they would be during non-test operation of the component. Where this is done, the design of the component must ensure that self-test executes only in the *Run-Test/Idle* controller state. However, the clock may be active in other controller states.

b) The on-chip system logic can be supplied with clock signals derived from the signal input at TCK in the *Run-Test/Idle* controller state. In this instance, in all other controller states, the clocks should not change state.

There are certain, additional, common sense requirements placed on an IC that implements RUNBIST and on such an IC's documentation. All stages of the test data register selected by the RUNBIST instruction must be set to determinate logic states (0 or 1) in the *Capture-DR* controller state (i.e., while the test result is loaded). A duration for the test executed in response to the RUNBIST instruction must be specified (e.g., by citing a number of rising edges of the signal at TCK or of a system clock). Because it is likely that self-test in one component may complete before the self-test in another, the test results produced by the execution of the RUNBIST instruction and deposited in the specified test data register ready for shifting out of the IC must be stable through the period of delay before the *Capture-DR* controller state is entered, which period is likely to be unpredictable at the time of the component's design. Use of the RUNBIST instruction must give the same result in all versions of a component. These requirements were included in the standard to ensure that the test for an assembled board is independent of the versions of components mounted on it. Such independence is an important consideration when working in a maintenance or repair environment, where the versions of components used on a board may very well be unknown. The Standard's requirement can be met by forming the exclusive-OR of the result from execution of the RUNBIST instruction with a fixed (version-dependent) pattern. The output from this function would become the result loaded into the boundary-scan register or other test data register connected between TDI and TDO during the time the RUNBIST instruction is current.

REFERENCES

[1] ANSI/IEEE Std 1149.1 (revision b) – "Standard Test Access Port and Boundary-Scan Architecture", *IEEE*, New York, 1994

[2] *C.M. Maunder and R.E. Tulloss*, "An Introduction to the Boundary-Scan Standard", *IEEE Computer Society Press*, Los Alamitos, CA, 1990

[3] *B. Konemann et al.*, "Built-in logic block observation techniques" Proceedings, *International Test Conference*, pp. 37-41, Los Alamitos, CA, 1979

[4] *Joint Electron Device Engineering Council*, "Standard Manufacturer's Identification Code" *JEDEC Publication 106-A*, July 1986. Obtainable from JEDEC, 2001 Eye Street NW, Washington DC 20006, USA

[5] *FPM. Beenker*, "Systematic and structured methods for digital board testing;" Proceedings, *International Test Conference*, pp. 380-385, 1985

[6] *C.W. Yau and N. Jarwala*, "A new framework for analyzing test generation and diagnosis algorithms," Proceedings, *International Test Conference*, pp. 63-70, 1989

[7] *N. Jarwala and C.W. Yau*, "A unified theory for designing optimal test generation and diagnosis algorithms for board interconnect", Proceedings, *International Test Conference*, pp. 71-77, 1989

[8] *H.N. Scholz et al.*, "ASIC implementations of boundary-scan and BIST," Proceedings, *8th International Custom Microelectronics Conference*, pp. 43.1-43.9, 1988

Chapter 3

IEEE 1149.4 Architecture and Instruction Set

Steve Sunter
LogicVision, sunter@lvision.com

BACKGROUND

Similar to the other chapters in this book, this chapter is written with readers in mind who may not be familiar with the 1149.1 digital test standard because they are more familiar with analog and mixed-signal ICs, which typically do not use boundary-scan. Also, these instruction and architecture descriptions should be read in addition to, and not instead of, the 1149.4 Standard itself. The instruction description here is intended to be more concise than that in the 1149.4 document so that readers can easily comprehend the full instruction suite without needing to understand every detail.

In the following sections, italicized instruction names (e.g., *EXTEST*) refer to instructions, and plain text instruction names (e.g., EXTEST) refer to the mode invoked by an instruction upon entering the *Update-IR* state of the TAP controller.

3.1 The 1149.4 instruction set

This section briefly reviews the 1149.1 test access port (TAP) communication protocol to show how instructions are defined and how they are conveyed into an IC. Then each instruction is described, starting with the mandatory ones followed by the optional ones. Lastly, each pin's test capabilities are described.

The 1149.1 TAP has a specific transaction protocol for conveying digital instructions to an IC. Figure 3.1 shows the state diagram for the TAP controller, which should be referenced to understand the next two paragraphs.

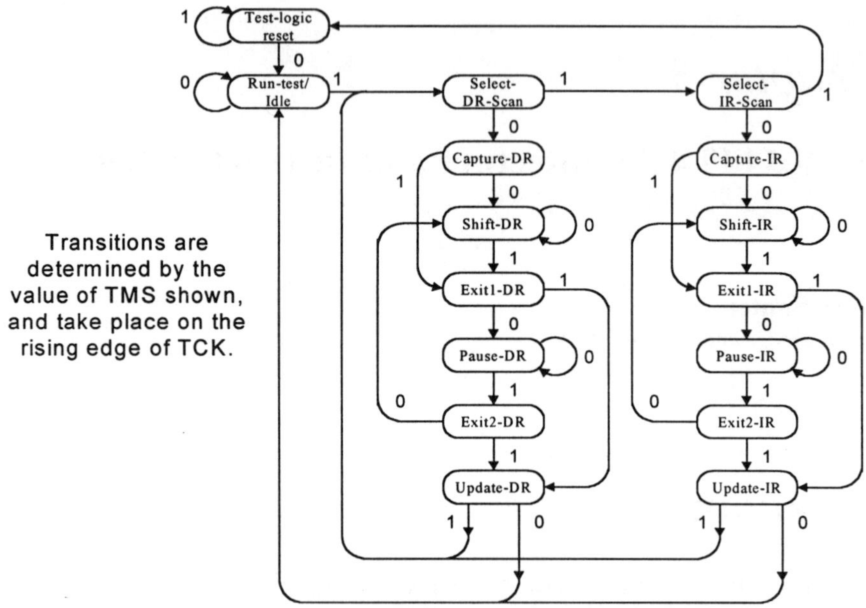

Figure 3.1. **TAP controller state transition diagram**

First, a sequence of bits is sampled at the TMS pin by the rising edge of the TCK pin signal. Consider a specific example: starting from the *Test-Logic-Reset* state, the sequence 01100 sets the register pointer to the Instruction Register. Subsequently, bits corresponding to an instruction are shifted into the TDI pin synchronously with TCK rising edges (while TMS is held at logic 0). The bits shift into the Instruction Register, whose length is chip-specific, until it contains the full instruction. Then, the sequence 110 is applied to TMS to return the TAP controller to the *Run-Test/Idle* state.

From the *Run-Test/Idle* state, the TMS bit sequence 100 will set the register pointer to the boundary-scan Data Register. Next, while TMS is held at logic 0, bits on the TDI pin are shifted into this register, which comprises all boundary module test registers for all the digital and analog function pins of the IC. In this way, each pin is programmed to perform a pin-specific test mode function later when the Update-DR state is reached. Lastly, applying the bit sequence 11 at the TMS pin advances the TAP controller to the *Update-DR* state, at which time the contents of the boundary-scan data register are loaded into the controlling latches of each pin. If an instruction such as *EXTEST* is in the Instruction Register, the pins of the IC will drive pin-specific test signals.

The bit sequence for each instruction that is to be shifted into an IC is chip-specific and documented in the BSDL description [1] for the IC, except for bit sequences which are mandated by 1149.1 (and which are mentioned later in this section). For example, any instruction could be represented by 10101010, or by 111000, or any other sequence of any length, as decided by the IC designer.

The set of instructions for an IC that is 1149.4-compliant is almost identical to that for an IC that is 1149.1-compliant. Only the *PROBE* instruction must be added. However, many of the 1149.1 instructions are interpreted differently for analog pins. For example, where 1149.1 indicates "drive logic 1" in EXTEST mode, 1149.4 effectively says, "drive the maximum voltage that the pin would normally drive in function mode."

3.1.1 Mandatory instructions

3.1.1.1 BYPASS

This instruction overrides all other test instructions to ensure the chip operates in normal function mode (mission mode), not test mode. The instruction also tells the IC to allow test data shifted into the TDI pin to bypass all registers in the IC and to be routed via a one bit register to the TDO pin. Each TDI bit value is sampled by a rising edge of the TCK signal, and shifted out at TDO synchronous with the next falling edge of TCK.

BYPASS can be used to allow test access to any IC in a group of ICs connected in a series, TDO of each IC being connected to TDI of the next IC. Without this instruction, it might require thousands of "padding" bits to precede and follow the bits destined for one specific IC in the middle of the chain. If it is desired to bypass an IC's boundary-scan register without causing the IC to leave test mode, then the optional instruction *CLAMP* or *HIGHZ* can be used.

As a fail-safe mechanism, all ICs must interpret all 1's in the instruction register as a *BYPASS* instruction. This increases reliability because the 1149.1 standard mandates an on-chip pull-up for the TDI and TMS pins, so that a board-level open circuit on the TDI signal to an IC will cause the IC to be bypassed and to remain in mission mode. The *BYPASS* instruction may also be represented by other bit sequences.

All aspects of this instruction are interpreted identically in 1149.1 and 1149.4, except that 1149.4 specifically mentions analog pins and analog test buses.

3.1.1.2 SAMPLE/PRELOAD

The primary purpose of this instruction is to allow data to be shifted into and out of the boundary-scan Data Register without affecting the normal function mode of the IC. Data can be shifted in to pre-load each pin's serial boundary-scan register in preparation for applying an instruction such as *CLAMP*. Data is shifted out, after sampling the signal at every pin simultaneously without interrupting mission mode for the IC, thus providing a 'snapshot' of normal operation.

In 1149.4, sampling the value at a pin is interpreted as sampling a digitized equivalent of the analog voltage at the pin and storing the result as one bit. The reference voltage for the digitizer (i.e., the voltage that distinguishes a logic 0 from a logic 1) is V_{TH}, whose value is pin-specific and is approximately mid-way between the maximum and minimum voltages seen at each pin.

3.1.1.3 EXTEST

This instruction puts all pins into test mode to allow testing of pin functions and testing of the board-level interconnect between pins. It is the key instruction for facilitating analog measurements of elements connected to the pins, typically by injecting stimulus current and monitoring the resultant voltage, both via the analog test busses.

This is the only mandatory instruction that takes an IC out of mission mode into test mode. When this instruction is in the Instruction Register, immediately upon entering the *Update-IR* state, all function pins become isolated from the core of the IC and their state is determined solely by the contents of the analog and digital boundary module register bits.

When in this test mode, data shifted into TDI is routed through the boundary-scan Data Register to TDO. To simplify coding, all ICs must interpret all 0's in the instruction register as an *EXTEST* instruction. The instruction may also be represented by other bit sequences.

3.1.1.4 PROBE

This is the only new instruction in 1149.4 relative to 1149.1, and its primary purpose is to allow continuous-time access to pin signals, without affecting mission mode operation. The instruction is similar to the *SAMPLE/PRELOAD* instruction, which stores one-bit digitized samples of pin signals and serially shifts the samples out. For both analog and digital signals, continuous-time access provides much more information about a single pin. In effect, the *PROBE* instruction allows 'virtual probing' of selected pins of an IC with minimal impact on their normal function.

While it is possible to also inject signal current or voltage during PROBE mode, this should only be done with care, as it will clearly affect mission mode operation. The amount of current that can be injected is minimal (<100 μA, as detailed elsewhere). The frequencies that can be accurately monitored may be quite low, e.g., less than 10 kHz, unless higher bandwidth is specifically designed into the IC's analog test buses.

3.1.2 Optional instructions

The following instructions are not required to be performed by every IC conformant with the 1149.4 or 1149.1 standards, but if the IC does respond to any of these instructions, then the IC must respond in the ways specified in the Standard.

3.1.2.1 INTEST

The primary purpose of this instruction is to facilitate testing of the interior (core) of the IC by controlling and observing all digital inputs and outputs of the IC's core via digital boundary module register bits. In this mode, upon entering the *Update-IR* state, all digital inputs (except clocks) and outputs must become isolated from the off-chip world and from the analog core. However, analog inputs and outputs must remain electrically connected to their respective function pins. This significant difference between digital and analog pins is necessary because most analog inputs cannot be meaningfully tested by a DC signal.

As is also true for EXTEST, the boundary-scan data register is connected to TDI and TDO during INTEST, which allows repeated shifting in of new stimulus vectors and shifting out of captured results without updating the Instruction Register. Similarly, in this mode, any analog pin signal can be probed via the analog buses to allow signal monitoring or signal injection. For an IC that supports *INTEST*, the boundary-scan register must include digital boundary modules to control and monitor signals between the analog and digital cores so that the tests of both cores can be deterministic (hence 'boundary' refers to each of the digital and analog cores, not the IC).

During INTEST for a totally digital IC, the digital input pins of the IC do not need to have meaningful signals applied to them because the boundary register overrides to supply values for all digital inputs to the core. However, during INTEST for an 1149.4-conformant IC, all analog input pins (and possibly some analog output pins) must have meaningful signals applied to them from off-chip.

3.1.2.2 CLAMP

This instruction is almost a combination of the *BYPASS* and *EXTEST* instructions. All pins are disconnected from the core and their state is controlled by data in the boundary-scan Data Register, but data shifted into TDI bypasses the boundary register and goes directly to TDO. As always, data is shifted into TDI synchronously with rising edges of the TCK signal, and shifted out on TDO synchronously with falling edges of the TCK signal.

This instruction can be used to allow rapid access to any IC in a group of ICs connected in series. The number of "padding" bits which precede the bits destined for a specific IC is equal to the number of ICs which come before the IC in the scan chain. If it is desired to bypass the test registers without causing an IC to enter test mode, then *BYPASS* can be used.

3.1.2.3 HIGHZ

This is the simplest instruction that places all function pins in test mode. All pins, including AT1 and AT2, are simply disconnected from the core and placed in the *Core-Disconnect* state (CD-state), which is preferably a high impedance state. This allows a board-level bed-of-nails tester to impose currents and voltages on the board-level interconnect wires, without contention from IC output drivers.

In 1149.1, digital pins are set to the high impedance state, but in 1149.4, it is recognized that this is not always practical without jeopardizing analog performance (especially for differential pins). A more detailed explanation of the CD-state is given later in this chapter.

This instruction also causes data shifted into TDI to bypass the boundary-scan Data Register and go directly to TDO.

3.1.2.4 RUNBIST

This instruction is quite different from all preceding instructions because the test stimulus is not scanned in, nor are detailed captured results scanned out. Instead, the instruction initiates a built-in self-test (BIST) when the TAP controller enters the *Run-Test/Idle* state. The BIST must run autonomously, without needing any further test data. When the BIST is complete, the test result signature must be in the data register selected by this instruction, so that the following *Shift-DR* state will shift out the signature via TDO. For an IC that supports RUNBIST, the boundary-scan register must include digital boundary modules to control and monitor signals between the analog and digital cores so that the tests of both cores can be deterministic

While in RUNBIST mode, the analog output pins of an IC must be inactive. They are either individually driven to states controlled by the boundary-scan data register, or all pins are simply placed in the CD-state (i.e., the core disconnect state, which is preferably high impedance).

It is quite practical to use BIST to test all random logic in an IC, especially if the logic comprises more than 100K logic gates and operates above 50 MHz. BIST typically uses a linear feedback shift register (or a more distributed circuit called cellular automata) to generate a pseudo-random bit stream which is scanned into the digital core's scan chains. One or two clock cycles after all data is scanned in, the parallel inputs to the scan chain are sampled, and then serially shifted out to a multiple-input signature register (MISR) which combines the outputs of each chain with previous results. After this sequence is repeated a few hundred times, the MISR will contain a digital signature, containing 100 bits or more. This signature can then be shifted out via TDO.

BIST for mixed-signal circuitry is relatively uncommon. Although it has been the subject of research papers throughout the 1990's, a commercial mixed-signal BIST was not announced until the International Test Conference in Oct. 1997 (some of the principles used are described in [2]).

Using BIST to test digital memories in ICs is increasingly common, and has become the most cost-effective way to test memories in ICs containing dozens of them. It is less common to see ICs that use RUNBIST to produce a single signature representing all random logic and all memories on the IC. If implemented, it is done by causing each BIST to be invoked by successful completion of the preceding BIST. Many ASICs permit BIST for the random logic and memories to be invoked with separate 'private' instructions. Similarly, it is more practical to implement a separate BIST for each analog circuit block of an IC because many condition-specific test limits will need to be shifted in, and many measurement values will need to be shifted out.

3.1.2.5 IDCODE

This instruction has no effect on the IC in its test mode or mission mode. Its sole purpose is to allow external reading of an optional identification register in the IC. This register contains the manufacturer's identity, the part number, and a version number. This data identifies the IC to ensure that it is the correct IC and to allow test software to use an appropriate version of the test.

This instruction causes data shifted into TDI to be routed through the identification register and then directly to TDO. When in IDCODE mode, upon entry into the *Capture-DR* state, the identification register is parallel loaded with the identification data. The data is then shifted out during the *Shift-DR* state. The *IDCODE* instruction will also be automatically loaded immediately following the *Test-Logic-Reset* controller state, if the IC supports this instruction.

3.1.2.6 USERCODE

This instruction is similar to *IDCODE*, in that it has no effect on the IC in its test mode or mission mode. Its purpose is to allow external reading of a second identification register in the IC, if the functions of the IC are user-programmable so that *IDCODE* does not provide sufficient IC identification to facilitate testing. For example, a field programmable gate array (FPGA) might have both an identification code to identify the manufacturer, and a user-programmable code to identify the gate interconnection pattern that was loaded during power-up.

Similarly to *IDCODE*, this instruction causes data shifted into TDI to be routed through the identification register and then directly to TDO. When in USERCODE mode, upon entry into the *Capture-DR* state, the identification register is parallel loaded with the user-programmable identification data. The data is then shifted out during the *Shift-DR* state.

3.1.2.7 PRIVATE and PUBLIC instructions

ICs that are compliant with 1149.4 or 1149.1 are permitted to respond to PUBLIC and PRIVATE instructions. PUBLIC instructions include all mandatory and optional instructions described in the preceding paragraphs, and include any TAP-loaded instructions that are suitably documented for users of an IC. Typically, PUBLIC instructions will provide test access to registers as an aid for diagnosing software or application problems.

PRIVATE instructions are typically intended for use only by the manufacturer or designer, and are not documented. Often, PRIVATE instructions will provide test access to core signals that aid in diagnosing manufacturing or design problems.

Table 3.1. **Summary of instructions**

Instruction	Requirement	TDI – TDO path	Mode of function pins
BYPASS	Mandatory	Direct	Function
SAMPLE/ PRELOAD	Mandatory	Bscan register	Function
EXTEST	Mandatory	Bscan register	Test
PROBE	Mandatory	Bscan register	Function
INTEST	Optional	Bscan register	Test – digital pins Function – analog pins, clock pins
CLAMP	Optional	Direct	Test
HIGHZ	Optional	Direct	High-Z – digital pins CD-state – analog pins
RUNBIST	Optional	Bscan register	Test – output pins Function – input pins
IDCODE	Optional	Ident. register	Function
USERCODE	Optional	Ident. register	Function

3.1.3 Analog pin test capabilities

Analog function pins are defined in 1149.4 as all pins that are not power pins and not digital pins. Test capabilities of the digital pins must conform to the specifications in 1149.1. In summary, test circuitry at digital input pins need be able to only monitor (observe) the logic values at the pins. The circuitry may also drive test signals into every digital pin of the IC, simultaneously, in place of the pins' logic values – this is not possible for analog pins in 1149.4. At each digital output pin, test circuitry must be able to drive a logic 1, 0, or high impedance, or the normal function mode signal.

Every analog pin must be able to be connected to the core of the IC during Function mode, and disconnected (in effect) during Test mode. "Disconnected" means that for all intents and purposes, activity in the core does not affect measurements at the pin, and *vice versa*.

In test mode, each analog pin (input, output, or bi-directional) must be capable of driving a high and a low voltage, denoted as V_H and V_L, respectively. These voltages may be unique to each pin or common to all pins. The voltages are the maximum and minimum voltages, respectively, that the pin normally handles in function mode.

Each analog pin must also have a CD-state capability (CD is derived from "core disconnect"). Preferably, a pin in CD-state has a high impedance output, but it may have any impedance provided the impedance can be modeled by a network of linear passive elements and independent DC voltages. There are several interesting cases where the CD-state is important. Bias voltage pins are an example of an analog input/output for which disconnecting the pin from the IC core might jeopardize all other analog functions. Since a bias voltage can be modeled as an independent DC voltage, it does not need to be disconnected. Similarly, low impedance pull-up resistances or inductances often cannot be disconnected because the isolation transistor would be impractically large or inaccurate.

It must be possible to digitize the voltage at each pin, relative to a reference voltage, V_{TH}, and the one-bit result is latched by the rising edge of TCK during the *DR-Capture* state. V_{TH} is mid-way between V_H and V_L, and may be unique to each pin.

Each pin shall also be able to drive an analog ground voltage, V_G, which must be a stable, low-noise DC voltage supplied to the IC by a pin, such as V_{SS}. Lastly, it must be possible to deliver the current at AT1 to any analog pin, via the on-chip bus AB1, and to transmit the voltage at any analog pin to AT2, via the on-chip analog bus AB2.

Several of analog pin's test mode capabilities can be activated simultaneously – for example, a pin can drive V_L while a current is also being delivered to it via AT1, and while the voltage is monitored via AT2. Any capabilities that are inherently contradictory are made impossible to achieve, e.g., it is not possible to set bits so that a pin is simultaneously driven to V_H and V_L, or to V_L and V_G.

Summary of analog function pin mandatory test capabilities

While connected to core (function mode):

- Compare pin's voltage to V_{TH}, and latch as a one-bit result
- Convey current to the pin from AB1
- Convey the pin's voltage to AB2

While disconnected from core (test mode):

- Drive the pin's maximum or minimum voltage (V_H or V_L)
- Connect the pin to an analog ground (V_G) pin
- CD-state (preferably high impedance)
- Compare pin's voltage to V_{TH}, and latch as a one-bit result
- Convey current to the pin from AB1
- Convey the pin's voltage to AB2

3.1.4 Analog Test Access Port (ATAP) pin capabilities

In this section, the capabilities of the analog test bus pins or ATAP are described. These pins are unique to 1149.4.

The analog pins connected to the test bus interface circuit (TBIC) are referred to as the ATAP. There are two mandatory ATAP pins: AT1 and AT2. AT1 is an analog input pin for providing a stimulus current, and AT2 is an output pin for monitoring the resultant voltage. These pins may also be bi-directional, and may convey both voltage and current. Optionally, when differential test signals must be conveyed, there may be two optional pins, AT1N and AT2N, where AT1N carries the inverse of the AT1 signal, and AT2N carries the inverse of the AT2 signal. Almost all rules that apply to AT1 also apply to AT1N, if it exists, and the inverting pins of differential functions, if they exist. There is a similar duality between AT2 and AT2N.

When suitably programmed, the IC must convey the current at its AT1 pin to the AB1 bus. Similarly, it must convey the voltage on the AB2 bus to the AT2 pin.

Two additional capabilities are mandated to facilitate *in situ* calibration of current and voltage offsets gains, and other analog transmission characteristics.

It shall be possible for the voltage at AT1 to be monitored by AB2/AT2 just like the voltage at any other analog function pin. By driving AT1 with an accurately known voltage, and monitoring the voltage at AT2 with AT1-AB2 and AB2-AT2 paths enabled, the voltage characteristics of the paths and the AB2 bus can be deduced.

Similarly, it must be possible for the current delivered into AT1 and AB1 to be conveyed to AT2 just like the current from any other analog function pin. By driving AT1 with an accurately known current, and conveying the current to AT2 via enabled AT1-AB1 and AB1-AT2 paths, the current characteristics of the paths and the AB1 bus can be determined.

The primary purpose of the ATAP is to convey continuous-time analog test signals. However, to aid in rapidly testing the board-level interconnect, each ATAP pin must also have the same capabilities as any other analog pin. Specifically, each ATAP pin must be capable of driving a high and a low voltage (V_H and V_L), and a high impedance state, and be able to sample and digitize the pin's voltage. Whereas analog function pins must have a CD-state, which is preferably high impedance but could be low impedance, it is essential that the ATAP pins have a true high impedance state (e.g., 10 MΩ). This last capability ensures that when more than one IC are connected to the AT1 and AT2 wires at the board-level, uninvolved ICs will not interfere with an analog measurement via these wires.

Summary of ATAP pin mandatory test capabilities

While connected to the internal analog test buses:

- Convey current from AT1 to AB1
- Convey voltage from AT1 to AB2
- Convey current from AB1 to AT2
- Convey voltage from AB2 to AT2
- Compare AT1 voltage to V_{TH}, and latch as a one-bit result
- Compare AT2 voltage to V_{TH}, and latch as a one-bit result

While disconnected from the internal analog test buses:

- High impedance state
- Compare pin's voltage to V_{TH}, and latch as a one-bit result

3.2 The 1149.4 Architecture

The general architecture of an 1149.4-compliant IC is shown in the schematic of Figure 3.2. The architecture of a board containing ICs with 1149.4 and ICs with 1149.1 is shown in Figure 3.3. There are many features worth noting about these two figures, and they will be discussed from both the board-level view and on each IC.

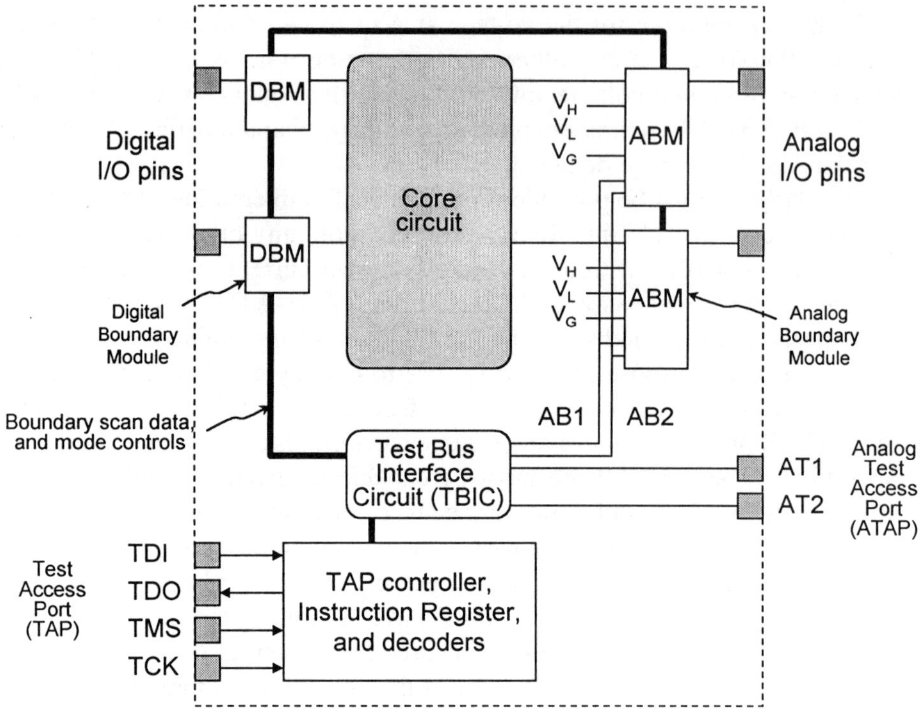

Figure 3.2. **Architecture of simple IC with 1149.4**

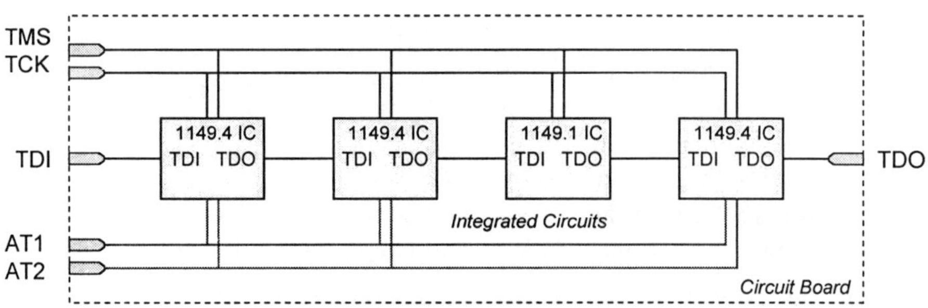

Figure 3.3. **Architecture of simple board with 1149.4 and 1149.1 ICs**

3.2.1 Board-level view

A single boundary-scan path can link any number of 1149.1 and 1149.4 ICs, with TDO of one chip connected to TDI of the next. Similarly, TMS and TCK can be common to both types of ICs. Test software that is designed for 1149.1 can be used for testing the wire interconnects between all these ICs. Of course, software and hardware for controlling the analog stimulus and measuring the analog response will need to be added.

Precisely two semiconductor switches are required between any IC function pin to be tested and the edge connector of a board containing the IC: one switch is in the pin's ABM and the other is in the IC's TBIC. Limiting the switches to two minimizes distortion and noise while allowing sufficient electrical isolation to accommodate many voltage and frequency ranges within a single board.

The capacitance and leakage current of the ATAP pin buses can be quite large if many ICs are connected to the buses. Various switches, calibration methods, and connection techniques are described in 1149.4 to address these parasitic effects. For example, the buses may be split into multiple bus pairs so that the number of ICs per bus pair is limited. Output drivers can be used for the AT2 pin (and AT2N) to isolate the board-level capacitance from the on-chip bus (AB2), so that accessed nodes are affected by only the on-chip bus capacitance and leakage. TBIC switches also isolate the on-chip buses from the off-chip buses, so that the capacitance and leakage of the on-chip buses of uninvolved ICs do not affect a measurement.

Capacitance between AT1 and AT2, and between AB1 and AB2, can significantly affect measurements of small impedances and should be minimized by placing a ground wire (or other DC voltage) between the bus wires. Of course, the same applies to the optional AT1N and AT2N buses.

When ICs operating at different power supply voltages are to be accessed, separate ATAP buses may need to be used on the board. Each bus can be accessed separately by the off-board tester, or on-board relay switches can be used to multiplex the buses to a single set of ATAP pins for connection to the tester. Alternatively, semiconductor switches could be used provided they do not introduce distortion and noise, or limit the bandwidth. These switches would obviously need to have a voltage range that encompassed the ranges of all the bus pairs.

When the ATAP pins are not driven by any IC on the board, and not driven by a tester, then these pins will be in high impedance state. Board designers need to make provision for driving or tying these buses to a DC voltage when they are not being used, otherwise, they may conduct noise or electro-static discharges (ESD).

3.2.2 IC-level view

3.2.2.1 Analog Boundary Module (ABM)

Unlike 1149.1, the test capabilities are the same for an analog input and output in 1149.4. However, the implementations may be quite different. A 4-bit register, which is part of the boundary-scan Data Register, is dedicated to each analog pin (digital pins in 1149.1 require a boundary module with between one and three bits).

The four bits, seen in Figure 3.4, are denoted C, D, B1, and B2 (for Control, Data, ABUS1 and ABUS2). Parallel update latches, optional in 1149.1, are mandatory in 1149.4. The latches prevent the serial shift of scan data from affecting pin voltages and possibly causing serious transients and long settling times.

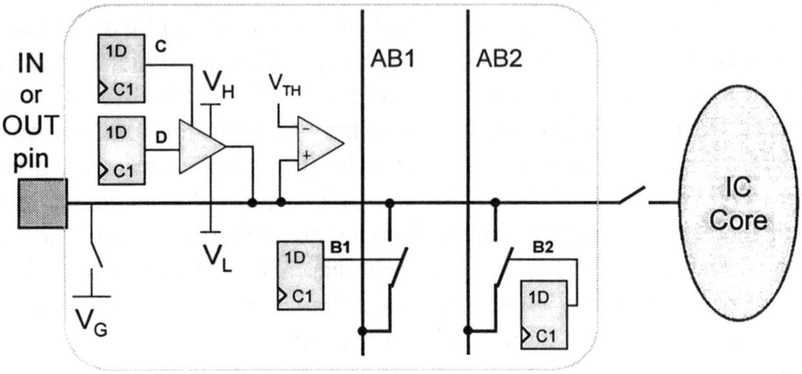

Figure 3.4. **Architecture of ABM**

In test mode, the C-bit acts as an enable for the pin, while the D-bit provides the pin's logic value (V_H or V_L). When the C-bit and the D-bit are both zero, the pin is in CD-state, however, when {C,D} = {0,1}, the pin is connected to V_G. This latter state exploits a pin state that is redundant in 1149.1. The B1-bit controls the switch connecting the pin to AB1, and B2 controls the switch connecting the pin to AB2. The boundary module register bits control the pin's state whenever it is in test mode, except when the *HIGHZ* instruction is in effect.

All four bits of an ABM should also be used to sample designer-selected signals in the ABM, for shifting out during the controller's *Shift-DR* state. The D-bit must be dedicated to sampling the pin's digitized voltage.

At least two mode lines (not shown in Figure 3.2) are needed to communicate the four instruction types to all ABMs. The first instruction type (EXTEST, CLAMP, and RUNBIST) puts all pins into test mode wherein the state of each pin is controlled by its 4-bit ABM register.

The second type of instruction (PROBE, INTEST) leaves all pins in function mode, but permits the pins' signals to be probed by the analog buses. The third instruction type (HIGHZ) puts all pins into CD-state. The fourth combination of mode line values is for instructions that must have no effect on function mode (BYPASS, SAMPLE/PRELOAD, IDCODE, and USERCODE). The mode lines are also used in the TBIC to control the ATAP pins.

The analog architecture of an ABM is equally important as the just described digital architecture. A correctly implemented ABM facilitates measurement accuracy limited only by external test equipment. Two aspects of the analog architecture are crucial to accuracy. First, all switches or other circuitry which drive a bond pad (or pin) of an IC must be connected via the same resistive electrical path to the pad. Second, all switches and other circuitry, which sense the voltage on the pad, must be connected to a second resistive electrical path. With this architecture, any amount of current flowing to the pad will have no effect on the accuracy of the voltage detected. Typically, the most significant source of resistance in the paths to and from a bond pad is the electro-static discharge (ESD) protection circuitry.

Figure 3.5 shows a schematic of an example ABM which implements all test capabilities required by 1149.4.

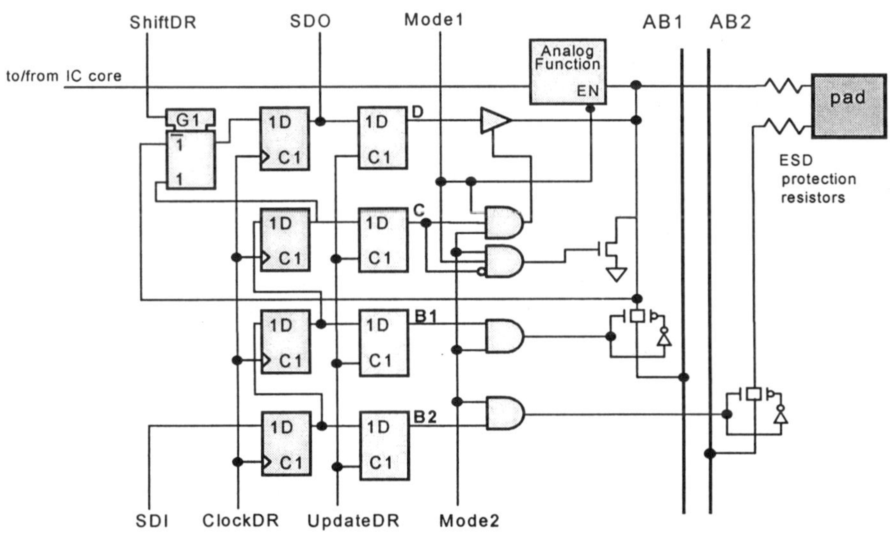

***Figure 3.5.* Example of an ABM schematic**

3.2.2.2 Test Bus Interface Circuit (TBIC)

The TBIC is the interface between the on-chip and off-chip analog test buses, and is the key circuit for enabling calibration and testing of the whole analog test bus infrastructure. The TBIC comprises four essential analog switches (S5-S8), shown in Figures 2 and 6, plus a variety of other switches, shown only in Figure 3.6. Switches S1-S4 and the two comparators provide the same digital test facilities that are present in every ABM to allow simple interconnect testing – they drive V_H or V_L, and monitor the pins' voltage. Switches S9 and S10 are optional but are recommended – they clamp the on-chip analog buses to a DC voltage when the buses are not in use.

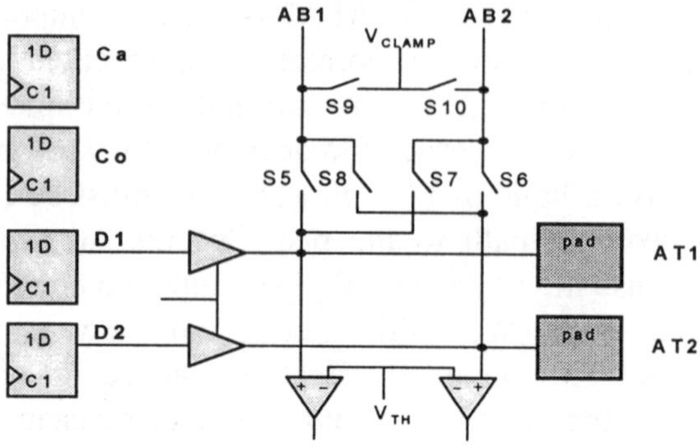

Figure 3.6. **TBIC switching architecture**

Like an ABM, the TBIC has its own 4-bit scan register, but the bits serve different purposes. The four bits are denoted Ca, Co, D1, and D2 (for Calibrate, Control, Data1 and Data2). The Co and D bits are similar to the C and D bits of an ABM, in that they provide enable and data for digital drivers connected to the pins. Unlike an ABM, there is no provision for connecting V_G when Co=0. Instead, when Co=0 (and Calibrate=0), the data bits D1 and D2 control the AT1-AB1 switch S5 and AB2-AT2 switch S6, respectively. When Calibrate=1, D1 and D2 enable the AT1-AB1-AT2 and AT1-AB2-AT2 paths, respectively.

All four bits of the TBIC may also be used to sample designer-selected signals in the TBIC, for shifting out during the controller's *Shift-DR* state. D1 and D2 must be dedicated to sampling the digitized voltages of AT1 and AT2, respectively.

The ATAP may be fully differential, comprising AT1 and its inverse AT1N, and AT2 and its inverse AT2N. To control these pins and their connections to their respective on-chip buses, a second TBIC is used, and it is a duplicate of the TBIC shown in Figure 3.6. It is the responsibility of the IC user to scan in bits that enable the inverting buses when accessing differential function pins.

There is only one TBIC per IC, however, it can be a multi-faceted structure for some ICs. For an IC with more than a few dozen analog pins, it is quite likely that there will be several pairs of AB1/AB2 buses, and possibly several pairs of AB1N/AB2N buses. Each section of these partitioned buses requires a partial TBIC, each containing two control bits (instead of four) to enable calibration and isolation. The primary reasons for partitioning are preventing excessive leakage and isolating incompatible operating voltage ranges or frequency ranges. Figure 3.7 shows a simplified view of a multiple analog bus architecture.

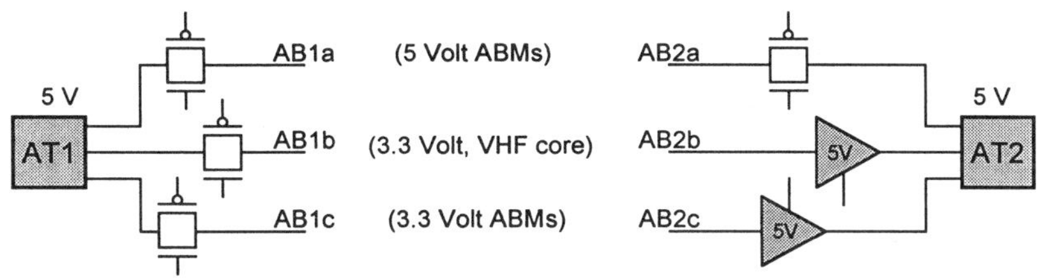

Figure 3.7. **Simplified example of partitioned analog buses**

The analog architecture of the TBIC is less critical than it is for the ABM. Only a single resistive path to each bond pad is necessary. Analog design choices include whether to use a voltage buffer for driving AT2, choosing the buffer's gain (less than unity), whether to allow the use of AT2 as a stimulus source (requires bypassing output voltage buffer, and a control bit for the bypass), and how to ensure ESD/latch-up robustness. Using a current buffer at AT1 provides no measurement accuracy advantages in CMOS, but might be needed to prevent ESD susceptibility or latch-up.

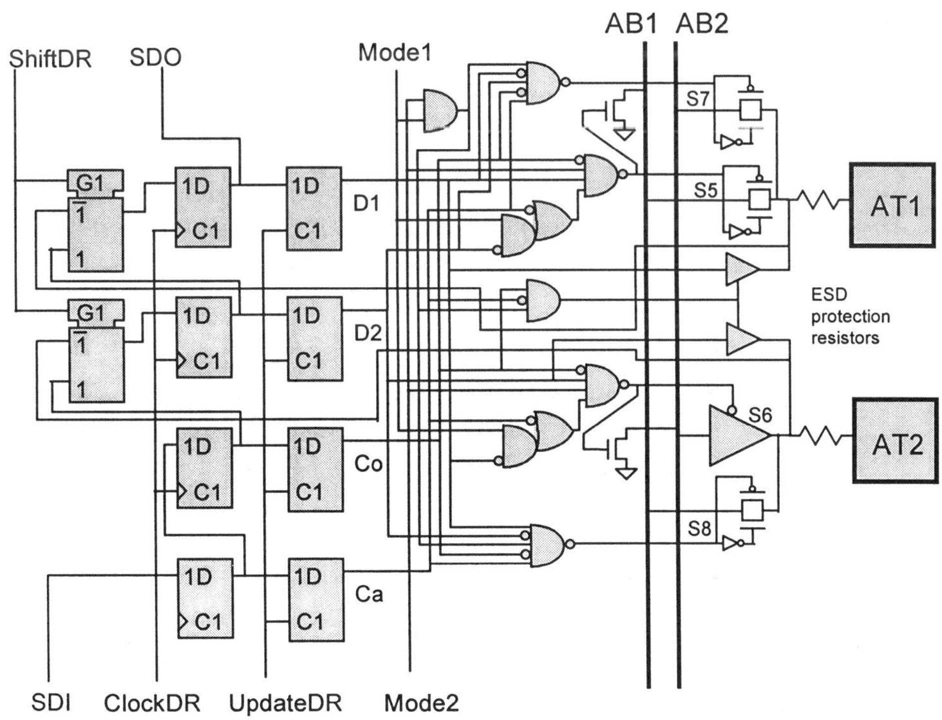

Figure 3.8. **Example TBIC schematic**

Figure 3.8 shows a schematic of an example TBIC which implements all capabilities required by 1149.4.

3.2.2.3 Connections between the TAP controller, TBIC, and ABM

The area for wires connecting circuit blocks on an IC can be just as significant as the area for the circuit blocks. Implementing 1149.4 on an IC typically requires only two or three more global wires beyond those required by 1149.1. The following is an overview of the wires typically required by 1149.1 and 1149.4. Both standards do not specify how many wires must be used to communicate between the TAP controller and the function pin test circuitry. This is intentional so as to allow innovation, considering that users of an IC typically care only about the ICs function and cost, and not how the functions were implemented. In any case, users are not able to easily verify how on-chip test facilities are implemented.

The eight wires listed below comprise a typical *minimum* number of on-chip global wires required by 1149.4, for an IC design using edge-triggered D-type flip-flops. The name and ABM function are listed for each wire. All these wires are also used in the TBIC.

1149.1 global wires (that are also used by 1149.4)

ShiftDR – selects data source for scan register, i.e., serial scan or parallel capture
ClockDR – clock input to all boundary-scan flip-flops
UpdateDR – clock input to parallel update latches, whose inputs come from scan register
Mode_1 – selects data source for all pins, i.e., test data or function signal
Mode_2 – (optional) forces HIGHZ for all digital output pins
SO/SI – scan-out to scan-in: the serial connection between boundary modules

Additional global wires needed by 1149.4

AB1 – analog stimulus bus or buses, optionally used for response too
AB2 – analog response bus or buses, optionally used for stimulus too
Mode_2 – (mandatory) enables analog access; when Mode_1=test, forces CD-state

3.2.3 Calibration Facilities

Various elements of the 1149.4 architecture facilitate calibration of the analog test buses so that measurements can be more accurate. The most important principle associated with accuracy is that of measuring a known quantity so that when an unknown quantity is measured, the two results can be compared. Calibration is the procedure of adjusting measurement equipment so that when the equipment measures a known quantity, it reports the correct result. In practice, this is often accomplished by simply subtracting the measurement of the known quantity from that of the unknown quantity.

To prepare for an accurate measurement via the 1149.4 analog buses, it is necessary to first measure the characteristics of the analog buses themselves, isolated from any function pins and from the on-chip buses of any uninvolved ICs. So the TBIC provides switches to connect the AT1 to AT2 via either on-chip bus, and to disconnect the on-chip buses from the off-chip buses. Typically, a known current would be applied to AT1, a known load temporarily connected to AT2, and the voltage at AT2 measured (via AB1) – this calibrates the offset, gain, linearity and bandwidth of the current delivery. Next, a known voltage would be applied to AT1, and monitored at AT2 (via AB2) to calibrate the offset, gain, linearity and bandwidth of the voltage transmission. All these measurements can be performed on the IC at the board level.

If any differential pairs of analog bus wires are used (AB1N and AB2N), or if the buses are partitioned into multiple pairs, then each pair will have its own set of four switches to allow calibration and isolation.

Unfortunately, it is difficult to calibrate the exact characteristics between a function pin and an ATAP pin at the board level. This is because the exact value of any load connected to the function is not known and it is not possible within an ABM to bypass the off chip connected load. Fortunately, the simplest way to implement switches in ABMs is the CMOS transmission gate, and its transmission characteristics will match sufficiently well with transmission gates in the TBIC, which can be calibrated. The most important elements to be calibrated are the capacitance and leakage current of the analog buses, the resistance of the switches, and the buffer amplifier recommended for driving AT2. If voltage or current buffers are used as switches in an ABM, then the IC manufacturer must test for the gain of each function pin to ATAP path to ensure that the gain variation meets the 1% overall accuracy objective of 1149.4.

3.2.4 Differential Access

As the power supply voltage of ICs decreases with each new technology, and as transmission speeds increase, differential signal transmission becomes more important and is used more frequently for both digital and analog signals. The 1149.1 standard admittedly devotes little attention to differential signals, in anticipation of a complete solution in 1149.4. For these reasons, extensive consideration was devoted to differential signals in the development of 1149.4 (in fact, there are a few rules in 1149.4 devoted exclusively to digital differential pins, which are not discussed here).

The ATAP comprises two single-ended analog buses or two differential analog buses (it is not permitted to have a 3-pin ATAP, i.e., a single-ended bus and a differential bus). A differential ATAP permits use of a differential stimulus and simultaneous differential observation of the response. It is worth noting that every switch connected to the buses is controlled independently, which allows single-ended use of a differential bus and differential use of two single-ended buses. AT1 and AT2 can be used to differentially drive or monitor the signal across any two pins, including both pins of a differential input or output. To support this application, it becomes important that the AT1/AB1 and AT2/AB2 paths are reasonably well matched in terms of switch impedances, connections to the functions pads, and wire routes.

For differential outputs and inputs of an IC, 1149.4 requires that all test facilities be implemented at each pin, however, some relaxation of the rules is permitted to ensure that differential function-mode performance is not jeopardized. Most significantly, the pins are not required to drive the same voltage at the same time. Also, the CD-state, because it is not required to be high impedance, permits low impedance resistors to remain connected between a pin and a power rail.

An extra test capability is required for all differential pins – the differential voltage must be captured by the boundary-scan register as a single bit, in addition to capturing the two single-ended voltages. For differential inputs, this is typically accomplished by capturing the single-ended output of the differential receiver. For a differential output, an additional comparator must be connected across the pin pair. For both inputs and outputs, the extra differential bit capture is the only way to determine if a differential signal path is functioning correctly in a noisy environment.

3.2.5 Accessing core signals

Although 1149.4 is aimed primarily at facilitating board testing, it can also provide a gateway to the core of ICs. After the introduction of 1149.1 into the marketplace, the TAP became the most flexible, low cost way for digital core tests to be controlled. The TAP is also evolving into a convenient port for in-system programming of ICs such as FPGAs (and is being addressed by the IEEE P1532 Working Group). Similarly, testing and diagnosis of mixed-signal systems might be more efficient if 1149.4 access to on-chip signals is provided, and non-test uses for the continuous time and unlimited resolution of the ATAP buses might evolve.

There are no rules for accessing core signals in an 1149.4-compliant IC. As long as the 1149.4-defined test modes are not affected, any core access circuit can be used. In practice, a few structures are most practical, and they are described next.

Switches to access nodes internal to an analog or digital core must be controlled by a data register that is *not* part of a boundary module. It is possible to include the data register as part of the boundary-scan register if "safe" values are declared for each bit so that boundary-scan software does not affect the core functions. An update latch will typically be necessary to avoid transients in the core when the scan register is being serially shifted. Note that the boundary-scan Data Register will be serially shifted twice for each board-level measurement (one shift for each node voltage to be monitored), and this shifting operation must not cause core circuitry to behave in a way that can affect a measurement in progress.

To observe a core signal, a CMOS transmission gate switch can be enabled by a bit in a scan data register. Typically, this is only practical if the AT2 pin is driven by a voltage buffer, and if the signal node being monitored has a low impedance. Otherwise, the capacitance and leakage of the analog buses can affect the signal, making any measurement inaccurate and affecting the circuit's performance.

To stimulate a core signal, a scan register-enabled transmission-gate multiplexer can be used to disconnect the mission mode input to a circuit, and connect the AB1 bus. Alternatively, the normal driver can be put into its high impedance state, and a transmission-gate enabled to connect AB1. In either case, AT1/AB1 will typically convey voltage instead of current.

For high frequency circuits, buffering will be needed as part of the switch that connects the analog circuit node under test to the bus [3]. Even so, using chip-level buses to convey high frequency analog signals is not desirable. An alternative is to use various conversion circuits, such as under-sampling [4] or strobed capture [5], which only convey very low frequencies on the bus.

3.2.6 Support for BIST

BIST is another technique for testing on-chip circuitry, as described briefly in the *RUNBIST* instruction description. The AB1 and AB2 buses that connect to all analog boundary modules and possibly some differential boundary modules, also provide means for on-chip BIST circuitry to access pin signals and core signals. The isolation switches in the TBIC allow the measurements to be performed entirely on-chip, without involving the off-chip buses.

It is also possible for the BIST circuitry on one IC to test circuit blocks on other ICs via AT1 and AT2, though this will need to be orchestrated by a board-level boundary-scan controller. The concept of an IC being placed on a board solely to test the other ICs has been implemented by a few companies for 1149.1, it was discussed in P1149.4 Working Group meetings [6], and it is discussed extensively in IEEE 1149.5 (an IEEE standard for system-level test access).

REFERENCES

[1] *K. P. Parker, S. Oresjo*, "A Language for Describing Boundary-Scan Devices", *Journal of Electronic Testing: Theory and Applications,* pages 43-55, vol. 2, n°. 1, March 1991

[2] *S. Sunter & N. Nagi*, "A Simplified Polynomial-Fitting Algorithm for DAC and ADC BIST", Proceedings, *International Test Conference*, pp. 389-395, November 1997

[3] *S. Sunter*, "A Low Cost 100 MHz Analog Test Bus", Proceedings, *VLSI Test Symposium*, pp. 60-65, May 1995

[4] *R. Mason*, "Mixed Signal DFT at GHz Frequencies", Proceedings, *VLSI Test Symposium*, April 1998

[5] *K. Lofstrom*, "Early Capture for Boundary-scan Timing Measurements", Proceedings, *International Test Conference*, pp. 417-422, Washington DC, October 1996

[6] Meeting minutes of P1149.4 Working Group, October 1995, http://grouper.ieee.org/groups/1149/4/min1095.html

Chapter 4

System Test Methodologies Using IEEE 1149.4

Mani Soma
Univ of Washington, soma@ee.washington.edu

BACKGROUND

This chapter discusses applications of the IEEE 1149.4 Mixed-signal Test Bus Standard in system testing, chip design, and in-situ measurement. The reader is assumed to:

- understand the basic issues in mixed-signal system test (see Chap 1),
- be familiar with IEEE 1149.1 Boundary-scan standard ([1] and Chap 2),
- understand the key features of the architecture and instruction set of IEEE 1149.4 ([2] and Chap 3).

This chapter illustrates these key features by applications and examples of 1149.4 in structural test, parametric test, and internal test. The capabilities of the digital boundary modules (DBM) and analog boundary modules (ABM) are reviewed briefly to set the stage for their usage in these test applications. For each type of test (structural, parametric, and internal), examples are presented to show how the IEEE 1149.4 instructions are applied to perform the test. These applications also identify where the 1149.4 instructions, in extending the IEEE 1149.1 instructions, have provided additional tools to the system design and test engineers to reduce the mixed-signal test difficulty.

The approach chosen in this chapter is top-down and brings the reader to the level of using the IEEE 1149.4 infrastructure to apply tests. Subsequent chapters extend the coverage to the physical level as follows:

- Chapter 5 covers examples of the physical design of the boundary modules (digital and analog) and provides guidelines for design-for-test using IEEE 1149.4, and
- Chapter 6 covers the measurement issues using IEEE 1149.4, especially in conjunction with calibration, parasitics, and accurate parameter extraction from measurements.

4.1 Capabilities of digital and analog boundary modules

The digital boundary module (DBM) and analog boundary module (ABM) have been described in Chapter 3 and are briefly reviewed in this section with regard to their operations during test application, and wherever applicable, with regard to differences between IEEE 1149.1 and IEEE 1149.4.

4.1.1 Digital Boundary Module (DBM)

A digital boundary module (DBM) in IEEE 1149.4 is identical in function and operations to a boundary-scan register in IEEE 1149.1. During test execution, the DBM at a digital pin is capable of:

- setting the pin to logic 1 or 0.
- capturing the logic state of the pin.
- putting the pin in a high-impedance condition (for tri-state pins with 2 associated DBMs).

For digital pins, the fundamental differences between the two standards involve:

- how the DBM is specified for differential logic pins and for the on-chip analog/digital interface,
- the possible use of an ABM at a digital pin instead of a DBM. The use of an ABM in this manner will be explored in more depth when the capabilities of an ABM are discussed below.

4.1.1.1 Differential-logic DBMs

A key difference between the two standards with respect to differential logic pins is that each digital pin, including differential logic pins, must be connected to either a DBM or an ABM (Rule[1] 7.2.1 a, IEEE 1149.4). If the designer chooses to connect a DBM to a differential logic pin, the standard does provide two types of DBMs (control-and-observe and observe-only) to reduce possible performance impact while still meeting the test objectives.

[1] This Rule and others below are described in their context in the IEEE 1149.4 standard document [2]

A control-and-observe DBM places circuitry between the core and the system pin, and this circuitry may be disabled by the TAP controller during the test execution. An observe-only DBM is non-invasive and does not disable the path between the core and the system pin.

A digital differential driver to two system output pins (Figure 4.1) may be associated with DBM in several ways:

1. if it is possible to use the two outputs independently, each output pin shall be connected to a control-and-observe DBM (Rule 7.2.1 b). There is no requirement that the differential driver input be connected to a DBM even though this connection is recommended and permitted by the standard. Figure 4.1 (a) illustrates this configuration.
2. if it is not possible to use the two outputs independently, each output pin shall be connected to an observe-only DBM (Rule 7.2.1 c). In this case, the input to the differential driver shall be connected to a control-and-observe DBM (Rule 7.2.1 d). Note that the two system output pins actually require a total of 3 DBMs. Figure 4.1 (b) illustrates this case.

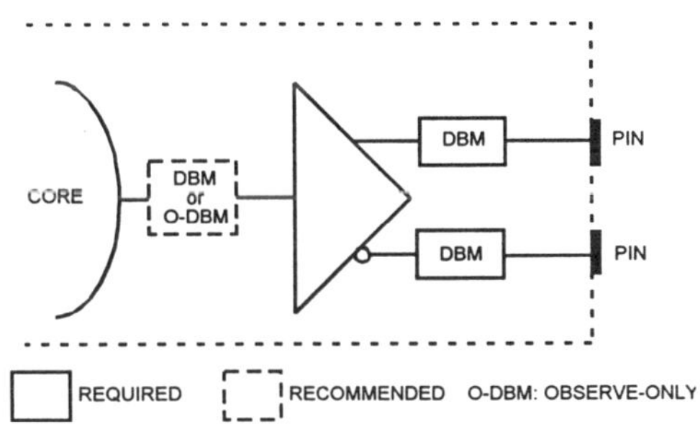

(a) DBM for driver with possible independent outputs

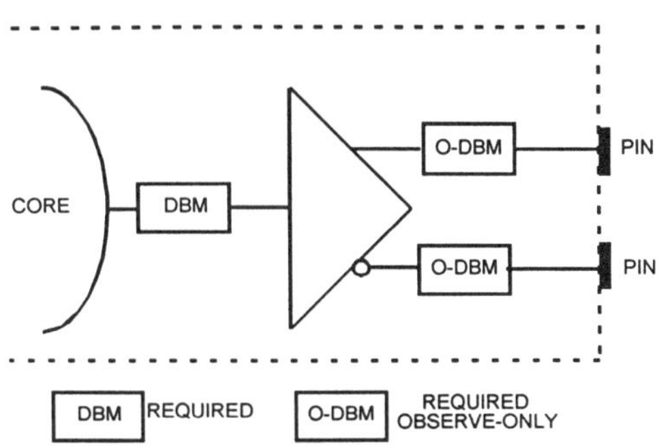

(b) DBM for driver with dependent outputs

Figure 4.1. **DBMs for a differential logic driver**

A digital differential receiver accepting signals from two system input pins (Figure 4.2) must be associated with a total of 3 DBMs:

a) Each input pin must be attached to one DBM. This DBM may be control-and-observe or observe-only. Figure 4.2 illustrates this configuration.
b) One DBM must be attached to the differential receiver output (even if it is an internal node and not a system pin) if its two inputs are system function pins (Rule 7.2.1 e). This DBM may be control-and-observe DBM or observe-only DBM. Figure 4.2 illustrates this configuration.

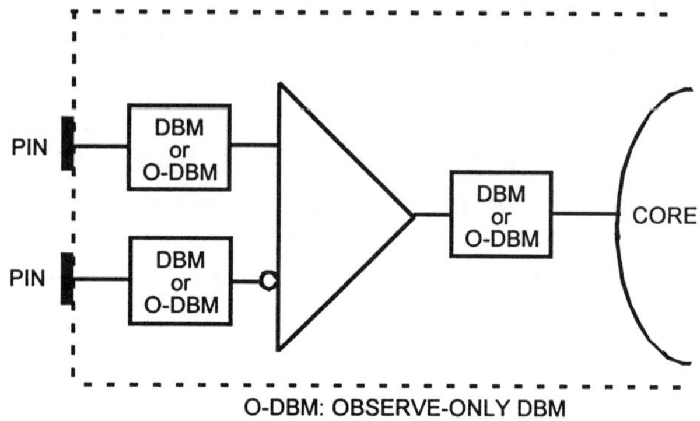

Figure 4.2. **DBMs for a differential logic receiver**

These requirements permit the testing of the differential interconnects as single interconnects or as a functional block consisting of both interconnects and the driver/receiver circuits.

4.1.1.2 On-chip analog/digital interface

Another essential difference between the two standards relates to the specification of the on-chip analog/digital interface. IEEE 1149.4 requires a control-and-observe DBM connected to each pin at this interface only if INTEST or RUNBIST is supported. Otherwise, this DBM is optional.

4.1.2 Analog Boundary Module (ABM)

The ABM is an innovation introduced by IEEE 1149.4 and with respect to capabilities, can perform more functions than a DBM. During test execution, an ABM attached to a system pin is capable of:

- Delivering either of two voltage levels (V_H or V_L) to the pin. This capability makes it possible for the ABM to set the pin to logic 1 or logic 0.

- Capturing a logic 1 or logic 0 by digitizing the voltage at the pin. The pin voltage is digitized with respect to a threshold V_{TH} whose value is in the range

$$\left(\frac{V_H + V_L}{2}\right) \pm \left(\frac{V_H - V_L}{4}\right) \qquad \textit{Equation 4.1}$$

- Putting the pin in a high-impedance condition. This is usually possible and implemented by the Core Disconnect facility. In the very rare cases where such a disconnect function is not implemented due to critical performance requirements, the designer must ensure that no part of the core circuitry will be adversely affected by any signal present at the pin during test execution.
- Transmitting a current from the test bus AB1 to the pin.
- Transmitting a voltage from the pin to the test bus AB2.
- Setting the pin to a relatively stable ("reference quality") voltage V_G.

Note that the first three capabilities of the ABM are exact duplicates of the capabilities of the boundary-scan circuit or DBM attached to a digital system pin. The last three capabilities are the extensions defined by IEEE 1149.4 to permit a wider range of tests required in mixed-signal systems. We will have opportunities to examine how these extended capabilities are used in this and the next two chapters.

The ABM is required to be connected to every analog system function pin. Its capabilities make it suitable for digital pins as well. Fundamentally, a digital circuit is also an analog circuit whose signals are discretized to two usable voltage levels. Thus a system pin, be it a digital pin or an analog pin, may be considered as an analog pin and connected to an ABM instead of a DBM. The standard does permit this connection and in fact, encourages this use of the ABM at a digital function pin if the designer or test engineer wants to take advantage of these extended capabilities to measure, for example, a pull-up resistor or the characteristics of the digital bus attached to the pin.

4.1.2.1 Differential analog circuit ABMs

Similar to the case of differential-logic DBMs, system pins connected to differential analog circuits need some special consideration. Each pin must be connected to an ABM as usual. The two ABMs are subject to additional control and have additional capabilities as discussed next.

A differential analog driver to two system output pins (Figure 4.3) has one ABM at each pin:

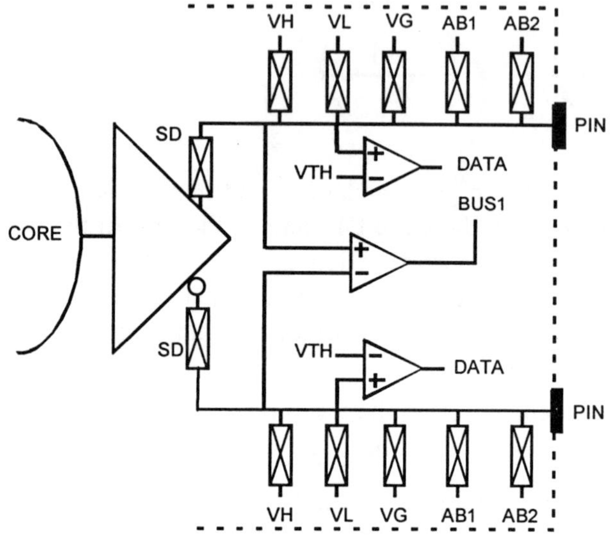

Figure 4.3. **ABMs for a differential analog driver**

- Either, neither or both pins can be totally isolated from the core circuit and from all DC test signals. This capability stems from the individual ABMs and independent controls.
- Either pin can be connected to V_G while the other pin is isolated from the core. This capability stems from the individual ABMs and independent controls.
- Either pin can be connected to V_H while the other pin is connected to V_L. This capability stems from the individual ABMs but control is now exercised jointly to set two pins to two opposite voltage states.
- Either, neither or both pins can be connected to the on-chip buses AB1 and/or AB2. This capability stems from the individual ABMs and independent controls.
- An additional differential comparator is required whose inputs are the two pins and whose logical output is stored in the BUS1 bit of the ABM control register attached to the non-inverting output pin. Note that BUS1 bit is thus the digitized voltage difference of the two analog outputs. This differential comparator is in addition to the two comparators already included in the ABMs at the individual pins.

Note that, in contrast to DBMs for differential logic drivers (Section 4.1.1.1), only two ABMs (not three) and one additional comparator are required for differential analog drivers.

A differential analog receiver attached to two system-input pins (Figure 4.4) has one ABM at each pin:

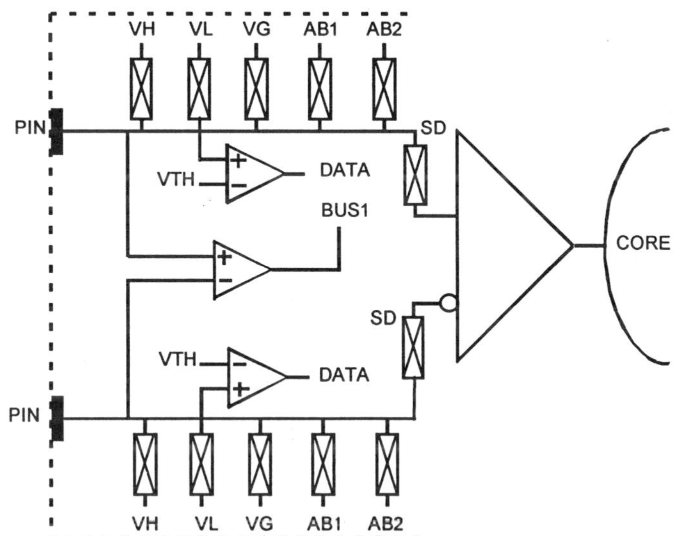

***Figure 4.4.* ABMs for a differential analog receiver**

- Either, neither or both pins can be totally isolated from the core circuit and from all DC test signals. This capability stems from the individual ABMs and independent controls.
- Either pin can be connected to V_G while the other pin is isolated from the core. If core isolation is not implemented due to critical performance requirements, either pin must be capable of being isolated from all DC test signals. This capability stems from the individual ABMs and independent controls.
- Either pin can be connected to V_H while the other pin is connected to V_L. This capability stems from the individual ABMs but control is now exercised jointly to set two pins to two opposite voltage states.
- Either, neither or both pins can be connected to the on-chip buses AB1 and/or AB2. This capability stems from the individual ABMs and independent controls.
- If the differential receiver on chip generates a logic bit, this bit must be captured in the BUS1 bit of the ABM control register attached to the non-inverting input pin. Otherwise, an additional differential comparator

is required whose inputs are the two system input pins and whose logical output is stored in the BUS1 bit of the ABM control register attached to the non-inverting input pin. In any case, BUS1 bit is the digitized voltage difference of the two analog inputs.

Note that, in contrast to DBMs for differential logic receivers (Section 4.1.1.1), only two ABMs (not three) are required for differential analog receivers, with the probable addition of a differential comparator.

The standard does permit a more extensive set of test bus interface circuits (with two more on-chip bus lines AB1N and AB2N and two more test access ports AT1N and AT2N) but for the purpose of understanding and using the boundary modules in test application, the ABMs as described in this section are sufficient. Interested readers should consult Section 7.4.3 of the standards document [2].

4.1.2.2 On-chip analog/digital interface

Another essential difference between the two standards relates to the specification of the on-chip analog/digital interface. IEEE 1149.4 requires a control-and-observe DBM connected to each pin at this interface only if INTEST or RUNBIST is supported. Otherwise, this DBM is optional.

4.2 Integrity test of the IEEE 1149.4 hardware

The boundary modules, together with the TAP controller and associated control registers described in Chapter 3 are dedicated to the system test applications. However, before these modules are used in test executions, they themselves need to be verified. This section covers the integrity test of the IEEE 1149.4 infrastructure and relies on structural test methods to check the various analog switching functions and the digital registers. This integrity test plan can be easily converted into a tester-specific language for execution.

Two assumptions are made before testing the integrity of the IEEE 1149.4 facilities:

1. The power supply nets on the various chips and boards in the system have been tested and verified.
2. The ICs have been individually tested before assembly on boards or on Multi-Chip Modules (MCM).

4.2.1 Testing the TCK and TMS nets

If there is a stuck-at fault in the TCK net, TCK will be all '1' or all '0', and no boundary-scan facilities work. The detection is simple. The same remark applies to stuck-at fault detection in the TMS net.

Bridging faults involving TCK and/or TMS are somewhat more difficult to detect [3, 4]. A possible test procedure includes these steps to check for various types of bridges:

- TCK bridged with another signal, TCK being the weaker signal. The transitions of the other signal are not always in steps with TCK, thus this bridge means that the TCK timing will be incorrect. Detection is accomplished by checking the state diagram of the TAP controller for incorrect state transitions. A simple way to do so is to operate a small segment of any boundary-scan register and verify the shifting of this register based on the TAP controller states. The same test procedure and fault detection method applies to the case of TMS bridged with another signal, when TMS is the weaker signal.
- TCK bridged with another signal, TCK being the stronger signal. A very specific test plan needs to be developed based on the characteristics of the signal suspected of being bridged. Testing the IEEE 1149.4 structure by itself probably would not detect this fault unless the suspected signal is a part of subsequent integrity tests. The same remark applies for the case of TMS bridged with another signal, when TMS is the stronger signal.
- TCK and TMS are bridged together, possibly due to adjacency of nets in the layout or of pins on chips. The two signals are synchronized and this can be detected by operating a small segment of the boundary-scan register. The test sequence TMS = {010101...} would result in test data shifted out at a rate four times lower than expected [4], and this is easily detectable.

Open faults in TCK or TMS are easily detected by various existing procedures for continuity tests.

4.2.2 Testing the TDI and TDO nets

TDI and TDO nets are connected via registers in the boundary-scan structure. The integrity of both nets can be partially tested by configuring the instruction registers from all chips to be the only registers on the TDI-TDO serial path. Since the last two LSB of each instruction register are required to be '01', a serial shift will produce this pattern (first bit is the first serial value shifted out): 10xxxx10xxxxxx10xxxx10, where the x's denote the bits from the other positions in the instruction registers. The first string of '10' shifted out via TDO is from the chip nearest to the board TDO, and the last string of '10' shifted out is from the first chip connected to TDI on the board. Other test sequences to check for stuck-at faults, fault locations, and bridges between TDI and TDO are possible and easily implemented [3, 4].

4.2.3 Testing the registers

Once TCK, TMS, TDI, and TDO have been checked for correctness, it is simple to verify all the registers: instruction registers, bypass registers, boundary-scan registers associated with DBMs, test bus interface circuit (TBIC) control registers, and boundary-scan registers associated with ABMs. The registers under test are configured to be connected between TDI and TDO, and various shift sequences (e.g., all 0's, all 1's, 01010101..., 001100110011..) are available to verify the registers' integrity and shift operations.

4.2.4 Testing the AT1 / AT2 nets

The AT1 and AT2 nets in IEEE 1149.4 are topologically different from the serial TDI/TDO net in IEEE 1149.1. Assuming that the mixed-signal test bus on the same board is not segmented, AT1 and AT2 are broadcast nets connected in a *star topology*, i.e., all AT1 pins on chips are connected to the board AT1 pin and all AT2 pins on chips are connected to the board AT2 pin. If various test bus segments are used to reduce the capacitive loads of the bus lines, the star topology still holds within each segment. Despite the broadcast nature of this test bus topology, parallel tests of AT1/AT2 are possible due to the requirement that the DATA1 and DATA2 stages of the TBIC control register must capture the 1-bit digitized values of the voltages on AT1 and AT2, respectively.

Assuming that the digitizing threshold V_{TH} for AT1 and AT2 pins (Figure 4.5) is set half way between the two voltage levels V_H and V_L, the integrity of AT1 and AT2 can be verified as follows:

- Test for AT1 stuck-at-0 and AT2 stuck-at-1:

1. Apply a voltage to AT1 higher than V_{TH} and preferably lower than the V_H of all chips.
2. Apply a voltage to AT2 lower than V_{TH} and preferably higher than the V_L of all chips.
3. Exercise Clock-DR to capture AT1=1 into the DATA1 bit and AT2=0 into the DATA2 bit.
4. Configure the TBIC control register between TDI and TDO. Use repeated Shift-DR to shift out the TBIC bits and examine for the sequence (first bit is the first serial value shifted out): 01xx01xx01xx... assuming that DATA2 bit is physically closer to the TDO pin than DATA1 bit. The x's denote the other two bits of each TBIC control register.

- Test for AT1 stuck-at-1 and AT2 stuck-at-0: the above test can be repeated, setting voltages on AT1 and AT2 so that the captured bits are DATA1=0 and DATA2=1.
- The above two tests taken together also cover open faults on and bridging faults between the AT1/AT2 nets.
- Testing for bridging faults between the AT1 net and another system signal net is more difficult. Depending on the characteristics of this system signal net, specific test plans and detection procedures need to be established and applied during system test, which may include functional tests. The same remark applies to testing for bridging faults between the AT2 net and another system signal net.

Even if different chips have different V_H and V_L values, as long as there exist two test voltages V_{t1} and V_{t0} such that

$$V_H(\mathit{any\ chip}) \geq V_{t1} > V_{TH}(\mathit{any\ chip}) > V_{t0} \geq V_L(\mathit{any\ chip})$$

Equation 4.2

All the AT1/AT2 nets to all chips can be tested in parallel. This condition is easily satisfied in many mixed-signal systems.

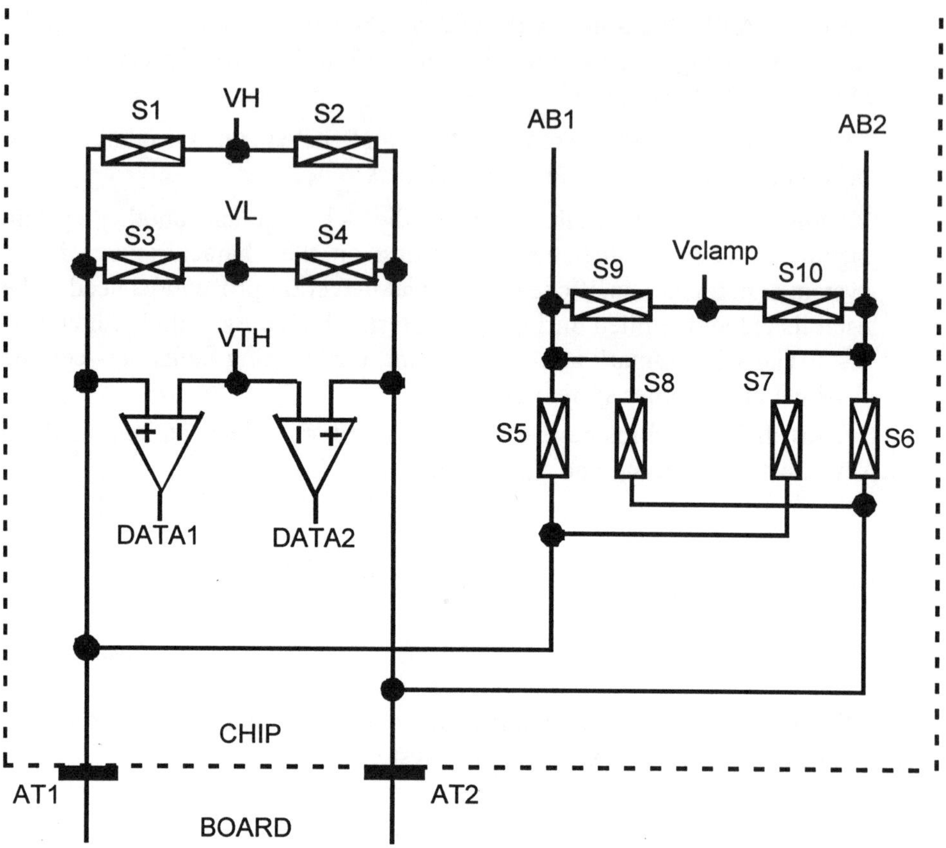

Figure 4.5. **TBIC switching structure**

4.2.5 Testing the TBIC switching structure

Before we describe a plan to test the Test Bus Interface Circuit (TBIC) switching structure, it should be emphasized that in the IEEE 1149.4 context, the term "switch" is used to describe the function of the device (to connect or disconnect between two data lines) without mandating a specific physical design. Chapter 5 and previous works [5, 6] discuss various possible physical designs of a switch and their impact on mission circuit performance. For testing purpose described in this and subsequent sections, we only need to specify the state of a switch ("open" or "closed") in a test; the detailed control signals are easily derived based on the IEEE 1149.4 control descriptions and the specific physical design implementations.

The switching structure connecting the on-chip buses AB1/AB2 to the AT1/AT2 ports on each chip can be tested as follows (Figure 4.5):

- Test for S5 and S8:

1. Close S5 and S8 to connect AT1 and AT2 together.
2. Close S10 to clamp AB2 (for safety and noise reduction).
3. Run any continuity test to verify that AT1 and AT2 are connected as a single net.

- Test for S6 and S7:

1. Close S6 and S7 to connect AT1 and AT2 together.
2. Close S9 to clamp AB1 (for safety and noise reduction).
3. Run any continuity test to verify that AT1 and AT2 are connected as a single net.

- Test for S9 and S10:

1. Close S5 and S9 to connect AT1 to V_{clamp}.
2. Close S6 and S10 to connect AT2 to V_{clamp}.
3. Measure the DC voltage on AT1 and verify that it is V_{clamp}. Repeat the measurement on AT2.

Since there is one switching structure per chip, the above three tests need to be repeated for all chips implementing IEEE 1149.4. Valuable information can be collected during these tests to save subsequent test time and improve parametric measurement accuracy. For example, when AT1 and AT2 are connected via S5 and S8, calibration can be performed both to verify continuity and to provide the physical data of this bus path (capacitive load, resistance, etc.) to be used in the future measurements to reduce errors due to bus parasitics.

4.2.6 Testing the ABM switching structure

Testing the integrity of the Analog Boundary Module (ABM) switching structure (Figure 4.6) is similar to testing the integrity of the TBIC switching structure, except for the core disconnect (CD) switch function. At the pin, the CD function must be tested first (for open and close operations) before any other test so that voltages placed on the pin in subsequent tests would not back-drive or damage the system circuit connected to the pin. To verify whether the CD function opens or closes the signal path from the core to the pin, a signal needs to be propagated from the core to the pin, which represents a problem since integrity tests are preferably executed first before the core is exercised in other functional or parametric tests. Given this test sequence constraint, there are three possible procedures to deal with the CD function at each pin:

- If the CD function is implemented as an integral part of the core circuit (e.g., part of a tri-state input or output buffer), it is usually safe to assume that the buffer can be disabled to disconnect the core from the pin. The probability of a defect affecting this buffer function is usually very low in state-of-the-art manufacturing processes. The closed condition of this buffer can be tested for as part of subsequent system tests.

- If the CD function is implemented as a separate physical switch on the signal path between the core to the pin, it is impossible to verify the open condition without exercising the core. Again it could be argued that the probability of a defect affecting this switch is so low that the open condition may be safely assumed and verified in subsequent system tests. Since a physical switch might adversely affect the system performance anyway, alternative implementations are preferred, which also facilitate integrity tests [6].
- If the CD function must be absolutely verified before further tests, a procedure to verify all CD pin functions of all chips in parallel will need a minimum of two DC tests. Core circuits need to be exercised as part of these two tests.

1. Use the core function to set the pin to a DC voltage lower than V_{TH} of the pin. With the CD function in the closed state, the digitizer at each pin will produce logic 0. Enter the test mode and exercise Clock-DR to capture this logic 0 in the DATA bit of each ABM control register. Configure the ABM control registers to be in the TDI-TDO serial path and exercise Shift-DR repeatedly to shift out the contents of these registers. For each ABM, the pattern (first bit is the first serial value shifted out) is xxx0 where the DATA bit is assumed to be physically the farthest from the TDO pin. The x's bits are the BUS2, BUS1, and CONTROL bits of each ABM, assumed to be don't care in this test.
2. With the CD function in the open state, use the core function to set the pin to a DC voltage higher than V_{TH} of the pin. The pin digitizer output should still remain at 0. Enter the test mode and exercise Clock-DR to capture the digitizer output in the DATA bit of each ABM control register. Configure the ABM control registers to be in the TDI-TDO serial path and exercise Shift-DR repeatedly to shift out the contents of these registers. For each ABM, the pattern (first bit is the first serial value shifted out) should still be xxx0 where the DATA bit is assumed to be physically the farthest from the TDO pin. The x's bits are the BUS2, BUS1, and CONTROL bits of each ABM, assumed to be don't care in this test.

There are cases where DC tests are not applicable (e.g., a pin is the output of an oscillator), and the integrity tests need to be developed for these specific cases. Note that there are also cases where the CD function is not implemented at a sensitive pin or the performance impact of such a function is not acceptable. In these cases, there is no need to perform the integrity tests for the function.

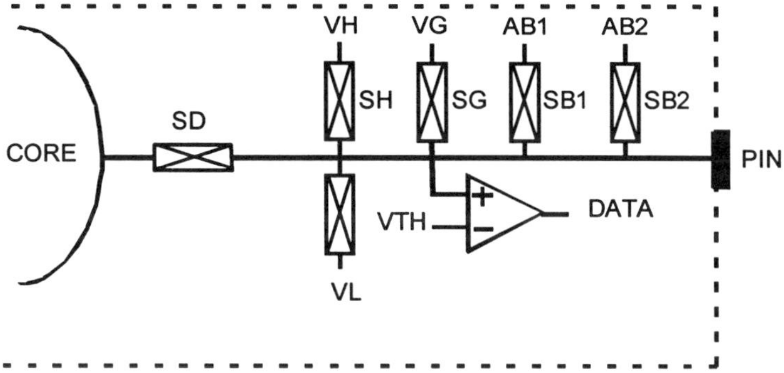

Figure 4.6. **ABM switching structure**

Once the CD function is verified, the remaining test of the ABM switching structure is straightforward. All CD functions are in the open condition for the following tests:

- Testing the SH switching function:

1. Activate the SH switching function to connect each pin to the V_H voltage at the pin. The pin digitizer output should be at logic 1.
2. Exercise Clock-DR to capture this logic 1 in the DATA bit of each ABM control register.
3. Configure the ABM control registers to be in the TDI-TDO serial path and exercise Shift-DR repeatedly to shift out the contents of these registers. For each ABM, the pattern (first bit is the first serial value shifted out) is xxx1 where the DATA bit is assumed to be physically the farthest from the TDO pin. The x's bits are the BUS2, BUS1, and CONTROL bits of each ABM, assumed to be don't care in this test.

- Testing the SL switching function. The procedure is identical to the test for SH, with the pin set to V_L and the digitizer output expected at logic 0.
- Testing the SG switching function. The procedure is identical to the test for SH, with the pin set to V_G. The digitizer output at each pin, captured in the DATA bit, is either 0 or 1 depending on the relationship between V_G and V_{TH} at each pin. The pin logical output is completely predictable and the correct bits can be checked for in the serial TDO output string.
- Testing the SB1 switching function:

1. Close S5 connect the AT1 pin to the AB1 on-chip bus.
2. Activate the SB1 switching function to connect the AB1 bus to all pins.
3. Apply a test voltage V_{t1} to AT1 such that:

$$V_H(any\ pin) \geq V_{t1} > V_{TH}(any\ pin) \qquad \text{Equation 4.3}$$

The digitizer output at each pin, captured in the DATA bit, should be 1. Alternatively, a test voltage V_{t0} may be applied to AT1 such that:

$$V_H(any\,pin) \geq V_{t0} > V_L(any\,pin) \qquad \text{Equation 4.4}$$

and the digitizer output in this case should be 0.

4. Exercise Clock-DR to capture this logic value in the DATA bit of each ABM control register.
5. Configure the ABM control registers to be in the TDI-TDO serial path and exercise Shift-DR repeatedly to shift out the contents of these registers. For each ABM, the pattern (first bit is the first serial value shifted out) is xxx1 where the DATA bit is assumed to be physically the farthest from the TDO pin and V_{t1} was applied in step 3. The x's bits are the BUS2, BUS1, and CONTROL bits of each ABM, assumed to be don't care in this test.

- Testing the SB2 switching function. The procedure is identical to the test for SB1, using the AB2 and AT2 bus connections to set the logical state of the pins. If the AT2 bus is designated for monitoring only, the AB2-AT1 connections can be used instead.

4.3 Simple interconnect test using IEEE 1149.4

A simple interconnect is a conductor on circuit board or MCM or other types of assembly, connecting two or more pins of the same or different integrated circuits. Manufacturing test of simple interconnects checks for stuck-at faults, opens and shorts on these nets. While a simple interconnect has an intrinsic resistance and capacitance, the simple interconnect tests are not intended to measure these parameters. Parametric tests are described in a later section.

Both IEEE 1149.1 and IEEE 1149.4 provide complete facilities and instructions for simple interconnect tests, with IEEE 1149.4 covering both analog and digital pins. To illustrate the use of IEEE 1149.4 in this application, we will first consider a very simple interconnect with only two pins on the net in Figure 4.7. This net may connect two digital pins, two analog pins, or one analog pin and one digital pin. Test procedures for more complicated nets with multiple fan-in and fan-out will be described later.

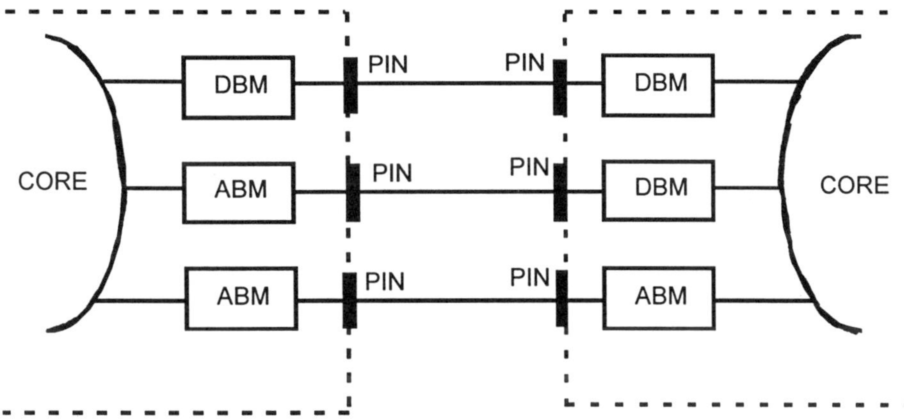

***Figure 4.7.* Simple 2-pin nets and associated boundary modules**

The fundamental use of the test bus is to provide a logic signal to drive one pin of a net (the driving node) and to sense the logic signal received at another pin of the same net (the sensing node). The ABM at a pin mimics logic 1 by driving a voltage V_H to the net, logic 0 by driving a voltage V_L to the net. The digitizer of the ABM converts the voltage at a sensing node to a data bit, which can be scanned out. The test bus facilities and the two instructions SAMPLE/PRELOAD and EXTEST are exercised to run the test as follows.

4.3.1 PRELOAD the driving nodes and setting up the sensing nodes

If the logic states or expected voltages at the driving nodes are unknown before the interconnect tests (e.g., at power-up of the chips or the board), the SAMPLE/PRELOAD instruction is executed to set the driving nodes into known states.

1. Enter the test mode and load the SAMPLE/PRELOAD instruction into the instruction registers of all chips containing the driving nodes of the nets to be tested. Update the instruction registers. Note that the instruction decoder will generate two logic signals Mode1=0 and Mode2=0 so that each TBIC is properly set for the normal system operation (all pins connected from chips to boards, all test circuits and test voltages disconnected from pins). For pins connected to DBMs, only one Mode signal is necessary and Mode1=0 can be used in place of the Mode signal shown in the IEEE 1149.1 standard document.
2. Load the data registers of the pins driving the nodes with the logic values to be used in subsequent EXTEST:

a) for a pin connected to a DBM, the driving logic value (1 or 0) is loaded from TDI serial input path to the data bit.

b) for a pin connected to an ABM, there are two possible data patterns to be loaded from the TDI serial input path:
   i) If the driving logic value during subsequent EXTEST is 1, the data to be loaded are DATA=1, CONTROL=1, B1=0, and B2=0. This will ensure that the pin, during subsequent EXTEST, will be connected to V_H, equivalent to logic 1.
   ii) If the driving logic value during subsequent EXTEST is 0, the data to be loaded are DATA=0, CONTROL=1, B1=0, and B2=0. This will ensure that the pin, during subsequent EXTEST, will be connected to V_L, equivalent to logic 0.

The values of the other two bits on the ABM control register (B1 and B2) are set to 0 so that the internal test buses are not connected to the system analog test buses during EXTEST.

The data registers of all driving nodes are loaded by repeated Shift-DR.

3. Set up the data registers of the sensing nodes in subsequent EXTEST:

a) For a pin connected to a DBM, it is not necessary to preload the node even though a known state is helpful for fault detection in subsequent EXTEST. Preloading a logic value is accomplished by scanning that value from the TDI serial input path.

b) For a pin connected to an ABM, the data to be loaded are DATA=0, CONTROL=0, B1=0, and B2=0. This will ensure that the pin, during subsequent EXTEST, will be connected to the net on the board and not to any test bus facilities at the pin.

4. In the subsequent *Update-DR* state of the controller, a falling edge of TCK is used to parallel load the output registers (or update registers) of all DBMs and ABMs from the data registers in step 2 above. This is the PRELOAD phase of the SAMPLE/PRELOAD instruction.

At this time, the set-up of all driving nodes is completed. No logic signal or voltage is driven to the pins yet, until the next instruction, which is EXTEST. Note that this preload phase only needs to be executed once since subsequent repeats of the EXTEST instructions will start with known logical and control states of all pins.

4.3.2 Propagating signals through the interconnects: EXTEST

1. Load the EXTEST instruction into the instruction registers. Update the instruction registers with Update-IR. Note that the instruction decoder will generate two logic signals Mode1=1 and Mode2=1 so that the following actions will take place:

a) For a driving node whose pin is connected to a DBM (Figure 4.8a): the output of the update register is connected is connected to the net on the board. Note that the IEEE 1149.1 Mode signal is 1, since this Mode signal is the same as the Mode1 signal of IEEE 1149.4. The test stimulus is now applied to one pin of the net under test. The core is isolated from all the pins.

b) For a driving node whose pin is connected to an ABM (Figure 4.8b): the voltage applied from the pin to the net is either V_H or V_L depending on the preloaded registers during the PRELOAD operation above or from a previous EXTEST operation. With Mode1=Mode2=1 in EXTEST, the voltage applied is V_H if previously DATA=1 and CONTROL=1. The voltage applied is V_L if previously DATA=0 and CONTROL=1. The test voltage is now applied to one pin of the net under test. The core is isolated from all the pins.

2. At the sensing nodes of the nets, the following actions take place:

a) For a sensing node whose pin is connected to a DBM (Figure 4.8c): during the time when Shift-DR is low, the logic signal from the net is stored in the input latch in the *Capture-DR* state, which the TAP controller enters after the *Update-IR / Select-DR* states. Note that since Mode1=1 (or Mode=1 in the 1149.1 standard), the pin is disconnected from the system logic.

b) For a sensing node whose pin is connected to an ABM (Figure 4.8d): the pin is now disconnected from the system analog core and from all test bus facilities. The voltage driven to the pin is digitized and with Shift-DR=0, stored in the DATA bit of the ABM in the *Capture-DR* state.

At this point, the data has been captured in the boundary-scan registers of the sensing nodes and is ready for scanning out.

3. Repeated *Shift-DR* states result in the data from one execution of EXTEST scanned out for observation while new data is scanned in to set up the next interconnect test. At the end of the scan-in of a new test vector, the *Update-DR* state is entered and the EXTEST cycle is repeated until all interconnects are tested.

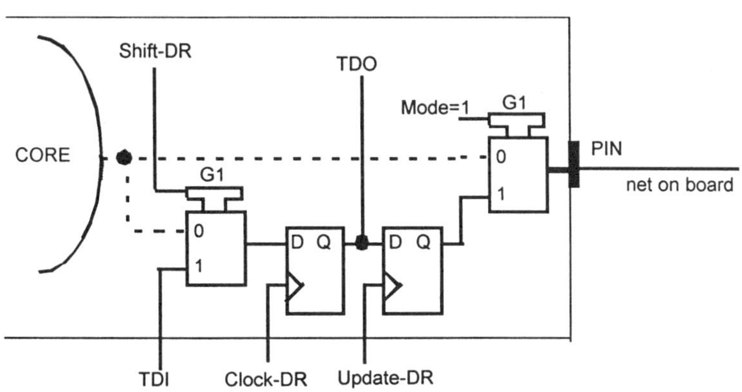

(a) Active datapath (solid line) of a driving DBM

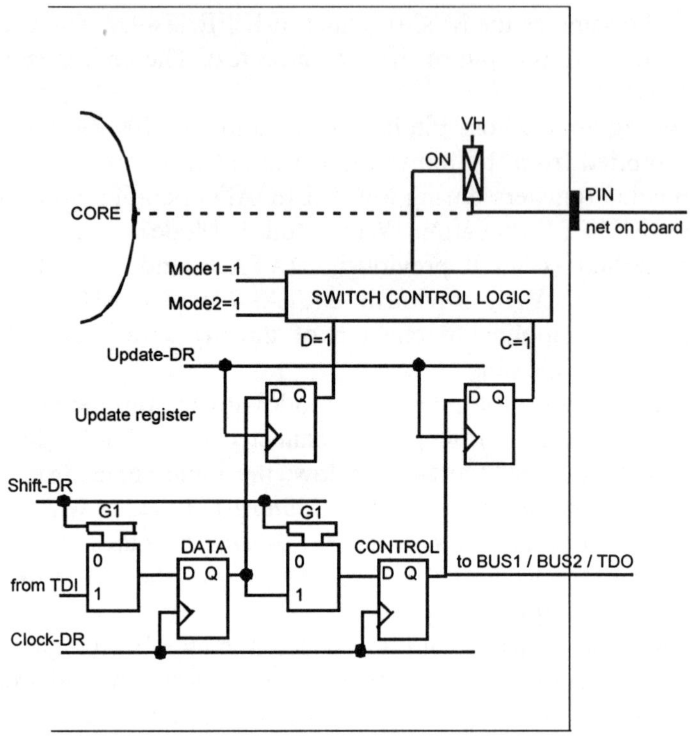

(b) Active datapath (solid line) of a driving ABM (assumed V_H drive)

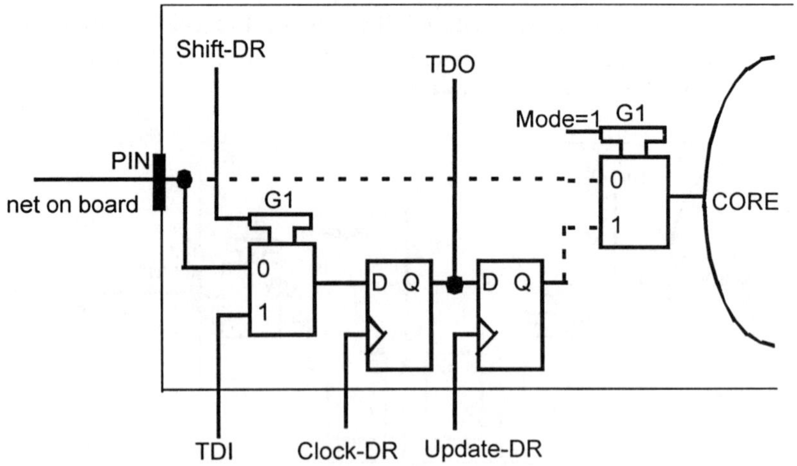

(c) Active datapath (solid line) of a sensing DBM

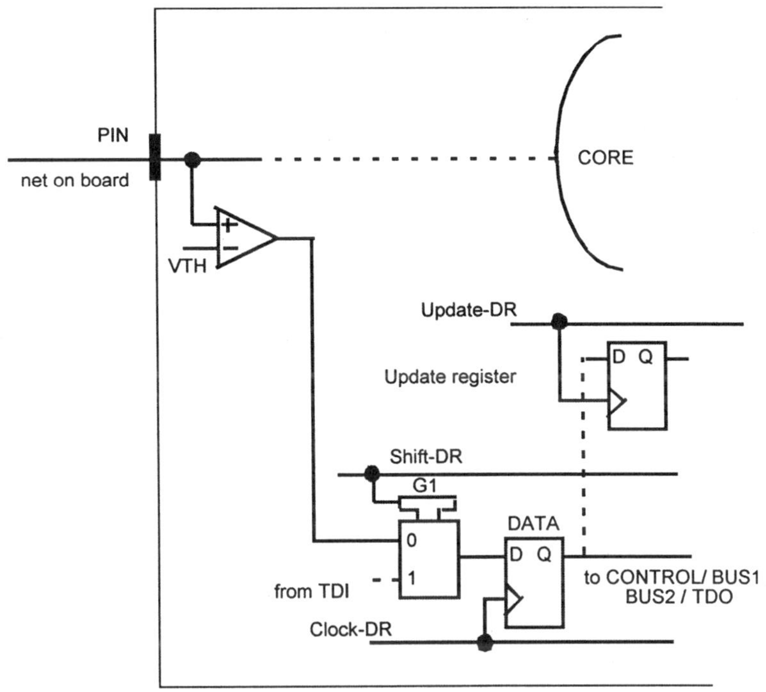

(d) Active datapath (solid line) of a sensing ABM

***Figure 4.8.* EXTEST data flow through DBM and ABM**

4.3.3 Interconnects with fan-in and fan-out

The procedure using EXTEST illustrated using a simple two-node net in the above sections can easily be extended to cover more complicated interconnects with fan-in and fan-out on the board:

1. For interconnects with fan-in, only one input is selected to drive the net in each test. All other inputs (analog or digital) to the net are placed in the high-impedance state or in the CD state. The ABM control register bits for the CD state are DATA=0, CONTROL=0, B1=0 and B2=0 to disconnect the pin from all test functions. Note that during EXTEST, Mode1=1, thus the SD switch is open as well. The test sequences are repeated for each input to the net.
2. For interconnects with fan-out, all the sensing nodes can be activated simultaneously to test all the fan-out branches of the net.

4.3.4 Test sequences for interconnects

Test sequences to detect stuck-at faults, shorts and opens have been extensively published in the literature [3, 7], together with diagnosis procedures to locate specific faults on the board. The reader is encouraged to consult these references. Automatic test systems supporting IEEE 1149.1 also have test programs developed for this purpose. Since IEEE 1149.4 provides exactly the same capabilities for interconnect tests, these test programs can be easily modified to perform the same task for mixed-signal systems. The main difference, as illustrated above, involves the loading of the ABM control registers during EXTEST.

4.4 Parametric interconnect test using IEEE 1149.4

The power of IEEE 1149.4 is in parametric testing of interconnects. In this context, interconnects, or "extended interconnects" as illustrated in Figure 4.9, may contain:

- Passive elements such as resistor, capacitor, inductor, crystals, etc. These passive elements (e.g., resistance and capacitance) may also be intrinsic to the interconnect.
- Pull-up or termination resistors for digital circuits and buses.
- Active components, e.g., diodes, sensors, etc.
- Impedance networks.

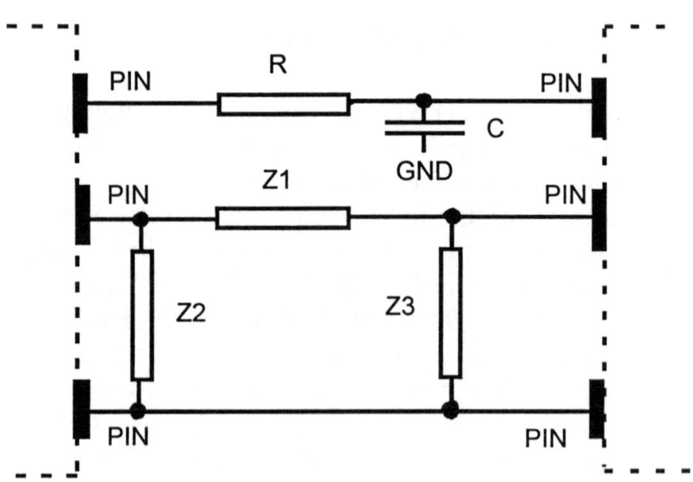

Figure 4.9. **Examples of extended interconnects**

The ABM structure provides facilities to measure the parameters of an interconnect using the EXTEST instruction together with external instruments connected to the two system analog test buses AT1/AT2. Fundamentally, the impedance of an element or an interconnect may be measured as illustrated in Figure 4.10:

1. Inject a current (DC or AC) through the element to ground (GND) or a stable DC voltage (V_G).
2. Measure the voltages V_1 and V_2 at the two end points of the element.
3. Compute the impedance:

$$Z = \frac{V_1 - V_2}{I}$$

Equation 4.5

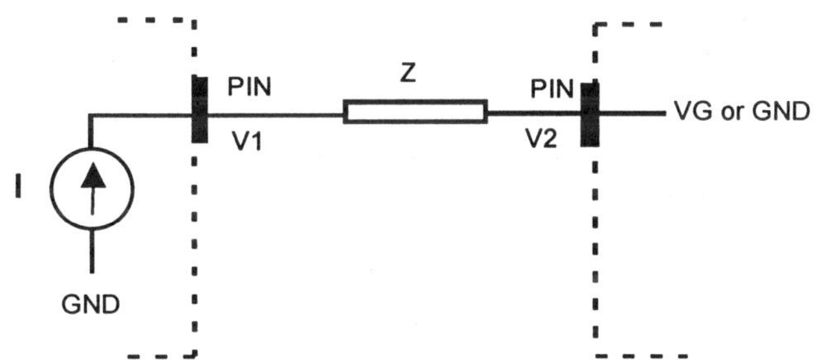

***Figure 4.10.* Conceptual technique to measure an impedance**

This section, describes how the test bus facilities are set up to inject current through an interconnect and to measure the voltage at a pin. Once these two procedures are understood, it is easy to use them in various sequences to perform a specific parametric measurement. The details involving guarded measurements, complex impedance measurements, network measurements, and other fine points in parametric tests and accurate data extraction are covered in Chapter 6.

4.4.1 Injecting a current through an interconnect

The fundamental instruction to be executed is still EXTEST but the set-up of the TBIC, the ABM and the pins is different. Using Figure 4.11 for topology, the injection procedure is as follows:

1. The control register of the TBIC of the chip containing pin 1 is loaded with CALIBRATE=0, CONTROL=0, DATA1=1, and DATA2=1. During subsequent EXTEST (Mode1=Mode2=1), this control ensures that the on-chip bus AB1/AB2 is connected to the system analog bus AT1/AT2 respectively (e.g., AB1 connected to AT1 via S5, AB2 connected to AT2 via S6).

2. If pin 2 is on the same chip as pin 1, the TBIC register is already loaded. Otherwise, the control register of the TBIC of the chip containing pin 2 is loaded with CALIBRATE=0, CONTROL=0, DATA1=0, and DATA2=0. This control ensures that the AB1/AB2 bus on the second chip is clamped to reduce noise, and are also disconnected from the AT1/AT2 system bus.
3. The control register of the ABM of pin 1 is loaded with CONTROL=0, DATA=0, B1=1, and B2=1. This control ensures that the pin will be connected to AB1 (which is connected to AT1 via the TBIC) for current injection and AB2 (which is connected for monitoring the voltage V_1. During EXTEST (Mode1=Mode2=1), the pin will be connected to the net without any other signal from V_H, V_L, V_G, or the core.
4. The control register of the ABM of pin 2 is loaded with CONTROL=0, DATA=1, B1=0, and B2=0. This control ensures that the pin will be connected to V_G during EXTEST. V_G, which may be GND or a reference-quality voltage to be used during test, provides a DC path for the current to flow through.

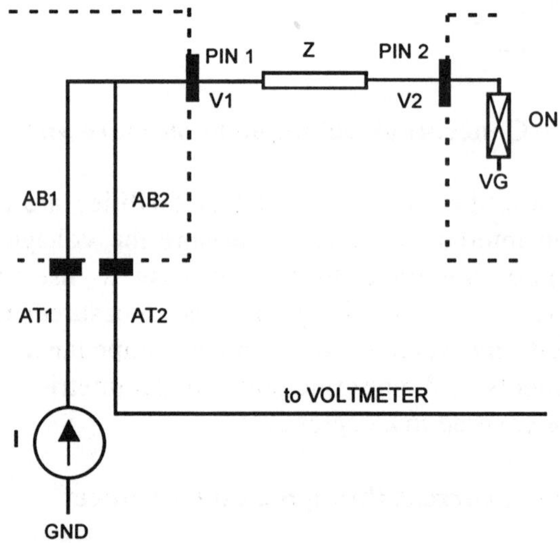

Figure 4.11. **Current injection using IEEE 1149.4**

When EXTEST is executed, an instrument (e.g., external ATE or some other off-board facility) injects a known current I (DC or AC, see Chapter 6) through the system analog bus AT1. The current return path is via V_G. A voltage-measuring device (e.g., external ATE or some other off-board facility) is connected to the system analog bus AT2 to measure the voltage V_1 at pin 1. The procedure to inject current is complete.

4.4.2 Measuring a voltage at a pin

The procedure to measure the voltage V_2 at pin 2 involves connecting the monitoring bus AT2 to pin 2. To continue the example illustrated in Figure 4.11, the second test to be applied to measure V_2 is as follows (Figure 4.12):

1. The control register of the TBIC of the chip containing pin 1 is loaded with CALIBRATE=0, CONTROL=0, DATA1=1, and DATA2=0. During subsequent EXTEST (Mode1=Mode2=1), this control ensures that the on-chip bus AB1 is connected to the system analog bus AT1 via S5, and the on-chip bus AB2 is clamped to reduce noise. If pin 1 and pin 2 are on the same chip, DATA2 must be loaded with DATA2=1 since the AB2 bus must be connected to AT2 for measurement at pin 2.
2. If pin 2 is on the same chip as pin 1, the TBIC register is already loaded with CALIBRATE=0, CONTROL=0, DATA1=1, and DATA2=1. Otherwise, the control register of the TBIC of the chip containing pin 2 is loaded with CALIBRATE=0, CONTROL=0, DATA1=0, and DATA2=1. This control ensures that the AB1 bus on the second chip is clamped to reduce noise, and is also disconnected from the AT1 system bus. The AB2 bus is connected to AT2 for voltage measurement.
3. The control register of the ABM of pin 1 is loaded with CONTROL=0, DATA=0, B1=1, and B2=0. This control ensures that the pin will be connected to AB1 (which is connected to AT1 via the TBIC) for current injection, so that the current path is not disrupted for the second measurement. During EXTEST (Mode1=Mode2=1), the pin will be connected to the net without any other signal from V_H, V_L, V_G or the core.
4. The control register of the ABM of pin 2 is loaded with CONTROL=0, DATA=1, B1=0, and B2=1. This control ensures that the pin will be connected to V_G during EXTEST and also to AB2 (which is connected to AT2 via the TBIC). The current path is maintained as previously.

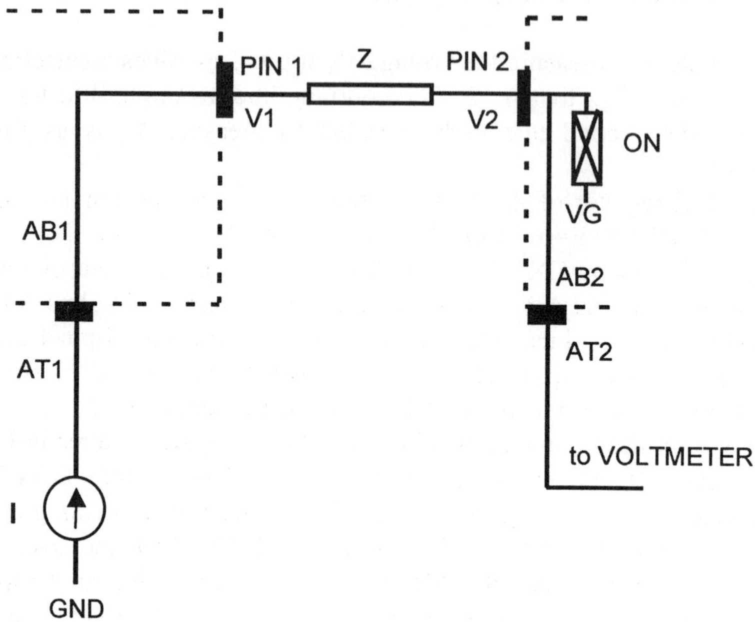

Figure 4.12. **Measuring voltage using IEEE 1149.4**

When EXTEST is executed, the current injection is as described above through the system analog bus AT1. The current return path is via V_G. A voltage-measuring device (e.g., external ATE or some other off-board facility) is now connected to the system analog bus AT2 to measure the voltage V_2 at pin 2.

At this point, the impedance Z is readily computed based on V_1, V_2 and I. Real-life measurements of course are much more complicated due to parasitics, multiple branches of an interconnect, errors due to voltage drops on buses, etc. These factors are examined in Chapter 6 with procedures for accurate parameter extractions.

Parametric testing is one of the crowning achievements of IEEE 1149.4. Even for strictly digital systems, the use of ABMs at digital pins connected to pull-up or termination resistors permits the characterization of these resistors. For high-frequency systems where it is important to verify or measure interconnect parameters, IEEE 1149.4 provides a complete facility for this task. For mixed-signal or analog systems, this is the first time a cohesive and well structured method is presented to standardize these critical measurements. With software tools and design support expected in the near future, these tasks will be automated and relieve a major system design and test problem for manufacturers.

4.5 Internal test using IEEE 1149.4

The infrastructure of IEEE 1149.4, while developed to address primarily the structural and parametric test problems provides high flexibility in system testing and performance verification. We will describe in this section two aspects of the standard for this purpose: probing nodes and extending test bus functions to use in other test applications.

4.5.1 PROBE a signal

The only additional mandatory instruction required by IEEE 1149.4, as compared to IEEE 1149.1, is PROBE. As the name implies, while PROBE is executed, the system functions normally, and the pin to be probed is connected to an analog test bus (e.g., AT2) so that some parameters of the signal at the pin can be measured. The bus AT1 can also be used to provide a stimulus to the pin simultaneously if necessary. All pins are connected to the core. The PRELOAD phase of the SAMPLE/PRELOAD instruction can be used to set up the pin for probing, similar to its usage in EXTEST.

During the PROBE instruction execution, Mode1=0 and Mode2=1. The TBIC configuration for a chip on which a pin is to be probed can be loaded in onc of thcsc configurations:

a) Connect AT2 to AB2 for monitoring, and clamp AB1 to reduce noise on chip (Figure 4.13). The control bits are: CONTROL=0, CALIBRATE=0, DATA1=0, DATA2=1, to close S6 and S9.
b) Connect AT1 to AB1, and clamp AB2 to reduce noise on chip. The control bits are: CONTROL=0, CALIBRATE=0, DATA1=1, DATA2=0, to close S5 and S10.
c) Connect AT1 to AB1, and AT2 to AB2. The control bits are: CONTROL=0, CALIBRATE=0, DATA1=1, DATA2=1, to close S5 and S6.

The ABM of the pin to be probed can be loaded in one of these configurations:

a) Connect pin to AB1. The control bits are: CONTROL=X, DATA=X, B1=1, B2=0, to close SB1. It is advisable to set CONTROL and DATA in a known state during test.
b) Connect pin to AB2 (Figure 4.13). The control bits are: CONTROL=X, DATA=X, B1=0, B2=1, to close SB2. It is advisable to set CONTROL and DATA in a known state during test.
c) Connect pin to AB1 and AB2. The control bits are: CONTROL=X, DATA=X, B1=1, B2=1, to close SB1 and SB2. It is advisable to set CONTROL and DATA in a known state during test.

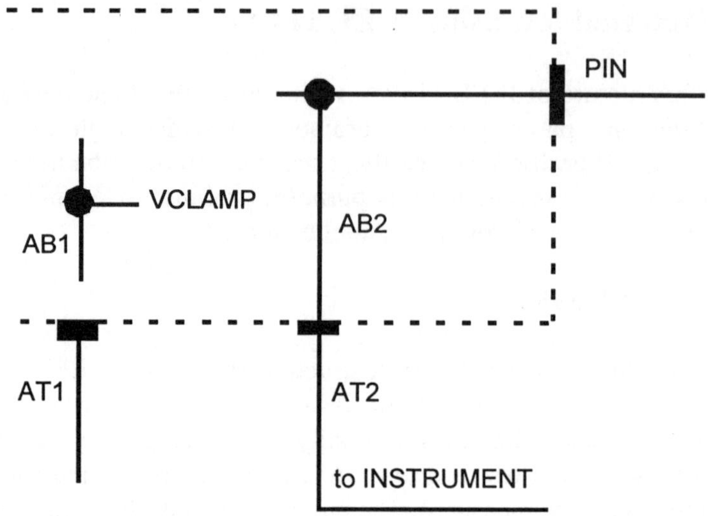

Figure 4.13. **One topology to PROBE a pin using AT2/AB2**

The characteristics of the signal at the pin can be measured in various ways. The simplest is to use an external instrument to monitor the waveform or to extract the desired parameters. The electronic probe is extremely useful in *in-situ* measurements in analog and mixed-signal systems. This capability is also another major advance in system test provided by IEEE 1149.4.

4.5.2 Extending test bus functions

The IEEE 1149.4 infrastructure brings test access and measurement capabilities to the pins of all integrated circuits. These test access and measurement capabilities are easily extended to on-chip test by using them in conjunction with test and design-for-test features already implemented by designers. Some examples of useful test bus extensions are:

1. Improved designs of the pin digitizer permit the use of equivalent time sampling to capture high-frequency signals using low-frequency techniques. This sampling method is widely used in oscilloscopes and has been implemented for mixed-signal test using IEEE 1149.4 [8]. "Early capture," as the implementation is called, was shown to work in the 100 MHz frequency range even though the test bus itself operates at low frequencies. The frequency range can be extended based on design and sampling techniques.
2. Many analog and mixed-signal systems use a multiplexer-based bus to monitor critical signals. This on-chip bus can be easily linked to either AB1 or AB2, and thus accessible at all system levels during functional tests and manufacturing tests. Instructions may be added to perform the

desired connections and control of the multiplexed bus. An on-chip signal may be probed using this extended bus system.
3. Several reconfiguration-based techniques for mixed-signal test, e.g., [9-12], can be controlled via the IEEE 1149.4 structure.
4. Additional waveforms can be simultaneously captured into the ABM control register bits CONTROL, BUS1, and BUS2 by additional 1-bit digitizers and techniques similar to early capture [8]. Since each pin has three such bits, the number of captured waveforms should be more than sufficient in many mixed-signal test applications. For each chip, the two bits CALIBRATE and CONTROL in the TBIC control register may be used in a similar manner to capture two more signals.
5. A serial analog scan bus [13, 14] may be connected to the IEEE 1149.4 structure. This scan bus has one analog scan-in input and one analog scan-out output, and a controller to perform scanning. Since analog scan employs sampling principles, samples of signals may be scanned to and from the on-chip bus using the AT1/AT2 access port. Many on-chip signals may be captured and analyzed by this extended bus system.
6. On-line test techniques and myriad *adhoc* techniques in mixed-signal system test may be implemented in a more structured manner using IEEE 1149.4 for standardized access and as a framework to build on.

4.6 Conclusion

This chapter describes the implementation of various test methodologies using IEEE 1149.4. The standard provides a structure for interconnect test, parametric test, and other functional characterizations of mixed-signal systems. The fundamental and most straightforward applications include interconnect tests and parametric measurements using EXTEST, and monitoring waveforms using the new instruction PROBE. The capabilities of the new ABM are described and compared with the DBM in IEEE 1149.1, and for the first time, test bus structures for differential signals (logic or analog) were consistently and unambiguously incorporated. While the standard is a major advance in analog and mixed-signal system test, its development has also focused on reducing the impact of the test bus circuits on system performance (Chapter 5) and on accurate measurement techniques using the test bus structure (Chapter 6).

IEEE 1149.4 builds on the solid foundation provided by IEEE 1149.1 and has been driven by intense efforts of many capable participants from industry and academia all over the world. The credit for the technical accomplishments described in this chapter belongs to the members of the IEEE 1149.4 Working Group, and to other members of the test community who have directly or indirectly assisted in the standard development.

REFERENCES

[1] *IEEE*, "IEEE Std. 1149.1 - Standard Test Access Port and Boundary-scan Architecture (and 1149.1b-1994 Supplement)," *IEEE,* 1990

[2] *IEEE*, "IEEE P1149.4: Standard for a Mixed-Signal Test Bus (D26)," *IEEE,* 1999

[3] *F. de Jong*, "Testing the integrity of the boundary-scan test infrastructure", Proceedings, *International Test Conference*, pp. 106-112, Nashville, TN., 1991

[4] *H. Bleeker, P. van den Eijden, and F. de Jong*, "Boundary-scan test: A practical approach", Norwell, MA, *Kluwer Academic Publishers*, 1993

[5] *K. P. Parker, J. E. McDermid, R. A. Browen, K. Nuriya, K. Hirayama, and A. Matsuzawa*, "Design, fabrication, and use of mixed-signal IC testability structures" Proceedings, *International Test Conference*, pp. 489-498, Washington DC, 1997

[6] *K. Lofstrom*, "A demonstration IC for the P1149.4 mixed-signal test standard" Proceedings, *International Test Conference*, pp. 92-98, Washington DC, 1996

[7] *W. K. Kautz*, "Testing of faults in wiring interconnects" *IEEE Trans. Computers*, pp. 358-363, vol. C-23, 1987

[8] *K. Lofstrom*, "Early capture for boundary-scan timing measurements" Proceedings, *International Test Conference*, pp. 417-422, Washington DC, 1996

[9] *K. Arabi and B. Kaminska*, "Oscillation-test strategy for analog and mixed-signal integrated circuits", Proceedings, *VLSI Test Symposium*, pp. 476-482, Princeton, NJ, 1996

[10] *M. Renovell, F. Azais, J.-C. Bodin, and Y. Bertrand*, "Design-for-testability for switched-current circuits" Proceedings, *VLSI Test Symposium*, Monterey, CA, 1998

[11] *M. Soma*, "A design-for-test methodology for active analog filters" Proceedings, *EE International Test Conference*, pp. 183-192, Washington DC, 1990

[12] *M. Soma and V. Kolarik*, "A design-for-test technique for switched-capacitor filters", Proceedings, *VLSI Test Symposium*, pp. 42-47, Cherry Hill, NJ, 1994

[13] *M. Soma*, "Structure and concepts for current-based analog scan" Proceedings, *Custom Integrated Circuits Conference*, San Jose, CA, 1995

[14] *M. Soma, T. M. Bocek, T. D. Vu, and J. D. Moffatt*, "Experimental results for current-based analog scan," Proceedings, *International Test Conference*, pp. 768-775, Washington DC, 1997

Chapter 5

Peripheral Cell Design for IEEE 1149.4

Guide for the Integrated Circuit Designer

Keith Lofstrom
KLIC, keithl@klic.rain.com

BACKGROUND

IEEE 1149.4 can help improve system testability, time to market, and even performance through the miniaturization of interconnects and improved characterization. The main cost for a mixed signal design is the addition of some switches and their associated parasitics. This chapter describes how an integrated circuit designer can add those switches, or the functional equivalent, without compromising performance.

At first, IEEE 1149.4 seems like just another nuisance for the mixed-signal integrated circuit designer. With all the other conflicting needs, why add a new one? In today's highly compartmentalized organizations, it can be professional suicide for a chip designer to complicate and delay their design in return for some hypothetical downstream cost savings or quality improvement.

Although intended for production test, 1149.4 can also be used to access pin signals for debugging. Once the overhead of boundary-scan has been paid for, the test infrastructure can also support ad-hoc debugging structures useful to the design engineer. For example, a designer can use IEEE 1149.4 to control and access specialized debug circuitry, perhaps used only for prototype debugging and removed before production. As circuitry shrinks, test logic and latches are becoming much cheaper and less intrusive than probe pads.

As circuits and pad counts grow, boundary-scan is a viable alternative to probe cards with more than a thousand probes.

A significant part of any integrated circuit life cycle is application support. Integrated circuit designers often become involved in this activity, and the plethora of board designs and instrumentation setups at customer sites can make application support troublesome and time consuming. As a standard technique for accessing components instantiated in boards, IEEE 1149.4 provides designers and application engineers with a standard technique for debugging their components in third-party environments. One can even imagine small test instruments that connect to a test access port, to be accessed over the internet, allowing chip designers to help their customers perform sophisticated measurements without ever leaving home.

Ultimately, IEEE 1149.4 can form the nucleus of a set of standardized, reusable test techniques that a designer reuses from design to design. Standards foster design reuse and interchange, and should help enable the import or export of design blocks to other organizations. The standard will change the landscape of analog design, freeing the designer to concentrate on interesting new circuits rather than endlessly rehashing old ones.

5.1 Goals and Usage of IEEE 1149.4

5.1.1 Introduction

IEEE 1149.4 can be used for many things. The test buses and digital control form a very general "analog data highway" of great versatility, designed to be inter-operable with other chips, standardized test hardware, and test software. However, the intent of the standard itself is strictly oriented towards board test, for the purpose of locating shorts, opens, and defective components between chips. The first test hardware and software will be focused on this task. The analog test buses are ***not*** very fast, and without special provisioning are not likely to move high bandwidth signals on and off chips. Typical bandwidths are less than 1 MHz. Typical component test frequencies are in the 10 kHz to 100 kHz region. Surprisingly, these frequencies are adequate for the testing of very small capacitances and inductances, when measured properly. Many of the rules of IEEE 1149.4 are intended to facilitate proper measurement of these components.

IEEE 1149.4 probably will not be used for directly measuring signal quality in even moderate speed systems. The switches are too slow and non-linear. More importantly, test hardware and software will not support such measurements "out of the box". However, these and many other measurements can be performed with built-in-test hardware, supported by such instructions as RUNBIST, and IEEE 1149.4 is excellent for calibrating this built-in hardware. In the longer term, measurements may involve greater interaction between external and internal test facilities, but without some way of standardizing all the different test hardware and software that may be used with a given component, this type of "ad-hoc" test will probably be confined to single-use ASICs and companies large enough to control their own test software and hardware.

In time, continued work will lead to more general and transportable test languages, supporting complicated mixed-signal tests on a variety of test platforms. This will make it possible for ad-hoc test schemes designed for on-chip test circuits to be used by most board test operations in an automated way. That will probably not occur until long after IEEE 1149.4 is well established for its primary goals.

5.1.2 Staying out of trouble

The IEEE 1149.4 standard mandates capabilities and behavior, not structure. The intention of the standard is to insure that test capabilities are available and describable to automatic test generation software. Test software is not intelligent; it cannot follow the reasoning of a pin designer or construct ad-hoc tests. While a test designer can append ad-hoc, one-of-a-kind tests to the automatically generated tests, this will delay production and eliminate one of the big advantages of 1149.4.

Ad-hoc tests may be very useful within an organization for debugging and internal chip testing, but they should not be treated as substitutes for standard 1149.4 tests. If such tests prove generally useful to the entire test community, they may be suitable as public extensions of future versions of the 1149.4 standard, and worthy of inclusion in future versions of test tools. This will not be common, and a designer probably should not attempt to invent new test techniques until the techniques supported by current standards and tools are fully exploited.

The mandated behaviors in the standard describe minimum acceptable behaviors for most situations. Given the broad range and unpredictable nature of analog design, it is impossible to anticipate many different ways an analog pin may be implemented or used. This makes it possible for an integrated circuit designer to meet the letter of the IEEE 1149.4 standard, while designing test circuits that are useless in practice. The standard assumes some maturity and creativity on the part of the integrated circuit designer, as well as some understanding of the test processes the test circuitry is intended to facilitate. Hopefully, this chapter can contribute to that understanding.

One of the worries that often surfaced during the formation of the IEEE 1149.4 standard is the tendency of some designers to look for loopholes in the standard, and to design circuits that meet the letter of the standard but create nightmares for the test engineers who eventually have to design them into boards. The writers of the standard acknowledged that there were plenty of opportunities to abuse the standard. We chose to make the standard relatively unrestricted, to allow analog designers the freedom to make the wonderful and unpredictable design advances we find so marvelous. This prevented us from closing all the loopholes. However, we ask that designers use this freedom wisely, and avoid circuits that make testing difficult. Again, we hope this book will explain enough about the testing process envisioned for IEEE 1149.4 that a good designer can make their components easy to test. Treat the limitations of the standard, not as barriers to push against, but as hazards to be avoided.

The ultimate proof of an adequate IEEE 1149.4 pin design is its acceptance in the marketplace and its compatibility with the software tools that emerge to support the standard. New technologies will certainly modify how the standard is applied to real world problems, and future revisions of the standard will reflect these technologies. But some things will not change; software will stay stupid, board designers will use components in unpredictable ways, and testing will remain a difficult and time-consuming chore. Standards and automated test generation software can reduce the test problem to manageable proportions, freeing designers for more interesting and profitable tasks. Pin designs that are not compatible with emerging test software, are intolerant of widely ranging board designs, or have ad hoc behaviors unsuitable for automation, will not help automate test and are likely to fail in the marketplace.

5.1.3 The Black Box Model

The IEEE 1149.4 standard goes into great detail describing the behavior of pins in terms of switches, internal buses, logical models, and so forth. These are all conceptual models, however, and only describe the black box behavior of a chip that is 1149.4 compliant. A chip may actually contain complicated networks of switches, buffer amplifiers, charge pumps, and bias generators, and still be compliant with the standard. However, the external behavior must be such that the chip *appears* to be a collection of simple switches and buses.

For example, bipolar transistors and MESFETs do not make good transfer gate switches, compared to insulated gate MOS transistors. However, current amplifiers can be used to transmit current from AT1 to the ABM pin, and voltage buffers can transmit the measured result from the

ABM pin to AT2. Although the standard is easiest to implement in modern CMOS, that is certainly not the only technology it can be applied to.

Later, we will discuss residual elements. These may be complicated circuits built out of many transistors, emulating a linear resistor with square-law devices and amplifiers, for example. However, as long as these complicated circuits can be approximated at the pin as resistors, capacitors, and other simple elements, that are independent of bias voltages on other pins, they are allowed by the standard.

The logic behind the switches can be implemented in many different ways, as long as the switches make their externally observable transitions on the falling edge of TCK during the UpdateDR state. Like IEEE 1149.1, designers can use very different circuits and cells inside the chip than the examples indicate, as long as the logical behavior at the pins is the same.

The common theme is that the standard describes a black-box model. The circuitry the designer uses to implement that model is irrelevant to the standard. However, great care must be taken to insure that the black box model is accurate; even small deviations may be detectable from the pins and cause tests to give poor results. As always, when deviating from the recommended cell topologies, the prudent designer will carefully test alternative circuits to make sure they behave exactly the same from the pins outwards.

5.2 Designing for IEEE 1149.4

5.2.1 The Analog Boundary Module

The IEEE 1149.4 standard adds test functions to pins. The intent is to provide test systems with a predictable set of responses to testers, so test software can be automatically generated. This is relatively easy for IEEE 1149.1 digital systems. In that case, the behavior is purely digital, and the logic circuits of the pin driver can be modified without significant effects on the normal mission operation. Outputs remain outputs and inputs stay inputs, and the modifications can be made on internal logic without concern for drive levels or ESD effects. Speed and size remain issues for IEEE 1149.1 digital boundary-scan, but clever design can reduce these effects to negligible proportions.

IEEE 1149.4 requires more care. The pin circuitry is modified, and performance can be degraded unless care is taken. Since IEEE 1149.4 compliant pins perform analog manipulations on external components, the test circuitry is exposed to ESD, leakage, and latchup. The standard mandates that the test circuitry is able to drive external impedances with significant currents and voltages. And the test circuitry itself may degrade sensitive analog performance if not properly designed.

A functional block diagram for a pin circuit with IEEE 1149.4 testability is shown in Figure 5.1. As long as the designer implements the functional equivalent in a black-box-equivalent way, great latitude is allowed in choosing switch sizes and implementations, and in merging switch functions with existing mission circuit structures.

The analog boundary module (ABM) shown is used on all input, output, and bias pins. It is not necessary for digital pins, which may be connected with simpler digital boundary modules (DBMs) instead. DBMs are essentially the same as the input, output, and bi-directional boundary cells of IEEE 1149.1.

Analog boundary modules are implemented with real transistors, with significant parasitic resistance and capacitance. Transistors have voltage and current limits that can put severe limits on testing. The transistors must also be protected from electrostatic discharge, and ESD protection circuits, if designed improperly, can cause IEEE 1149.4 metrology problems.

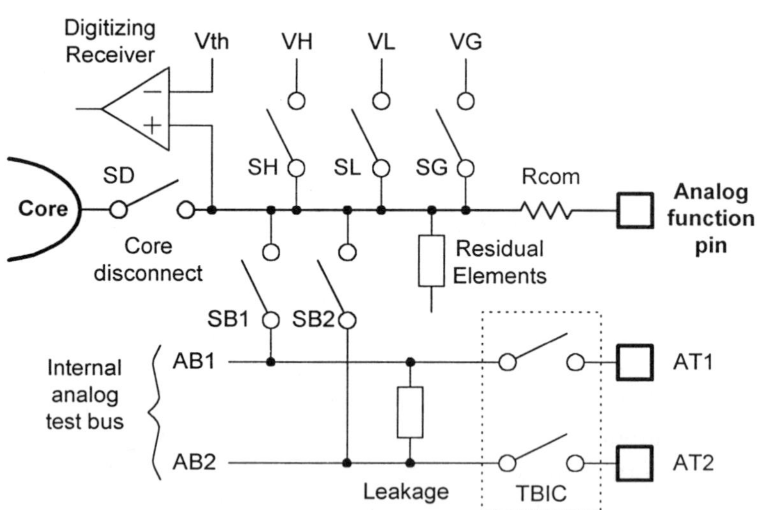

Figure 5.1. **The Analog Boundary Module and part of the Test Bus Interface Circuit**

An ABM is also not necessary for power supply pins, which must be directly wired through solid metal to the edge of the board or a test point. This relaxation of requirements is not intended to allow an integrated circuit designer to designate bias pins, jumper pins, or other signal-bearing pins as power supplies. The possibility that a pin *can* be connected to a hard power rail in some instances does not mean it will always be so connected. Power supplies are just those; sources of the operating currents used to power the chip.

A pin should not be connected with an ABM if a core disconnect on that pin disables or interferes with the operation of the test access port (TAP). For example, in some cases it may be possible to put an ABM on a logic bias pin, or on a pin controlling current sources or logic thresholds in ECL logic. However, if the TAP logic also uses this bias pin, then disconnecting the bias may render the TAP unprogrammable - the chip will go into a state it cannot escape from. More subtly, a bias pin may affect other ABM-equipped pins. This may cause interactions between pins that cannot be described to the test generation software. Later discussions of residual elements will describe which interactions are permissible and which are not.

The same ABM capabilities are specified for input, output, bias, or any other kind of analog pin. This simplifies the standard, the test, and the test generation software. While some analog pins can be easily characterized as input or output pins, others do not fit into these simple categories; for example, a pin connected to an external filter capacitor. While the capabilities are universal, the actual implementation may be different for every pin.

A slightly different ABM may be used for differential inputs and outputs; a differential ABM allows the partial merging of two single-ended ABMs. The standard recognizes that a differential structure may be precluded from implementing all possible pairs of switch closures, or generating all possible output states. More information on differential ABMs will be presented later in this chapter.

The integrated circuit designer is free to include portions of ABMs even on digital pins and power supplies. The ability to measure power supplies may enhance testability. Digital pins with ABMs can be tested for V_{OH} and V_{OL}, allowing the detection of subtle faults that pure digital methods cannot detect. ABMs and other switching structures may even be added to internal nodes of the integrated circuit. While such structures may not be usable automatic board test generation software, they may be valuable for integrated circuit debug or production test.

The ABM provides three functions:

- Disconnection of the mission function during testing.
- Test stimulus and measurement of external components.
- Test stimulus and measurement of external wiring.

These functions are described below.

The black box behavior is strongly constrained by the standard, so that the same control bits will cause the same results on all integrated circuits conforming to the standard. However, the designer has wide latitude in how to achieve this. This is analogous to the way that many different manufacturers can make different microprocessors in different technologies with different microcode, yet all will have the same basic instruction set.

Unlike the digital 1149.1 boundary cell, which must simply pass ones and zeros to predefined logic levels, an 1149.4 boundary module contains many design decisions about switch sizes, voltage ranges, thresholds, and so forth, and these decisions must be made on the basis of the test envisioned for the circuitry connected to that module. If the pin is capable of driving many watts into an 8 Ω load at audio frequencies, different design trade-offs will be made than for a pin that connects to a GHz RF input. It may be difficult to design boundary module switches for components with a wide range of uses.

5.2.1.1 Core Disconnect

Disconnection ("Core Disconnect") is the most misunderstood part of IEEE 1149.4. The purpose of core disconnect is to simplify and isolate the circuitry connected to the pins. This simplifies measurement as much as possible, and prevents damage to the circuit that test voltages might cause. Output circuits must be transformed into simple passive networks, and input circuits should be rendered immune to test voltages. However, much of the circuitry may remain connected to the pin, as long as its interactions are characterizable in specified simple terms.

Core disconnect eliminates the flow of signals through the chip. This simplifies automated test generation software, which assumes simple linear element networks without gain. The software also assumes that these networks robustly tolerate the voltages and currents placed on them by testing. The test signals are usually quite small, perhaps hundreds of millivolts and microamps. However, some analog integrated circuits will respond unpredictably to even small signals, and so they must be rendered quiescent by core disconnect.

Where possible, putting the pin into a high impedance state during core disconnect is ideal. This does *not* have to be done with an explicit switch as shown in Figure 5.1. Most input cells draw little current, and can be designed to tolerate the voltages encountered in test mode without compromise. Most output cells are equipped to exhibit a high impedance state for power-down operation, bus sharing, or other mission-related reasons; this high impedance state is perfectly adequate for core disconnect. In fact, the addition of an explicit series core disconnect switch can greatly reduce driver performance; such switches should be avoided if possible.

However, many analog pins cannot be reduced to a high impedance state. This is where "residual elements" are useful. The IEEE 1149.4 standard facilitates the measurement of external components; certainly on-chip circuit elements that can be modeled as external passives should be just as permissible. Residual elements are linear approximations of the circuit elements that remain connected to the analog function pins even after "core disconnect" has been asserted. Residual elements will be discussed in greater detail later in this chapter.

5.2.1.2 Test stimulus and measurement of external components

The main advantage of mixed signal IEEE 1149.4 over the digital-only IEEE 1149.1 standard is the ability to measure the external components that may surround and interconnect the IEEE 1149.4 chips. The off-chip force and sense buses AT1 and AT2 connect through the on-chip buses AB1 and AB2; these connect through switches SB1 and SB2 to the pin. The control of these switches will be discussed later. These two paths provide a "two-wire-Kelvin" connection to the components outside the chip. The currents and voltages involved in this component test are surprisingly small, in the hundreds of millivolts and hundreds of microamps, using AC stimulus with frequencies within a few decades of 10 kHz, and phase-sensitive measurements. These low frequencies are used for measuring component values when the normal operating frequencies exceed 1 GHz; accurate measurement can only occur if the test circuitry is well characterized, allowing these small reactances to be measured.

A complete current path through the component is required for testing. Test current flow in through AT1 and the AB1 bus. The other end of a component or network is often connected through the SG switch to the V_G supply. The V_G supply may be combined with V_H or V_L under some circumstances described later. The impedance of the SG switch and the V_G supply is subtracted by measuring the voltage on the pin with AT2.

A number of conditions must be met to allow accurate measurement of external components. First, the remaining on-chip impedances must be large or accurately characterized so their effects can be subtracted from the impedances measured. For this reason, although residual elements are allowed it is better if they are high impedance and well characterized during the core disconnect state in which measurements are performed.

Secondly, there may be some wiring shared between the SB1 and SB2 switches to the pad. This will have some resistance, shown as Rcom in Figure 5.1. An IEEE 1149.4 resistance measurement cannot differentiate between this resistance and the external resistance to be measured. More accurate measurement is possible if the common resistance is small, perhaps facilitated by running two different wires to the pad from the two different switches.

There may be leakage between the AB1 and AB2 buses on the chip. These leakages may be partially characterized and subtracted, but their net effect on measurement is to work in conjunction with parasitic switch resistances to produce voltage errors on AT2. These errors, and techniques to mitigate them, will be characterized later in this chapter.

While future CAD tools may be able to automatically deal with all these error sources, for now a competent analog designer must insure that errors are minimized so good measurements can be made.

5.2.1.3 Test stimulus and measurement of external wiring

While it is possible to use the external component measurement to verify that the wiring between chips is correct, current sources and voltmeters are relatively slow. The simple digital interconnect testing techniques used with IEEE 1149.1 are much faster. Therefore, where possible, these techniques should be used to look for shorts and opens between wires, while the more accurate component measurement tests should be used sparingly.

This kind of testing is called "pseudo-1149.1" testing. A typical pseudo-1149.1 test can be completed in $N \times \log_2(N)$ test clock cycles, where N is the number of boundary-scan pins. A board with 2000 pins and a 10 MHz-test clock rate can be tested for shorts and opens in under 3 ms. If analog parametric test is required on a node, many milliseconds may be required to make a single meter reading, taking many seconds per board.

Pseudo-1149.1 testing is performed by switching the analog pins to the V_H and V_L supplies through the SH and SL switches. The V_H and V_L "supplies" shown in Figure 5.1 may not be explicit "hard metal" power supplies. They represent the maximum positive and negative voltage levels the pin can reach under normal circumstances. They are used to drive the pin "digitally" to positive and negative "logic" levels. These levels are transmitted through the wiring between chips to be detected by "digitizing receivers" on the same or other chips. For an output driver, explicit switches may not be necessary, because the output drive devices of the output buffer itself can usually be taken over to do the task.

Typically the digitizing receiver threshold is set halfway between the maximum expected positive and negative levels. However, when analog circuits from a variety of sources are combined together, it is possible that one chip's levels may not be compatible with a different chip's expected thresholds. For example, chip A's output may be designed to produce levels of 0.0 to 1.0 V, while chip B's input is tolerant of voltage levels between 0.0 and 5.0 V, leading the designer to set the digitizing receiver threshold to 2.5 V - a level that chip A cannot produce. This will be rather common when mixing components from different vendors or different process generations, making simple interconnect testing unusable in many situations. Where it can be used, though, it can significantly save time and expense of test.

The analog wiring networks on IEEE 1149.4 boards are bi-directional. Every pin driving a wiring net may be either transmitting or receiving a logic level. Clever test software may some day be able to choose which pins on which chips to use as drivers and receivers to test the net, accommodating differences in drive voltage and digitizing receiver levels. However, for now, a chip designer should assume that the pin design should be able to drive distinguishable logic states into many different kinds of digitizing receivers. This suggests driving as large a swing during test as possible. This requires judgement, however, because large swings at low impedances may exceed the power supply range of another chip.

The digitizing receiver level of a pin design should be set where a number of other chips can drive it. Both drive voltages and digitizing receiver levels (and their variances) need to be specified so that automatic test software can predict their behavior.

5.2.1.4 Driving V_H and V_L levels

The SH and SL switches are use to put V_H high and V_L low "logic" levels on the pin, for simple interconnect test. These voltage levels need not be the same as any particular power supply level; they should be widely separated, though. Typically, V_H and V_L are the extreme voltages that an output buffer will produce, or the maximum range that an input buffer can be expected to see. Where the application permits, though, it is good to make the V_H and V_L separation as large as possible; if they can safely be switched to power supplies, one may do so.

The V_H and V_L levels are used to drive the connected net for simple interconnects test. An 1149.4 analog net will typically have more than one 1149.4 pin connected, and all 1149.4 pins are capable of both driving and receiving simple interconnect test signals. One of the drivers will be chosen by the test software to do the driving of V_H, and the same driver or another to produce V_L. The drivers will be chosen by two criteria: first, they must produce voltages compatible with the input swing limitations of all the other chips on the net, and secondly they should drive voltages capable of switching the digitizing receivers on all the other pins connected to the net. Sometimes these criteria cannot be met; in those cases we must apply the more complicated extended component tests using AT1 and AT2.

Different IEEE 1149.4 components, and even different pins on the same component, may produce different levels for V_H and V_L. Sometimes, a test may use the V_H from one component and the V_L from another to meet the voltage requirements.

On output pins, there are already large devices capable of driving the pin - the transistors in the Mission Mode driver. We can reroute the signals into these output transistors, and either disable the driver (in High Z mode) or use the pull-up and pull-down transistors to produce V_H and V_L. This is illustrated in Figure 5.2.

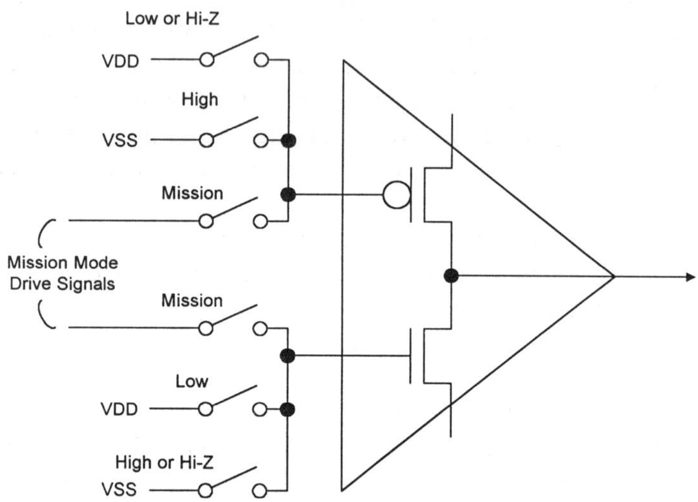

Figure 5.2. **Using the Mission Mode driver to produce V_H and V_L**

There is a subtle design concern when designing switches for V_H and V_L, or when adding V_H and V_L capability to an existing mission buffer. The purpose of simple interconnect test is to locate shorts and opens on a circuit board, so the board can be reworked and the faulty components or wiring repaired. A short on a pin should not destroy the components connected to it during testing, and the V_H or V_L switch could be turned on for many seconds driving a short while the tester isolates the failure. This requires the designer to make sure their design does not overheat or damage the devices connected to a pin when it is connected to a shorted net.

The V_H and V_L capabilities are also available for conditioning a network during external component measurement. We may use V_H and V_L signals to drive the gate of an external FET, for example, to measure its characteristics when turned on or off.

A third ABM switch, SG, connects a reference quality ground to the pin for external component measurement. This switch has relatively tight noise and offset characteristics, compared to the SH and SL switches; on the other hand, the SG switch is allowed to have relatively high impedance. However, where the application permits it and the supplies are clean, it is possible to combine the V_G voltage with either V_H or V_L supply voltage, and use one switch for two purposes. Designers choosing this option must think carefully about all the possible applications their chip will be exposed to, and keep in mind that the ground voltages used in some applications can differ from the voltages of the supplies.

5.2.1.5 Digitizing Receivers

The digitizing receiver is used to detect logic levels in pseudo 1149.1, simple interconnect test. This receiver may be as simple as a logic gate, or the input to a flip-flop, or it may be a full analog comparator to an internally generated reference. The digitizing receiver may be directly connected to the pin, or connected to the output of the mission mode input amplifier normally connected to the pin. However it is implemented, the pin designer should be aware of potential pitfalls in the design. One pitfall, coupling noise during mission mode operation, is illustrated in Figure 5.3.

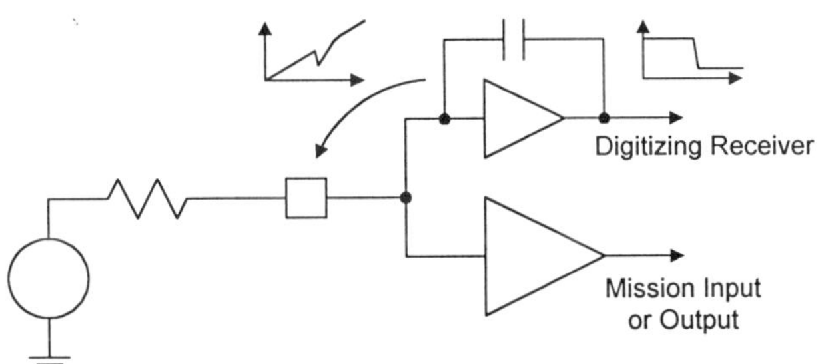

Figure 5.3. **A digitizing receiver with distorting feedback capacitance**

The pin designer should not forget that during normal mission mode, the digitizing receiver should not adversely affect normal analog signal transmission. There should be some way to disable it. The digitizing receiver is a type of comparator. The first gain stage after the pin may have parasitic feedback capacitance (Miller capacitance) between the output and the input. When the comparator switches, there is likely to be some charge injection back into the input node. While this is not a problem for simple interconnect test itself, such charge injection could add errors and distortion to the mission mode signal, unless steps are taken to disable the comparator during mission mode.

This may not be easy. The control signals typically provided to the scan cell are Mode1, Mode2, ShiftDR, ClockDR, and UpdateDR. There is no explicit control line for "digitizing receiver off" so a designer will typically use the Mode1 line to enable the digitizing receivers. This will eliminate noise during mission mode, but it will also make the digitizing receivers unavailable during PROBE and INTEST. Some testing or debugging schemes may attempt to use digitizing receiver during these instructions. Such schemes will not always work.

If the digitizing receiver is enabled by the OR of Mode1 and Mode2, it will be available for PROBE and INTEST, but the switching noise it injects in PROBE mode may cause unacceptable errors in test measurements. As always, judgement is required. The chip designer will have to choose the proper tradeoff between functionality, sensitivity, and digitizing receiver complexity.

As mentioned earlier in the chapter, a digitizing receiver may be implemented as a simple gate (such as a NAND gate), a transfer gate input into a mux or D flip-flop, or other scheme that minimizes the potential nonlinear feedback to the sampled node.

5.2.1.6 Residual Elements

Residual elements model the devices that cannot be disconnected during core disconnect. These were included reluctantly in the 1149.4 standard; test software would certainly be easier to write if it did not have to account for pin-specific component networks. However, sometimes devices cannot be switched off without severely compromising the speed or accuracy of an analog circuit. As an inferior fallback, a designer is allowed to include residual elements in a pin design, as long as they are meticulously characterized and described in copious detail. Any test should not include a residual element where one can be avoided. Including residual elements is risky, and may mean that an otherwise-compliant component may be difficult to test accurately in some circuits. If two otherwise-similar 1149.4 components are available for performing a given function, a board designer may choose the one with the simpler residual element behavior.

Residual elements are restricted to devices that can be modeled as networks of resistors, capacitors, inductors, and independent current and voltage sources from analog function pins to other analog function pins and to the supplies. Residual elements cannot connect to pins with DBMs, and they cannot connect to the core circuitry. The networks are allowed to connect to residual-only "internal" nodes, as long as those residual nodes do not in turn connect to the core. The residual element networks are not allowed to include dependent sources or amplification from other pins. While non-linear devices may be used in residual element networks, they should be close to linear over the range of test signals, so they can be modeled as linear elements with small errors. Thus, the source of a strongly turned on MOSFET may be modeled as a residual element for small signal voltages, while the emitter of a bipolar transistor with its much sharper non-linearity cannot be. It is best to de-bias these devices during core disconnect, so they do not swamp the measurement being performed. This can be done by dragging down the base of a bipolar transistor, for example.

Residual elements can be used to describe termination resistors, resistive ladders, and other high-current or high-precision networks. It is important to make the impedance of these networks reasonably high, however, because the values of the external components connected to these pins cannot be accurately measured if they are in parallel with smaller and less accurate impedance. Typical integrated circuit impedances are notoriously bad for absolute accuracy, and the 1149.4 standard does not presently set forth a method for characterizing the impedances on a particular chip. Thus, the accurate measurement of external devices can be compromised by low-impedance residual elements.

Terminating resistors on a bipolar-only chip may have typical impedances of 50 Ω, and there is no way to switch the device out of circuit. A termination resistor on an MOS chip does not face this restriction; typically the resistor can be disconnected by a large switch at the supply end without compromising performance, and a floating differential termination can be disconnected with a large switch at the center of the resistor. A subcircuit that is difficult to switch out of circuit would be the resistor ladder of a DAC or ADC; it may be necessary to leave the string of precision resistors connected between pads, without interrupting them with switches. For example, an on-chip R-2R ladder driven by on-chip current sources may reduced to a very simple residual element resistor by disconnecting all the current sources, as shown in Figure 5.4.

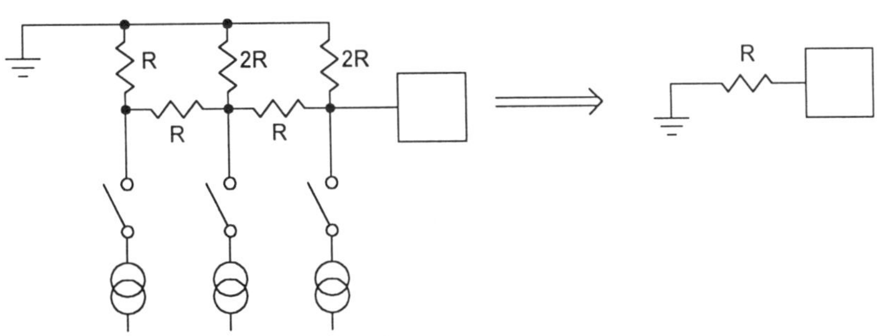

Figure 5.4. **A resistive ladder D/A converter becomes a simple resistor residual element**

5.2.1.7 Choosing Switches

MOS transistor conductivity is proportional to channel width, but so is the junction capacitance. Transistors robust enough to handle pin ESD often have longer channels and wider diffusions than transistors in the core, resulting in much higher capacitance-to-conductivity ratios at the periphery. Since ESD does not scale with technology, the capacitance-to-conductivity ratio of an advanced short-channel process may be the same or worse than the ratio for an older, longer channel process. As processes shrink, the available voltage for driving the switches, and the voltage back-biasing the nonlinear capacitance of the junctions, is reduced.

For example, the maximum typical drain capacitance divided by the minimum typical switch conductance for a PMOS/NMOS device pair constructed with the TSMC 1.2 micron N-Well process, at 25°C, Vdd = 4.5V, and mid-supply voltage, is typically about 70 ps, and is dominated by PMOS behavior. The TSMC 0.35 micron process at Vdd = 3V has a capacitance-to-conductivity ratio of about 70 ps as well. The device models yielding these results are proprietary to TSMC, but other manufacturer's data will probably have similar results.

These characteristic ratios can be reduced significantly by using N devices only for the switches. If the signal voltage and the test voltage are typically near the V_G ground voltage, and that voltage is near the negative supply for the chip, then test voltages will typically be near the negative rail. Conductivity is not needed near the positive rail, eliminating the need for a weak and highly capacitive P channel switch. The result is a factor of 3 to 5 improvements in the capacitance to conductivity ratio. However, this is a risky assumption to make; the chip designer usually cannot be absolutely sure that a board designer will not use the chip in some application where higher voltages are applied to the pin.

Another assumption is that the switches will be operated somewhere near 25°C. Board test environments are typically well controlled, and if the board being tested has proper heat sinking, the reduction of mobility at high device temperature should not be a problem. However, we may expect temperatures to be higher in the field, so while most of the switches can be characterized at low temperatures, a core disconnect switch (if explicit) will have to operate at the maximum device temperature. For a maximum junction temperature of 125°C, the core disconnect switch may need to be 1.6 times wider than necessary at room temperature, with a corresponding increase in temperature.

The characteristic peripheral capacitance-to-conductivity ratios for four TSMC processes are plotted in Figure 5.5. These ratios are approximations, and depend on switch design, junction shape, drive voltage. However, they can be used to estimate the capacitance of a switch given the desired resistance. For example, a rail-to-rail P+N channel switch with a maximum resistance of 50 Ω will have a capacitance of about 1.4 pF. An SB1 or SB2 switch, with a resistance of perhaps 1 kΩ, will have a capacitance of perhaps 70 fF. This smaller capacitance will be much smaller than the capacitance of a typical protection device, and the switch junction itself can be part of the necessary protecting junction.

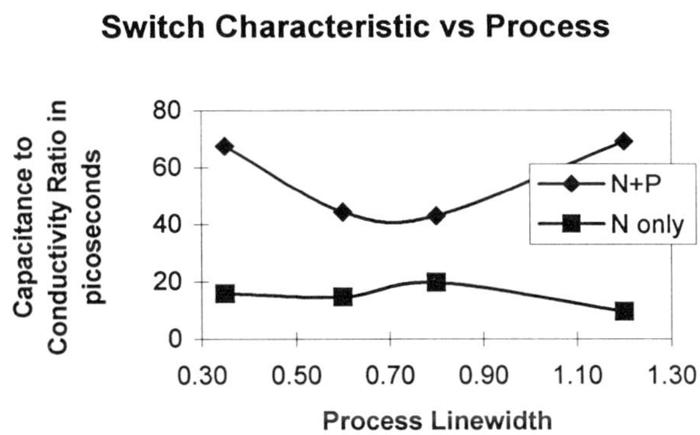

Figure 5.5. **Peripheral capacitance to conductivity ratios for four TSMC processes**

A common mistake in sizing switches for ABMs is making the switches connected to the pad minimum size. Unfortunately, small switches have small junctions and high fields at the corners, leading to preferential secondary breakdown. If small switches are desired, then series resistance is necessary to limit currents. Series resistors have their own problems; see below. Otherwise, the switches should be wide enough, and the "hot" drain junctions long and wide enough, to minimize electric fields during ESD events.

Chip designers must choose transistors with conductivity high enough to drive external circuitry, but capacitance low enough to avoid seriously degrading bandwidth. This makes the choice of transistor widths difficult. This probably also precludes using one "standard" analog boundary module for all the analog pins of a chip, unless those pins have similar impedances or very relaxed speed requirements.

The switches are specified in the standard to have resistances of less than 10 kΩ. While high resistance switches may enable marginally acceptable tests, lower values are typically preferred. CMOS switches are typically nonlinear, subject to self-heating, and the induced voltages can couple between test buses to create measurement errors. Where the application can stand the extra capacitance, lower switch resistances are better. Switch resistances usually do not need to be any lower than the minimum impedance found on a node, and typically ten times that impedance is quite good enough. Remember, though, that the impedance being tested is the AC impedance at perhaps a 10 kHz test frequency, so consider the low frequency reactances of capacitors and inductors when making this determination.

The SG switch may have special requirements. Typically, boards are loaded onto testers cold. As a chip heats up, the resistances of the switches increase. While the resistance of the SB switches is factored out by the nature of the Kelvin measurement, the increasing switch resistance of the SG switch can cause measurement errors, by increasing the test signal voltage across the switch. The voltage at the "ground" side of an external CUT (Component Under Test) returning test current through the SG switch is $V_{AT2\text{-}G} = I_{TEST} \times R_{SG}$. The voltage at the "top" of the external CUT is $V_{AT2\text{-}TOP} = I_{TEST} \times (Z_{CUT} + R_{SG})$. Since the AT2 voltage measurements are taken at two different times, the ground switch resistance R_{SG} will change and introduce an error in the impedance measurement of ΔR_{SG}. This error can be minimized by delaying the most sensitive device measurements to the end of the test, and doing interleaved averages of the two AT voltages, but this consumes valuable test time. It is better to minimize the SG switch resistance on components that can heat up significantly during test. SG switch heating is illustrated in Figure 5.6.

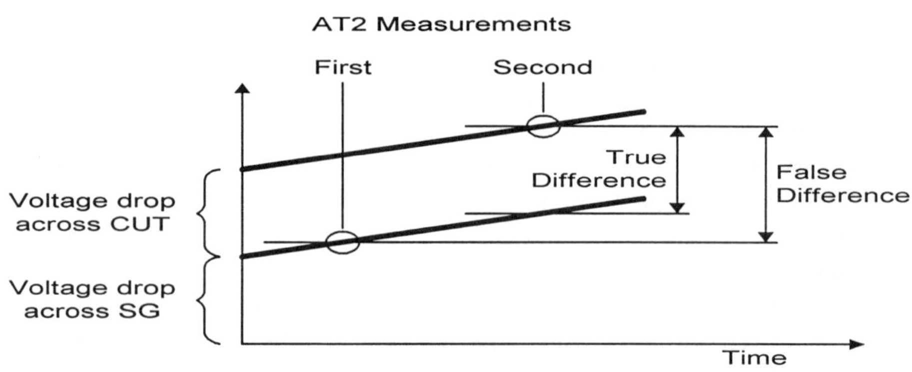

Figure 5.6. **The effect of device heating on SG voltage and measurement accuracy**

The purpose of board short circuit testing is to locate shorts, not to destroy the components connected to the short. This means that V_H and V_L switches (or driver transistors in a mission buffer reconfigurable for "virtual switches") may have to operate for considerable periods of time (perhaps minutes) with large voltages across them. This could lead to heating or hot carrier damage. This may require increased switch area and channel length to spread out device heating and short channel effects.

The AB1 and AB2 buses connect to many ABMs, and can form sneak leakage paths between pins unless care is taken. When signals or ground bounce induce large voltages in the pins, this may be enough to weakly turn on the normally off SB1 and SB2 switches of the ABM. This can affect both mission mode, probe mode, and test mode operations. In mission mode, the clamp switches in the TBICs should divert leakage currents to a supply. However, in probe mode, the AB1 and AB2 buses are carrying signal voltages and currents, and are not clamped, while the chip performs its mission function. Leakage from overshooting pins can leak through the SB1 and SB2 switches of supposedly off ABMs to affect sensitive probe signals from a different ABM. This is shown in Figure 5.7.

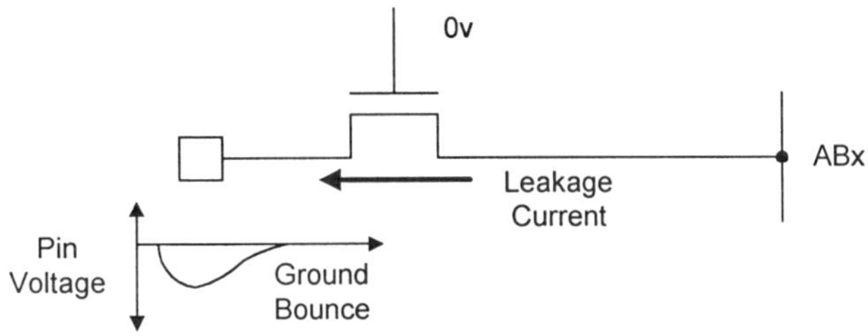

Figure 5.7. **Ground bounce causing switch leakage**

If sensitive measurements are required, and some analog pins may have bouncing or overshooting signals, the noisy pin leakage may be mitigated with the use of a "T switch", as shown in Figure 5.8. The switch is shown in the normally open position. Leakage currents through the first switch are diverted through the clamp switch to a supply, and do not leak into the AB bus. The T switch adds extra resistance and capacitance to an AB switch. However, the clamping switch can be of a minimum size, and both the clamp switch and the switch to the AB bus can be of minimum gate lengths, since these devices do not have to deal with ESD voltages.

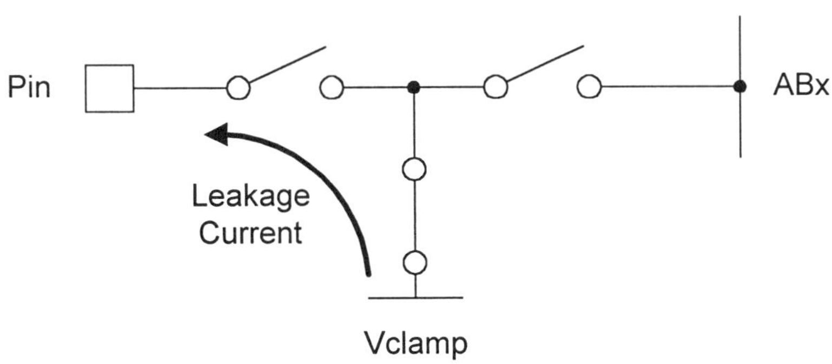

Figure 5.8. **The T switch minimizes leakage from noisy pins**

5.2.1.8 ESD Protection Resistors

A common form of ESD protection circuit is the resistor pi connection, shown in Figure 5.9. ESD currents flow through the outer diodes; the currents are large enough to put large voltages across the diodes. The resistor and the inner diodes limit the voltage on the inner node to levels that short-channel internal devices can tolerate.

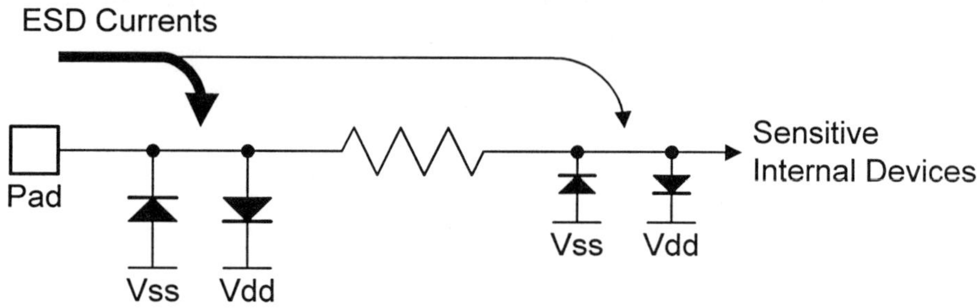

Figure 5.9. **An ESD protection circuit, not suitable for IEEE 1149.4**

The protection resistor can be a problem for metrology. If a measurement current is injected to the node through this resistor, the voltage drop across it will invalidate voltage readings at the protected end. Therefore, the AB2 voltage sensing switch, and the digitizing receiver, must be connected to a separate protection resistor and set of internal diodes, as shown in Figure 5.10. The common path through contacts, internal wiring, and the bond wire should be minimized as well, and may need to be characterized for the test measurement software.

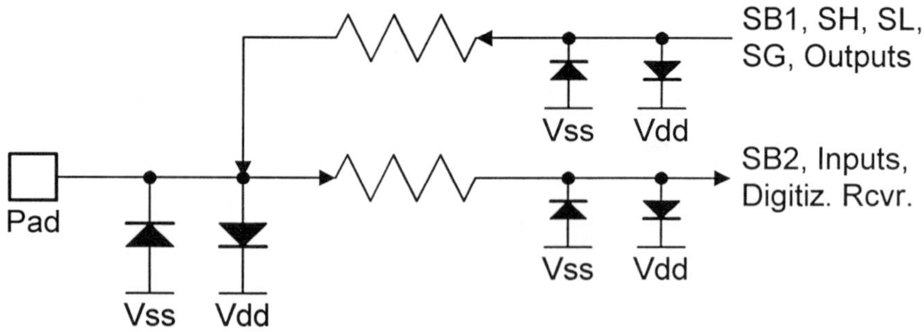

Figure 5.10. **Improved ESD circuit, compatible with IEEE 1149.4**

5.2.1.9 The AB buses and the TBIC

At the frequencies and impedances envisioned for IEEE 1149.4 - under 1 MHz, around 1 kΩ - switch junctions capacitances on test buses can be huge. If we were to connect directly to the AB1/AB2 side of all the pin switches on a board, we could be connecting to thousands of switches. By exposing these internal connections to the outside world, the internal connections now must be armored against ESD currents and voltages. Thus, large junctions with concomitant high capacitances are added with each switch. This could result in thousands of picofarads of loading.

For this reason, we partition the internal (ABx) and external (ATx) test buses with the Test Bus Interface circuit (TBIC). Only the external ATx pins of the TBIC need to be shielded against ESD, and so only one large junction capacitance is added per chip, rather than many. Further, the internal sides of the switches do not need to be ESD-proof and can use smaller, lower capacitance junctions. So even if the TBIC switches are closed, and the internal buses are connected to AT1 and AT2, the total connected capacitance is minimized.

Future mixed-signal VLSI chips may have hundreds or even thousands of mixed signal pins. When this happens, we may choose to partition the capacitance still further, by connecting many separate TBICs from one pair of ATx external pins to many pairs of internal ABx buses. This provides two further advantages - it allows the designer to prevent crosstalk from propagating across the chip, and it allows the combination of many different designs, each with its own TBIC, onto one "system on a chip". Note that the standard allocates four control bits to a single TBIC, while only two additional bits are necessary for each additional TBIC. Since the calibration and simple interconnect test capabilities do not need to be replicated, only two bits are needed to control the ABx to ATx switches for each partition. The standard needs these 2 bit extensions, because multiple and independent 4-bit TBICs give rise to some potentially nonsensical control combinations, such as V_H and V_L switches connected in parallel, or multiply-connected calibration paths. This may confuse test software, so it is best not to do it.

The switches in the TBIC are likely to be much larger than the switches in the ABMs. The TBIC switches are "reused" in series with all the different ABM switches on the chip, so the resistance in the TBIC is added onto every test switch used. On the other hand, the capacitance of all the little ABM switches, and the capacitance of the ABx bus, is likely to be much larger than the additional capacitance of the TBIC switch. As a rule of thumb, the optimum switch size for a TBIC switch is $\sqrt{N}$ larger than an ABM switch, where N is the total number of ABM switches. The optimum is broad, however, and not much is lost by making the TBIC switch larger or smaller than this rule of thumb.

When we are done minimizing the capacitance, we can still expect to see a capacitance in the tens of picofarads on each AB line. Typically, the AT1 and AT2 capacitance of each chip will be on the order of 2-10 pF, the capacitance of a bond pad and an ESD protection structure. If there are 20 chips on a board, all sharing the same AT bus lines, the total capacitance of each line may be on the order of 50 to 200 pF. This capacitance can be measured and subtracted out by the tester, but such subtractions are not perfect and there will be a residual error.

A subtle problem encountered when designing the ABx lines and ATx pin connections is crosstalk. Capacitive coupling between the lines causes errors in Kelvin measurement, appearing as an inductive series parasitic in the component under test. This is because the voltage on the AB1 bus is larger than the component voltage due to the IR drop in the SB1 switch of the ABM. This larger voltage can capacitively couple to the AB2 bus, and the current through the capacitance produces another IR voltage drop across the SB2 switch. This is illustrated in Figure 5.11.

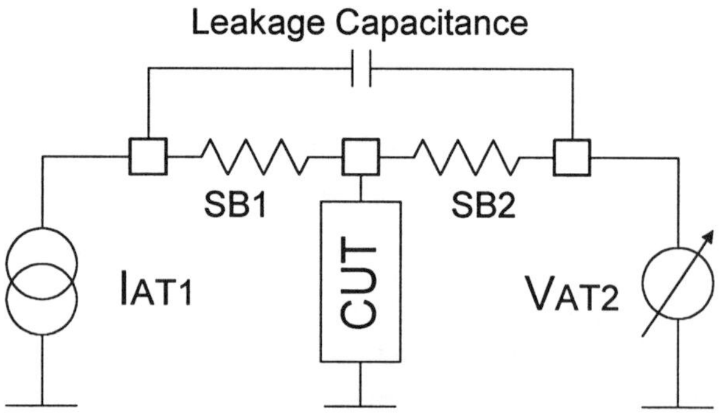

Figure 5.11. **The effect of leakage capacitance crosstalk on measurements**

The induced voltage across the SB1 switch is ($I_{AT1} \times R_{SB1}$). The current through the capacitor is $j\omega C \times (I_{AT1} \times R_{SB1})$, and the induced extra voltage across the SB2 switch is $j\omega C \times (I_{AT1} \times R_{SB1}) \times R_{SB2}$. This is equivalent to an inductance with a value of $C \times R_{SB1} \times R_{SB2}$, which appears to be in series with the component under test (CUT). While this can be partially calibrated out, it still limits the ability to measure small inductances.

The crosstalk can be minimized by running a ground wire between the AB1 and AB2 bus lines on the chip, and perhaps putting shielding metal layers above and below the lines in a multi-layer metal process. The AT1 and AT2 pins should be separated by a ground pin on the package if possible, but at least separated by other non-analog signal pins if a ground pin is too expensive.

When the AB1 and AB2 lines are not connected through the TBIC, they are connected to V_{CLAMP} (typically analog ground) through clamping switches. The intent is to keep these lines from carrying crosstalk signals from one part of a design to another. These clamping switches cannot be characterized though the TBIC interface - they are closed only when the switches to AT1 and AT2 are open. They can be closed in series with one of the ABM switches, however.

The AB1 and AB2 lines are permitted to connect to internal test circuitry, by switches controlled by either the boundary register or by optional user defined data registers. As described in greater detail later, these switches must disconnect themselves when the chip is in Test-Logic-Reset mode, and should not interfere with external component measurement.

5.2.1.10 Buffers and Bipolar Implementation of IEEE 1149.4

Passive switches and large test bus capacitance result in low bandwidth. This makes some component measurements difficult. Pure bipolar chips cannot implement passive switches at all. Recognizing these needs, IEEE 1149.4 permits the use of voltage and current buffers for driving internal buses and pins. These buffers are typically implemented in the TBIC, and controlled by the TBIC control bits. Buffers may permit operating frequencies above ten megahertz in some designs, and can minimize the loading on test buses.

A buffer on the AT1 current input to the chip might appear as a current-to-current or a current-to-voltage buffer, with a virtual ground as an input. When the AT1 switch is not selected, the buffer input connected to AT1 should become high impedance. With this buffer controlling the AB1 bus, it does not matter whether the internal AB1 signal is a virtually grounded current path, or a voltage path, as long as the signal reaches the selected ABM with the desired bandwidth. Differential line pairs are also possible, and may protect the system from noise and crosstalk. There can be signal gain, with the typical 100 μA test currents at AT1 converted to milliamps on chip. It is important that this gain and offset of the buffer be predictable and characterizable, using the TBIC AB1 calibration state (P9 in the standard).

When the AT1 current signal is delivered to the ABM on a pin, it goes through another selectable buffer, either a current buffer or voltage-to-current converter as appropriate, to make a current into the pin. This current buffer should have high output impedance, since that impedance is connected in parallel with the component under test. If it has a finite impedance, it needs to be characterized as a residual element. Note that the current buffer may be combined with the mission mode driver, just as we did earlier for the V_H and V_L switches. Unlike the V_H and V_L switches, the AT1 current buffer must be measurement quality.

Voltage buffers can replace the SB2 switch, either as voltage-to-voltage buffers or into virtual grounds as voltage-to-current converters. The signal may be transmitted differentially to reduce noise. Transmission as a current or as a differential signal implies that there is some measurement quality ground that the pin signal is differenced with, and that the very same ground signal is also available to the TBIC, where the modified signal is converted back to a voltage. Transmission as a unity-gain voltage avoids the need for this reference ground.

The TBIC contains two calibration paths, and these can be complicated when buffers are used. One calibration path measures parasitics on AB1, and with switches is used to transmit a calibration current from AT1, through AB1, to AT2. The other calibration path measures parasitics on AB2, and switches move a calibration voltage on AT1, via AB2, to AT2. When implemented with buffers, the AB1 calibration path should produce a current into AT2 that equals the current driven by the tester into the virtual ground of AT1. The AB2 calibration path should produce a voltage on AT2 that is the same as the voltage driven by the tester into a high impedance AT1.

Buffers can allow significant performance gains, but at the price of gain error, nonlinearity, and offset. Component measurement is not very sensitive to nonlinearity, assuming the tester is well designed, but square law errors can result in DC offsets, and cube law errors can result in incorrect amplitude measurements. Typical component measurements use low amplitude sinusoidal AC waveforms, and are less sensitive to DC offsets, but some residual errors may still occur. And since buffers are active, they are often dependent on bias voltages and power supplies that are shared between cells. Coupling through these lines can lead to the same sorts of errors that leakage capacitance does for passive switched lines, as previously described.

5.2.1.11 Decoding the Update Register

The data from the boundary-scan shift register is loaded into the update register on the rising edge of TCK during the Update-DR state. This information is combined with global signals from the instruction register to control the switches in the ABM and the TBIC. This is illustrated in Figure 23 of the 1149.4 standard [6]. It is important that the switches are able to immediately respond to changes in either the boundary update register or the instruction update register. On the other hand, it is possible to get decoding glitches when the boundary update register changes, and some designs may not tolerate these glitches.

In the ABM, for example, the SH, SL, and SG switches are all decoded from the C and D bits of the ABM's four bit Update register. A transition from {C, D}={0,0}, the high-Z state, to {1, 1}, the SH state, could temporarily glitch {1, 0}, the SL switch, or {0, 1}, the SG switch. The first could result in disturbances in a shared-function output buffer, while the second could inject noise into the ground line.

If either potential glitch is an issue, a cautious designer may chose to do the {C, D} decoding ***between*** the transport shift register (or control register) and the Update register, and using three bits rather than two to store the results. These bits are combined with global controls from the instruction register ***after*** the Update register, so the switches can immediately respond to Instruction changes.

The logic for the TBIC is more complicated, but it may prove less sensitive to glitches because it does not use modified mission mode drivers, and it does not contain a ground switch. If the designer chooses to connect the clamp switches to ground, it is possible to get clamp switch glitches when going between characterization mode and interconnect testing modes. However, this does not interfere with any other measurements being made.

As always, the "black box" model of the chip applies, and a designer may choose very different logic and decoding schemes than the standard suggests, while retaining external compatibility to the standard. When making such changes, the designer should be sensitive to the effects of control bit decoding on glitch creation, leading to ground noise or driver stress.

5.2.1.12 Differential Analog Boundary Modules

Differential interconnect is becoming increasingly common inside and outside of chips. All circuits are fundamentally differential; the "other side" of the differential signal for single-wire signals is usually a common ground wire, and such circuits are very sensitive to noise on that often-abused ground wire. Since single-ended circuits are so sensitive to supply noise, temperature, process variation, and the like, analog designers are making increased use of differential signaling. Differential signaling also offers double the dynamic range for a given signal amplitude, and decreases noise generation at the transmitter. Most offsets and square-law distortion errors are greatly reduced. On the other hand, differential circuits suffer from increases in the number of pins, wires, and devices. And differential circuits must often be tested differentially, which complicates the IEEE 1149.4 standard with the addition of the Differential Analog Boundary Module.

The standard does not require the use of Differential Analog Boundary Modules; in fact, if the design permits, it is often better to avoid them. Differential Analog Boundary Modules add some test capabilities, and remove others. Some early versions of test software may not accommodate these changes very well.

The removed capabilities reflect the nature of most real differential circuits. Modern differential signaling schemes use relatively small differential signaling levels, while tolerating a fairly wide range of common mode levels. The differential lines are often differentially terminated with a fairly small impedance. The mission mode drivers are often incapable of driving both lines low or high - there is only enough current in the cell to drive one side or the other of the terminator. Section 7.4.2 of the 1149.4 standard describes the issues quite well for the designer [6].

Since there is no definable threshold for a single-ended digitizing receiver, the differential ABM adds differential digitizing receivers, allowing a modified form of simple interconnect testing, treating wires as pairs. Unfortunately, shorts and opens may not be differential. Shorts between two differential pairs may result in strange behavior at high frequencies that do not show up with low speed differential testing. Parametric testing using the ATx lines is often required.

Our first case in the rogues gallery is a differential driver similar to a standard LVDS driver. LVDS is short for **L**ow **V**oltage **D**ifferential Signaling, and is described in the IEEE 1596.3 standard [5]. LVDS uses current drive, and forward and back terminated lines. Figure 5.12 shows a simplified version of such an interface.

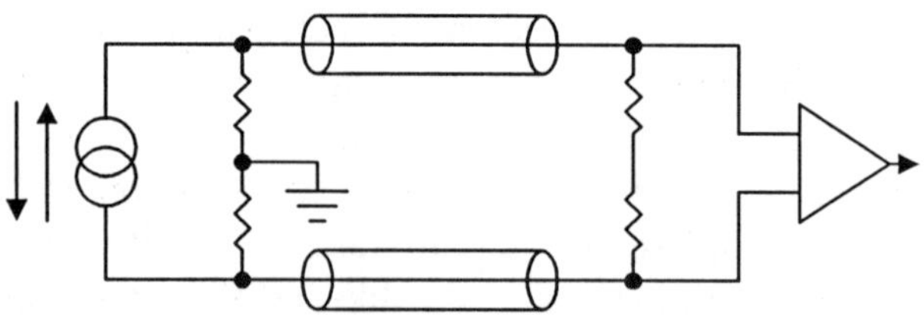

Figure 5.12. **A differential transmitter and receiver**

Assume each resistor is 1 Ω, and the characteristic impedance of the line is 1 Ω as well. Assume a drive current of ±1 A. This produces a received voltage of 1 V. Since the transmission lines are terminated at both ends, there should be no reflections or other signal anomalies. In reality, the impedances at either end of the line may vary in resistance, and the characteristic impedance on the board may also vary. Fortunately, impedance errors are attenuated by a double reflection before they can upset the receiver, making this doubly terminated scheme very robust against parametric variation.

But in the real world of short and open defects, all is not so well. Shorting or opening most single lines or resistors in the above diagram results in a system that *still works,* at least at low speeds. At high speeds, mis-terminations could result in excessive reflections, leading to degraded signal settling time. A differential low-resistance short, such as an extra 1Ω resistance between the differential lines, means the signal gets through, but with degraded amplitude and increased noise and pattern sensitivity. Even worse is a short between two pairs shown in Figure 5.13.

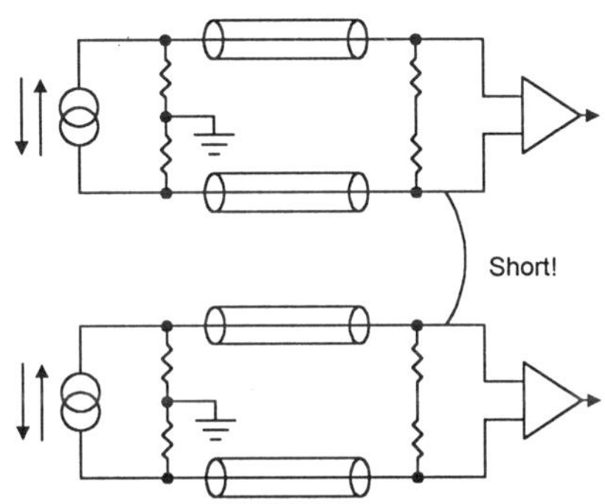

***Figure 5.13.* A short between two differential signals**

If the two lines happen to be carrying complementary data, and they have the same delay, there is no detectable change in the signals. When the signals are the same polarity, there will be appreciable signals flowing between the circuits, resulting in a 33% loss of signal but continued propagation of the DC logic. This may be just enough aberration to make the circuit fail.

There are many ways to find the errors. One method performs parametric measurements and looks for the small changes in received voltage - which, unfortunately, may be smaller than the expected range of device parametric variations. LVDS, for example, has a large permissible amplitude range. Another method is to look at the common mode point on the receiver and look for unexpectedly large variations in common mode signal. The recommended technique is to measure the resistance of each net - we should see a resistance of 0.75 Ω on each normal node and a resistance of half that on a shorted one.

Differential circuits are not as easy to deal with as single-ended ones. Fortunately, IEEE 1149.4 gives us additional tools to deal with them. In fact, most pure-digital differential systems will need to use IEEE 1149.4 to properly test for shorts and opens.

5.2.2 The Demonstration Chips

The test techniques described in IEEE 1149.4 were verified on two demonstration chips, while the standard was under development. IMP Semiconductor fabricated a chip in 1.5 micron CMOS, designed by KLIC, which tried various structures for ABMs, instruction decode, and ad-hoc circuit elements and tests [2]. Meanwhile, Matsushita designed and fabricated a 0.35-micron CMOS test chip with controllable ABM switch impedances. Both chips were designed using early revisions of the preliminary standard, and consequently do not conform to the released standard. However, both designs taught us about the trade-offs inherent in 1149.4 designs. Both devices also contained instructive design flaws.

The IMP/KLIC demonstration chip was a small die, packaged in a 40-pin DIP, containing four different ABM designs, a differential amplifier instrumented with digital ABMs, some digital boundary modules, virtual process monitors, other ad-hoc internal cells, and a test access port and instruction register decode. A design defect in the instruction register decode made an internal control/status register difficult to read back; fortunately, that register was rarely used, since it was there to enable various alternatives for ABM behavior.

The ABMs were implemented with different types of switches. One ABM used simple transistors with internal rules, protected by series protection resistors. These transistors were "series decoded", using multiple analog transistors in series rather than gates to decode the control logic. A similar ABM used ESD-rules transistors, without series protection resistors, also series decoded. These two ABMs demonstrated that it is a *really dumb idea* to use series gating rather than explicit gate decoding (as the rest of the design used).

A third ABM controlled switches directly from the update bits in the boundary register, and a fourth ABM used gate decoding from three control bits to produce control signals for single switch transistors. This last device is the closest ancestor to the ABM specified in the present standard. These cells were lower resistance, lower capacitance, and smaller. Gates are much smaller than even a single ESD-proof switch, and the advantage will increase as processes shrink. The main area costs of ABMs are the wires from the gates to the switches, which have to cross guard ring boundaries.

The chip demonstrated (as if proof is needed!) that explicit series core disconnect switches are expensive and unnecessary. A 50 Ω differential video amplifier on the chip had 140 MHz without disconnect switches. The frequency was reduced to 60 MHz with them. This amplifier also had a Disable input, a much more effective means of turning it off in core disconnect mode.

The demo chip had a diode-connected device, and a MOS threshold measurement circuit, used as a virtual process monitor. Virtual process monitors are described later in this chapter.

These ad-hoc circuits demonstrated the value of built-in circuitry for purposes other than interconnect test.

Among the lessons learned from the demo chip is the necessity of constructing accurate mixed-mode simulations of the chip. Few mixed-mode simulators were available at the time of the design. Other lessons learned involved the drift of the SG switch resistance, which had significant effects on the measurement of resistors. The resistance of switches must be harmonious with the current requirements of external components and of ad-hoc measurements; in some cases, current compliance limits were exceeded during testing.

The Matsushita demonstration chip [3] was equally instructive. Using multiple copies of a four-bit-control ABM cell (still not standard!), this chip had switch impedances controllable from 100 to 6400 Ω. This chip was very useful for simulating the addition of 1149.4 capability to regular analog components, and has been used in a number of research experiments. Matsushita performed experiments measuring signal propagation through the switches, demonstrating the distortion and bandwidth losses that switches can cause. The switches were controllable, but the junction parasitic was that of the largest, lowest resistance switch, so the extra capacitance per pin was on the order of 1 pF.

Due to an auto-routing problem, the Matsushita chip had a very large Rcom, on the order of 160 Ω. This was caused by the auto-router assuming that since two lines were running from the internal switches to the pads, they could be joined. This caused Kelvin measurement errors, of course. However, this resistance did not prevent them from accurately measuring capacitors and inductors. Even resistance measurements could be made after this error was characterized and subtracted. The Matsushita chip made a spectacular showing at the 1997 ITC, where it was demonstrated connected to a commercial Hewlett-Packard board tester. Both demonstration chips, and an increasing number of production chips, show that even with all the pitfalls inherent in early adoption of a standard, IEEE 1149.4 can successfully test boards and systems.

5.3 Extended uses of IEEE 1149.4

5.3.1 Accommodating non-1149.4 components

IEEE 1149.4 may never be used on every mixed signal component. Analog designers are a cranky bunch, and used to getting their own way. They are allowed this behavior because they solve some very difficult problems. Some of these problems are so difficult that less-gifted analog designers may not be able to solve them, and also meet the minimal requirements of the IEEE 1149.4 standard. Whether their arguments against 1149.4 are sensible or not, their designs must be accommodated in a scan testable system.

Nothing in particular has to be done to an 1149.4 component to tolerate its non-scan neighbors. With enough ad-hoc labor by a test engineer, many of these components can be partially tested with 1149.4. We can use AT1 to inject current stimuli into various connected nodes of an external circuit, and observe the results with AT2, where scan components interface with non-scan components. We can also expect "mixed signal scan aid" components to show up on the market, third party components capable of connecting otherwise untestable nodes to V_H, V_L, V_G, AT1, and AT2. What such components cannot do is verify that signals are actually reaching the pins of the non-scan components, and they cannot disable the outputs of non-scan components if those components have no controls for doing so.

To analog designers that insist that their components cannot be made scan testable (and for three terminal components like voltage references, they may have a point) they can still be good "scan neighbors" if they add a "disable" input to their components, allowing their outputs to go to a high impedance, or at least a core-disconnected condition, when the disable input is asserted. It also helps to characterize and document the residual elements that remain connected when the component is disconnected. While these simple requirements will not make their components compatible with the standard, they will make the components better neighbors on the scan testable boards of the future.

We should remember that one of the main advantages of boundary-scan is that permits pins to shrink way down, allowing more pins on smaller packages. It may well be that a chip scale package for a voltage reference, with scan and 16 pins, will be smaller than a naked three terminal voltage reference with three pins and a test land on the board. Reference circuits that can be characterized internally, via the 1149.4 test bus, are likely to evolve to tighter specifications and better performance than components that can only be characterized from three terminals.

5.3.2 Breaking feedback loops

When the IEEE 1149.4 chips go into EXTEST mode, which is necessary to enable V_H and V_L pin drive, all analog pins automatically go into Core-Disconnect mode. While residual elements may remain connected, these are simple networks without gain or signal differencing behavior. On a circuit board, this may break a control loop with feedback, eventually causing a regulator to fail. A board designer needs an overall strategy for dealing with broken feedback loops; designing circuitry to shut down or produce a low-but-acceptable unregulated voltage when EXTEST is asserted and the mixed signal components go out of circuit.

Of course, inadvertent feedback can be created if the mixed signal chip is involved in controlling its own power supplies, and one of those power supplies drives the TAP or the boundary register. While we can check the test program to never exercise the pins that turn off these power supplies, this may still occur inadvertently. While it may be better to make such control pins purposely un-scanned, there is usually a clever way to do partial measurements on such pins without bending the standard so much. For example, the V_H and V_L levels can be chosen so that interconnect test is still possible, but neither voltage will disable the power supply, with the voltage decision level for interconnect test well inside the input voltage margin of the power supply disable signal.

5.3.3 Ad Hoc Measurement Circuits

With the analog measurement system in place, it is tempting to use IEEE 1149.4 for other purposes besides board level testing. Switches, boundary bits, and extra registers can be added to the data-path hardware. User-defined instructions, either public (available to customers) or private (used only by the chip vendor) can be added. These options can be highly flexible, but the user must insure that the basic boundary-scan measurements and data interfaces are not jeopardized. Because these additional features are not described in the standard, it is not likely that they will be available from multiple test software vendors, and thus they will not be very useful for automated board level test. However, they may still prove useful for chip level testing and debugging. As long as the test pins are there, they might as well be used!

The two main rules for ad hoc measurement circuits are:

1) Do not compromise mission mode performance

2) Do not interfere with regular IEEE 1149.4 board test

These requirements are not easy to understand, but fairly easy to once they are understood. The first requirement can be subtle. When the Test Access Port is in Test-Logic-Reset mode, all testability on the chip should be turned off - including the ad-hoc circuitry. In other words, when the system is reset, the component should work normally. This means the ad-hoc circuitry should not require special setup to permit mission mode operation. Test circuitry should also not load or degrade signals, inject noise, or cause other signal integrity problems. Where possible, ad-hoc circuitry should be completely powered down.

These are simply good design practices, but it can be easy to forget them when focusing on the design of a test circuit.

The second requirement is difficult to grasp at first. The ad-hoc circuitry cannot require any special setup or programming that the test software is required to know about. Remember, the same component may go into multiple boards at multiple board houses. Those houses may be using different board testers or test generation software. Ad-hoc setups are not likely to be understood by all these different test tools. And even if ad-hoc setup wasn't just a lousy idea, it violates all the IEEE 1149.x standards.

There are "exceptions" allowed in the IEEE 1149.x standards. First, a chip may have "compliance pins". These test-related pins must be set up with specific DC voltage levels for board-level boundary-scan test to function. If they are jumpered or wired to other voltage levels, the boundary-scan functions may be altered or disabled to permit other kinds of testing or operation. The board test software will not know about these control pins - they will have to be jumpered or wired manually. And the pins should not be programmed sequentially - a straight DC pattern at power-up is all that is allowed. Pin patterns that require on-chip memory or state to determine whether the component is in boundary-scan or not, are ***not*** allowed.

Another "exception" is the use of non-standard, user-defined instructions, public or private. These instructions can initiate tests, target additional data registers, or use existing data registers in novel ways. There is no functional or design difference between public or private instructions - they are both optional sets of instructions, not specified in the standard that may access special data registers or cause temporary state changes in the circuitry. A public instruction is simply a private instruction that is documented in sufficient detail that a board level designer might use it. Do not hold your breath; it is unlikely that off-the-shelf test software will do much with the public instruction, anyway. But if user-defined instructions are designed into a component, it is neighborly to tell users about them. And if the instructions are very useful, perhaps the software vendors will make use of them in future versions of their test pattern generation programs.

When a user-defined instruction is loaded, many new capabilities are possible, but some are riskier than others. For example, a user-defined instruction may target the boundary register, but cause the bits to behave differently. When a transition is made from USER-DEFINED to EXTEST, it will cause the funny bits in the Update register, left over from USER-DEFINED, to take on their EXTEST meaning, with possibly unpredictable results. Since normal test programs are likely to be constructed entirely from the standard instructions like PRELOAD, SAMPLE, EXTEST, PROBE, and BYPASS, these unpredictable results should not occur in normal board testing. But they might - redefine the boundary register with great caution!

Another user-defined instruction might target a user-defined data register. This can be used to set up, control, and read back the digital values generated by the ad-hoc circuitry. The register may connect portions of the ad-hoc circuitry to the ATx or ABx buses, and are permitted to do so as long as those connections are ***only*** made when the user-defined instruction is asserted. The switches should disconnect in other modes, regardless of the state of the user-defined data register. Again, it would be best to shut down and disable the ad-hoc circuitry when user-defined instructions are not asserted.

The boundary-scan interface is useful for device programming. The IEEE P1532 preliminary standard [7] permits the use of the IEEE 1149.x test bus for programming of digital components. It can be used to program EPROMs, programmable logic, and switching fabric, among other digital functions. P1532 does not explicitly discuss mixed-signal components, the techniques may prove useful for programming or trimming analog features. A mixed-signal component whose behavior has been modified using P1532 techniques will have different behavior after those instructions have been run; how that changed behavior will be tested is beyond the scope of this book. Designers considering board-level adjustment of mixed-signal components should become familiar with P1532, then work with the test software vendors to come up with tools that merge programming and analog measurement. These two standards are inter-operable in theory, but it may be a long time before they are inter-operable in practice.

User-defined instructions, data registers, and the ad-hoc circuitry they support must not leave traces that affect the standard tests in any way. The test software should always see a repeatable environment - the same test should always have the same results. Ignoring this requirement will mean that every instantiation of a component on a board will require time-consuming, ad-hoc, and difficult manual test generation using disparate software and test systems. This is a costly mistake, since the available permissions in the standard allow just about any kind of ad-hoc chip test a chip designer might want.

What kinds of ad-hoc tests might a chip designer wish to include? V_{OH} and V_{OL} tests on digital pins, for example. Modern chips with hundreds or thousands of pins require expensive and unreliable probe cards that can reach all those pins. With boundary-scan test and IEEE 1149.4, we no longer need to make physical contact to all the pins on the die during wafer test. Instead, we can probe enough of the power supply pins to power up the chip, then program and observe the pins using IEEE 1149.4.

The AT1 bus cannot provide enough current to test a typical digital IO driver directly, but there are alternatives. First, we can use an ad-hoc on-chip current multiplier to take the 100 μA level currents from AT1, and force milliamp-level currents into the pin under test. Then we can measure V_{OH} and V_{OL} directly on the pins via AT2. This is not a standard test, but it can be set up from user-defined instructions and registers. Such behavior should be turned off during PROBE and EXTEST.

A more portable alternative performs resistance measurements with AT1 and AT2, then extrapolates driver behavior. A 20 mA, 500 mV driver should have an output resistance of 25 Ω or so, and this may be verified with the IEEE 1149.4 test features. This kind of testing replaces the digital boundary module on the pin with a full analog boundary module.

Other kinds of ad-hoc tests might include process monitoring, measurement of internal circuitry, oscillator frequency measurement, on-chip network analysis, speed testing, provision of special biasing, and Iddq testing of digital components.

These are all chip-related tests. Board-level tests might also be performed with ad-hoc tests, but there is that pesky test software compatibility issue, again. Some board-level components would greatly benefit from ad-hoc testing. For example, crystal resonators and SAW devices could be tested over the IEEE 1149.4 buses, with the aid of on-chip circuitry to do frequency conversion or oscillator biasing. Such ad-hoc tests may become common in vertically integrated organizations that can write their own test software and build their own specialized test circuits on-chip. However, it will take a while for these specialized techniques to migrate into general use.

5.3.4 Virtual Process Monitors

Part of the manufacturing of semiconductor wafers is the inclusion of process monitors, groups of transistors with separate pads that are probed on the wafer to characterize it. Process monitors usually consist of about a dozen transistors, some strings of contacts, resistors of various shapes and layers, and a small sea of probe pads so a tester can get at them all. The process monitor is either placed in the scribe grid and stepped with every die, or stepped into a wafer site instead of a mission circuit. In either case, the process monitor is a compromise between adequate process information and the large expanse of pads necessary to connect all the different devices.

IEEE 1149.4 can help eliminate separate process monitors, and can provide better wafer tests with significant wafer area savings. By placing test devices inside the chip and connecting them with AB1 and AB2, they can be significantly smaller than probe pads. Devices may be selected by decoding in an X-Y grid, and on-chip current sensors can measure the results. Small but sophisticated circuits can be designed to extract inferred process variables like gate threshold. Other test circuitry can mimic the critical portions of the mission circuitry on the die. On-chip measurement circuitry can be calibrated and controlled from off chip. Finally, all this circuitry can be disabled via a fusible link or other programming bit, allowing semiconductor foundries to protect their process data.

For example, the three-transistor cell shown in Figure 5.14 generates a voltage approximating the process threshold voltage. Other circuits, generating currents or voltages proportional to mobility, oxide thickness, and other parameters can be envisioned. With appropriate design of test circuitry, process characterization may be no more difficult than connecting an IEEE 1149.4 tester and measuring DC values.

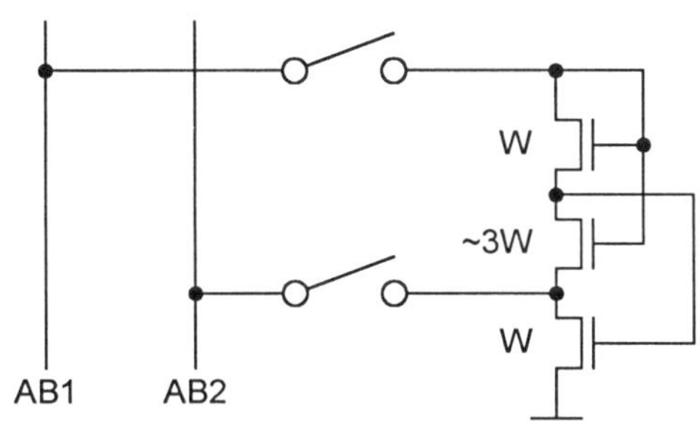

Figure 5.14. **V_T Measurement Circuit**

Such test monitors may be very useful to the fabless semiconductor manufacturer. With a standard test monitor accessible from the pins of a packaged component, the fabless vendor can track process variation without having access to the wafer vendor's separate process die. In fact, such process test information may prove very useful while the component is in the finished board, or after the component is removed from the board for failure analysis.

REFERENCES

[1] *J. E. McDermid, S. Oresjo,* "Structure and Metrology for an Analog Testability Bus", K. P. Parker, Proceedings, *International Test Conference,* pp. 309-322, Baltimore, MD, October 1993

[2] *K. Lofstrom,* "A Demonstration IC for the P1149.4 Mixed-Signal Test Standard", Proceedings, *International Test Conference,* pp. 417-422, Washington, DC, October 1996

[3] *K. P. Parker, J. E. McDermid, R. A. Browen, K. Nuria. K. Hirayama, A. Matsuzawa,* "Design, Fabrication, and Use of Mixed-Signal IC Testability Structures", Proceedings, *International Test Conference,* pp. 489-498, Washington, DC, November 1997

[4] *IEEE*, "IEEE Standard Test Access Port and Boundary-Scan Architecture", *IEEE Standard 1149.1-1990. IEEE Standards Board,* 345 East 47th St. New York, NY 10017, 1990

[5] *IEEE,* "IEEE Standard for Low-Voltage Differential Signals (LVDS) for Scalable Coherent Interfaces (SCI)"*IEEE Standard 1596.3-1996. IEEE Standards Board,* 345 East 47th St. New York, NY 10017, 1996

[6] *IEEE*, "IEEE Standard for a Mixed Signal Test Bus", *IEEE Standard 1149.4-1999. IEEE Standards Board,* 345 East 47th St. New York, NY 10017, 1999

[7][1] *IEEE*, "Draft Standard for Boundary-Scan-Based In-System Configuration of Programmable Devices", *IEEE Standard 1532-2000. IEEE Standards Board,* 345 East 47th St. New York, NY 10017, 2000

[1] Date expected to appear.

Chapter 6

Structural Testing

John McDermid
Hewlett-Packard, mcd@lvld.hp.com

BACKGROUND

The IEEE 1149.4 standard mandates no test methods or metrologies. Rather, the standard provides an architecture for controlling a set of minimum geometry switches within an IEEE 1149.4 compliant device. The issue of which test metrology and instrumentation to use has been left to the test engineer. This chapter focuses on one test method, structural testing, and the instrumentation needed for this method.

Two basic metrologies have emerged with the IEEE 1149.4 standard; functional test and structural test. Functional test methods have been the bulwark of test for many years. Functional testing is the preferred technique to determine whether a circuit will ultimately perform its mission. It directly measures the parameters needed by a circuit to complete the mission. Consequently, pass/fail functional tests must be manually generated from the parameters of the design and their generation requires skilled test engineers. Although the generation of functional pass/fail tests is time consuming, the generation functional test diagnoses is even more difficult. The success of functional test diagnoses is highly dependent on the mission of the circuit and its complexity. Although difficult, functional testing with IEEE 1149.4 is quite possible as long as the bandwidth limitations of the switches and test busses is observed.

6.1 Introduction to Structural Testing

Structural testing (verifying that the right components are connected in the right way) takes a different approach by making a key assumption[1]. Structural tests assume that designs are sufficiently robust to tolerate the worst case variation of all components within their specified tolerance range. Given this assumption, structural testing asks the question "Could the measured node voltages have been generated from any combination of components within their specified range?". Unlike functional testing, the stimulus for structural testing can be arbitrary so long as the expected response to the stimulus is computable. Stimulus can be inserted anywhere an IEEE 1149.4 compliant device is available and voltages can be measured at nodes connected to the compliant device. Structural test metrology begins with a linear (or piecewise linear) model of the circuit and uses the linear model to determine the limits of node voltage variation. Circuits that exceed these limits are judged to be faulty. No design knowledge is necessary to make this determination.

Structural testing has the advantage of producing *component level diagnoses* with an additional assumption, which in most applications does not limit the use of the technique. Structural testing assumes that a limited number of faults (less than the number of nodes minus two) exist in the circuit. When this assumption is true, each component (with the exception of like component types in parallel) affects the node voltages in a unique way. This uniqueness provides the diagnosis and can be obtained directly from automatically generated programs, which do not depend on the skill of the test engineer to extract diagnostics from functional test results. Automatic program generation eliminates much of the effort and expense intrinsic to a functional test.

6.2 The Structural Testing Method

The basic methods used to test individual components provide a foundation for the structural test of networks. The stimulus and measurement techniques used for structural testing are the same as for measurements of an individual component. The difference between individual components and networks occurs after the measurements are taken.

[1] The original proposal for determination of component values from node voltages appeared in [2]. A practical method appeared in [13].

6.2.1 Individual Component Measurement

Figure 6.1 illustrates how an individual component can be measured. The device being measured is *Zx*. The stimulus, *I*, is applied through AT1. The voltage across *Zx* is measured by a voltmeter (having a high input impedance) connected to AT2.

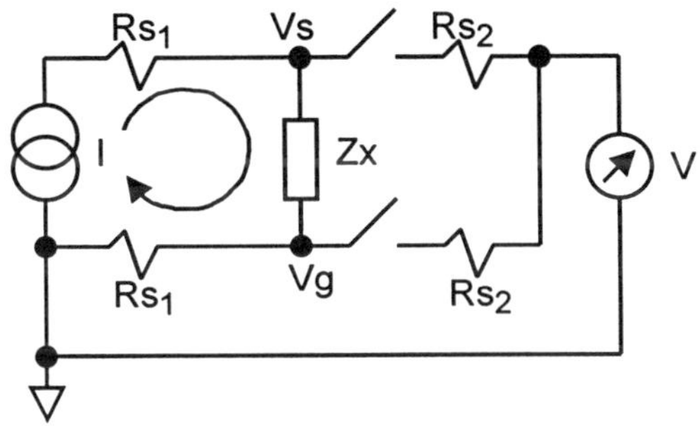

Figure 6.1. **Individual Component Measurement**

Rs_1 represents the equivalent resistance of all the switches in an 1149.4 device connected to *Zx*. Current, from the current source *I*, flows through *Zx* and the impedances labeled Rs_1 . Voltages appearing at the terminals of *Zx* are measured by the voltmeter, *V*. When the input impedance of a voltmeter is ideal, no current flows through the switch impedances Rs_2 in AT2. Because no current flows, the voltages *Vs* and *Vg* are directly determined from the measured values. The voltage across *Zx* is

$$Zx = \frac{(Vs - Vg)}{I} \qquad \textit{Equation 6.1}$$

where *Vs* and *Vg* are complex voltages. The impedance *Zx* is determined by

$$Zx = \frac{(Vs - Vg)}{I} \qquad \textit{Equation 6.2}$$

where it is assumed that *I* is known. The quantity *Zx* may contain both real and imaginary values when physical components are used.

Real physical components must often be modeled as two components in parallel for accurate results. A parallel model occurs, for example, when a resistor has some stray capacitance or a capacitor has a significant dissipation factor. Usually, it is the primary element (of the pair) that is specified and must be measured. Separate measurements of the elements in individual components are possible when the element types are distinct. The impedance, *Zx*, can be modeled as in Figure 6.2.

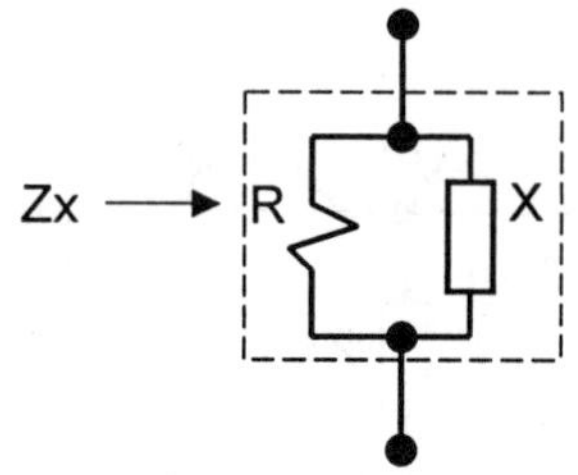

Figure 6.2. **The Parallel Model**

where the value of X is $\frac{-1}{2\pi fC}$ for a capacitor or $2\pi fL$ for an inductor. The current through the resistive branch, *Ir*, is $\frac{V}{R}$ and the current through the reactive branch, *Ix*, is $\frac{V}{jX}$.

These currents add to the total current such that

$$I = Ir + Ix = \frac{V}{R} + \frac{V}{jX} \quad \text{or} \quad V = RI + jXI \qquad \textit{Equation 6.3}$$

The real and imaginary values of V can be used to obtain the individual elements when complex voltages are available from the measurement instrumentation, as shown in Equation 6.4.

$$\begin{aligned} Re(V) &= Re(RI + jXI) = RI \quad \text{or} \quad R = \frac{Re(V)}{I} \\ Im(V) &= Im(RI + jXI) = XI \quad \text{or} \quad X = \frac{Im(V)}{I} \end{aligned} \qquad \textit{Equation 6.4}$$

6.2.2 Testing a Network

Testing a network with IEEE 1149.4 requires the same initial steps as testing an individual component. Current is applied to the network (via AT1) in the same way it was applied to the individual component. Voltages generated by the current in the network are measured at each node in the network (See Figure 6.3). Under the assumption of infinite input impedance, the voltage across switch impedances $Rs2$ is zero, making voltages $V1$, $V2$, ..., Vg directly measurable. The node voltages of the network, with Vg as a reference, are obtained by subtracting Vg from each of the other nodes.

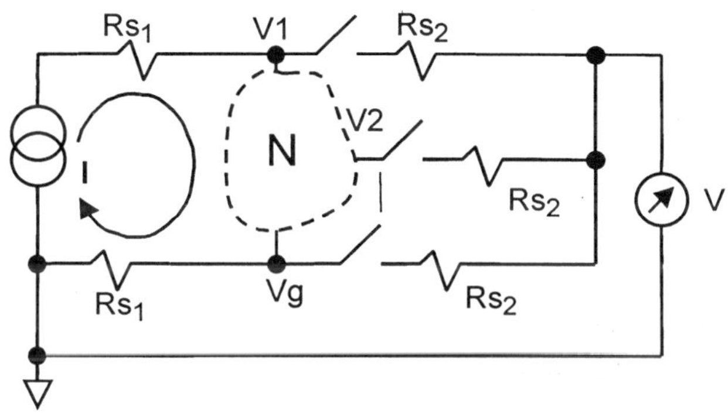

Figure 6.3. **Network Component Measurement**

The stimulus and measurement instrumentation required to determine the value of a single component is also applicable in a network. The difference between individual components and network measurements appears after the stimulus is applied and the voltages measured. The simple relationships shown in Equation 6.3 and Equation 6.4 do not apply in a network. To develop the required relationships for a network, a mathematical model of the network will be needed.

6.2.3 Circuit Models

Several circuit models have been used for circuit analysis. Perhaps the most widely known is the nodal admittance model. (See [3]) Any circuit model used must be linear system of equations having the form

$$MX = S \qquad \textit{Equation 6.5}$$

In Equation 6.5, M is a square matrix, X is a vector of currents and/or voltages, and S is a vector of stimuli. The choice of which circuit model used is arbitrary so long as one fundamental property is present. That property is directly tied to the parameters that can be measured with the IEEE 1149.4 standard and that parameter is voltage. Therefore, any model with this fundamental property has the form where V is the vector of node voltages.

$$MV = S \qquad \textit{Equation 6.6}$$

The response of the circuit to the applied stimulus, predicted by this model, is known as the nominal value. The problem is to determine whether the change in node voltages observed by actual measurement could have resulted from some combination of components within tolerance. Equation 6.6 can be modified to express the change in node voltage, ΔV, which occurs because of the change in component value. (See [7])

$$(M + \Delta M)(V + \Delta V) = S \qquad \textit{Equation 6.7}$$

Because S is the test stimulus, it does not change. When Equation 6.6 is subtracted from Equation 6.7 a relationship between the actual measured voltages, $V+\Delta V$, the range of component values, ΔM, the nominal characteristics of the circuit, M^1, and the change in node voltage, ΔV, is created. That relationship is

$$\Delta V = -M^{-1}\Delta M(V + \Delta V) \qquad \textit{Equation 6.8}$$

Note that M^1 depends *only* on the nominal value of the components. With all component values within their specified range, the range of ΔM is bounded. When the range of ΔM is bounded, ΔV is also bounded. The bounded property provides a connection between design and test.

6.2.4 The Connection Between Circuit Design and Structural Test

This bounded condition is the connection between a fundamental property of linear equations, known as a convex hull and the circuit design itself. The hull is a function of applied stimulus and the tolerances of components. Any particular stimulus will create a hull, but the shape of the hull is solely dependent on the component tolerance and their network interconnection. Thus, the shape of the hull depends on the very properties, which a design depends on to perform its mission. This property of the hull eliminates test dependence on circuit design objectives.

Convex hulls provide an important property needed for test generation. The hull is the surface of an object in ΔV space. In three dimensions that object might look like the outside surfaces of a box which is made with flat sides. With more dimensions, the surface is composed of hyperplanes. (Hyperplanes, in three dimensions, are simply flat sides.) The hull is made from the intersection of these hyperplanes. When measured node voltages are checked against each intersecting plane in the hull and found to be on the inside, the network is said to "PASS".

The hull of a circuit can be explored with Monte Carlo simulation (See the appendix). The circuit in Figure 6.4 is a simple example chosen because two dimensions are easily visualized and plotted. The values of *R1*, *R2* and *R3* are 20 kΩ, 10 kΩ and 10 kΩ respectively. All components have a tolerance limit of 10%. Nominally, V_2 will be one half of V_1.

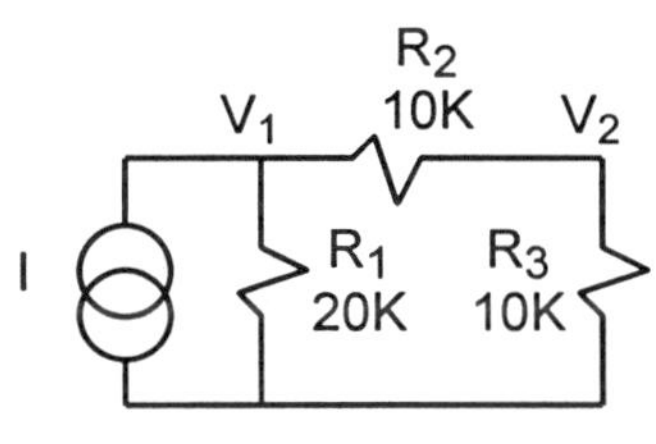

Figure 6.4. **A Simple Example Circuit**

Figure 6.5 plots the change in node voltage ΔV_2, with respect to the change in node voltage ΔV_1, for the circuit in Figure 6.4. The point (0,0) becomes the nominal value in ΔV space. Each point plotted in Figure 6.5 represents a change in voltage from nominal value, ΔV, due to some random combination of *R1*, *R2*, and *R3*. Two hundred points have been plotted. With such a small number of points there is no apparent pattern. The next figure, Figure 6.6, contains 2000 simulations of the same circuit. With 2000 simulations, a pattern emerges but the boundaries do not appear to be straight lines. The lack of straightness, however, is due to an insufficient number of plotted points. Figure 6.7 shows 20,000 simulations of the same circuit. Here the boundaries have become clear and are outlined with straight lines. No points extend beyond the boundary. With straight lines, testing becomes a matter of determining which side of the line a set of measured voltages falls. If any set of voltage measurements fall outside one line, the network is considered to have "FAILED".

More is visible in these simulations than just PASS/FAIL information. In Figure 6.8 one component is allowed to vary over approximately percent of its nominal value while the other components remain within tolerance. The passing region of the network is shown in the center of each sub-figure.

Measurements that fall within this region indicate a passing circuit. Suppose that a measurement (shown by an arrow in Figure 6.8) falls within the region indicated for *R1*. Since the arrow is outside the passing region but within the area defined by sweeping *R1* over percent of its range, *R1* is diagnosed as being faulty. Had the point fallen into the region for *R2*, then *R2* would be diagnosed as faulty. If a measurement falls outside all of the regions, multiple component failures would have been required to achieve those voltages.

In this example it is clear that node voltages determine both PASS/FAIL results and diagnosis when a single fault exists. When multiple failures exist in Figure 6.4, faults are detected but diagnosis is impossible. There must be more than three nodes to develop uniqueness for multiple faults. In general, there must be at least $n+2$ nodes accessible to determine n faults.

There are also measurements where the diagnosis does not indict a single component. An example of this can be seen in Figure 6.9, which is the composite of the graphs from Figure 6.8. The arrow in Figure 6.9 points to a region where a set of measured voltages could belong to either of the region caused by *R1* or *R3*. When a set of measured voltages is a member of more than one region, both components must be reported as possibly faulty since the results are indistinguishable.

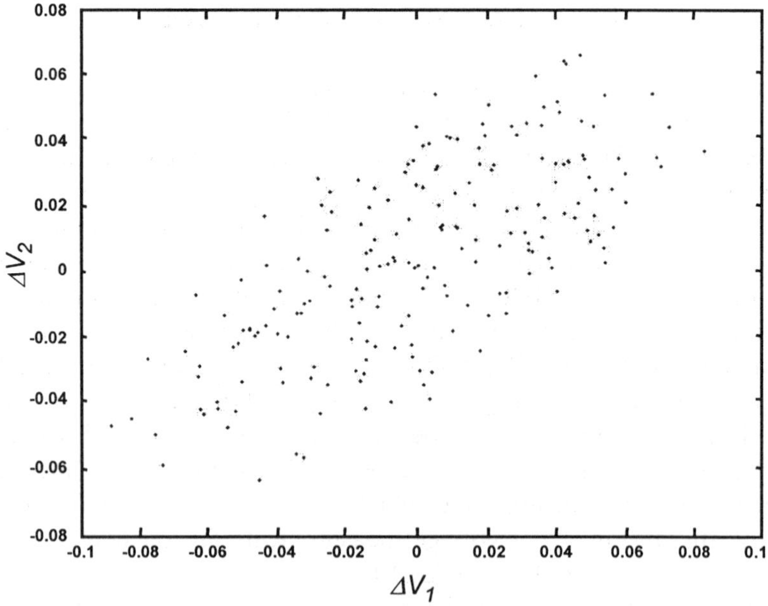

Figure 6.5. **Monte Carlo Simulation of the Example Circuit with 200 Points**

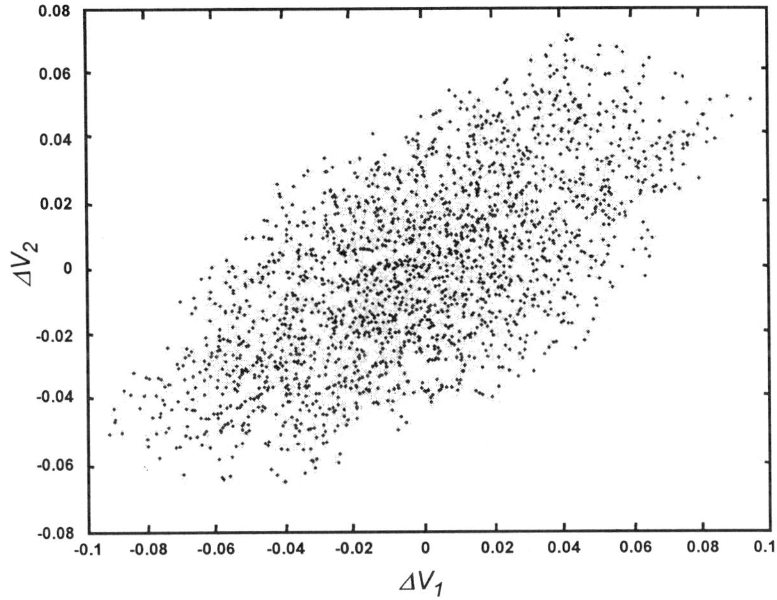

Figure 6.6. **Monte Carlo Simulation of the Example Circuit with 2000 Points**

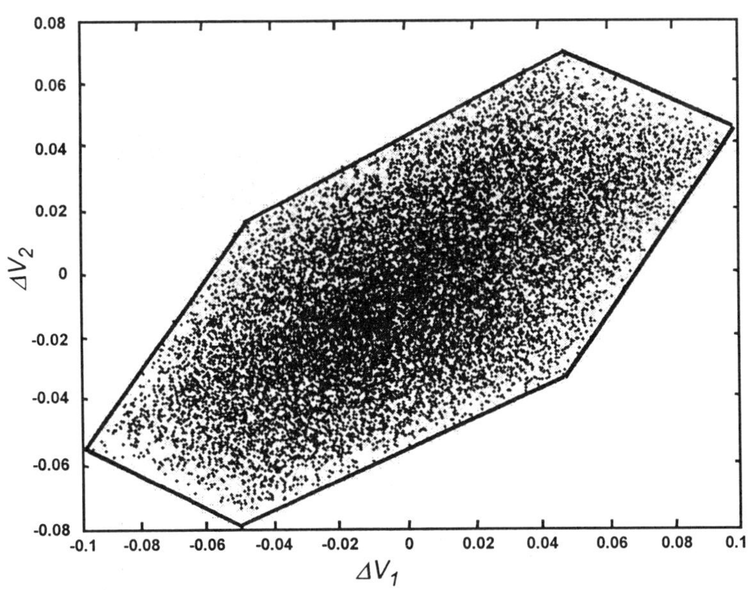

Figure 6.7. **Monte Carlo Simulation of the Example Circuit with 20,000 Points**

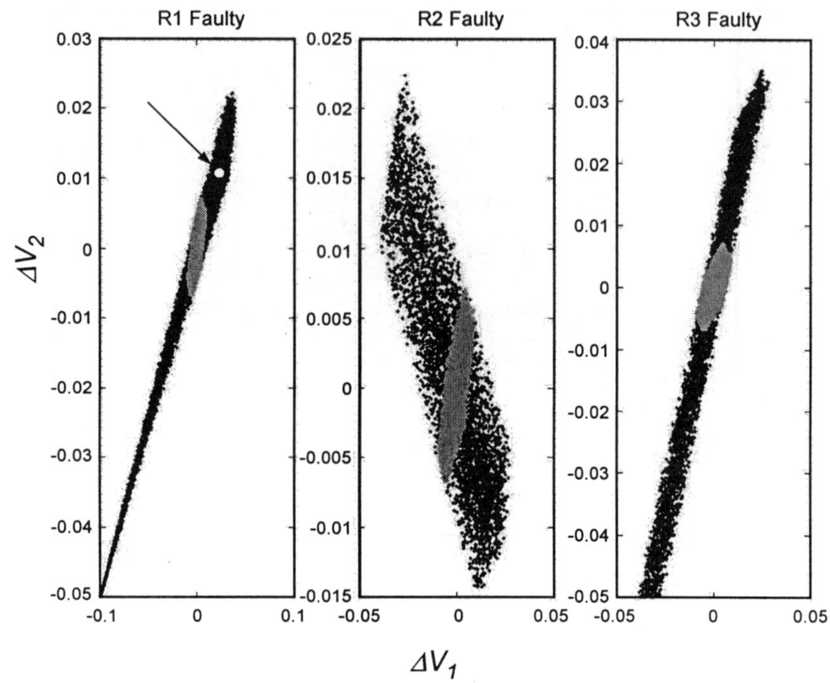

Figure 6.8. **The Fault Directions for *R1*, *R2*, and *R3***

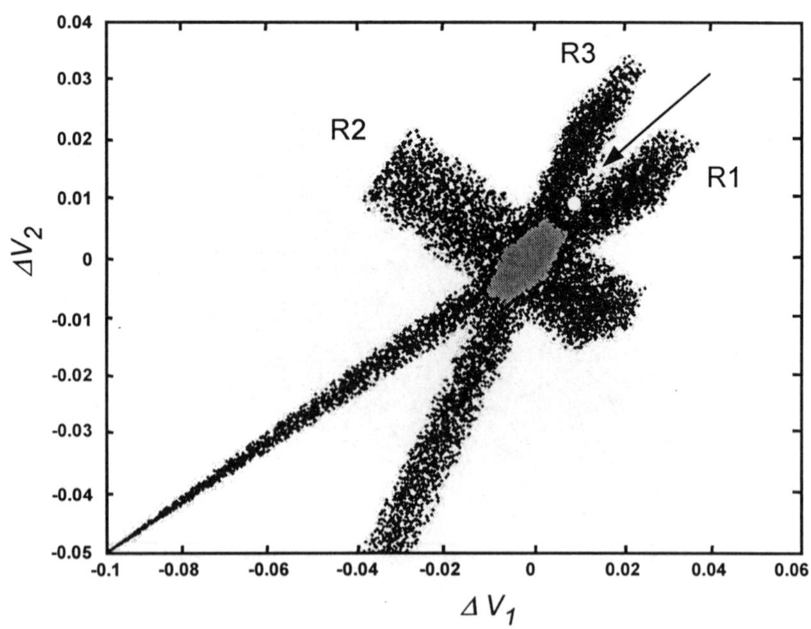

Figure 6.9. **Composite of the Faults**

An estimate of the failing component value(s) can be obtained from Equation 6.8. Once it has been determined which component is faulty, those locations in ΔM which are not faulty can be set to zero and the equation solved for the component value. Setting non-faulty values to zero approximates the measured value of the component. When all non-faulty parts are at their nominal value (that is, their locations in ΔM are set to zero) the estimate is exact. In most instances these errors are much smaller than estimated value of the component, making the estimate acceptable for diagnostic use. If there is concern, the error bounds can be computed by using the values of ΔM that create the maximum and minimum values of ΔV.

6.2.5 When Complete Nodal Access is not Available

The example above has full access to all the nodes in the circuit. Structural testing, however, does not need access to all the nodes to provide acceptable results. In the full access circuit in Figure 6.10, the change in *R3* has exactly the same effect on node voltages as a change in *R4*. Therefore, it is only possible to diagnose a failure as belonging to the set of components {*R3*, *R4*}. Component groups that affect the node voltages in a circuit in the same way are known as equivalence classes.

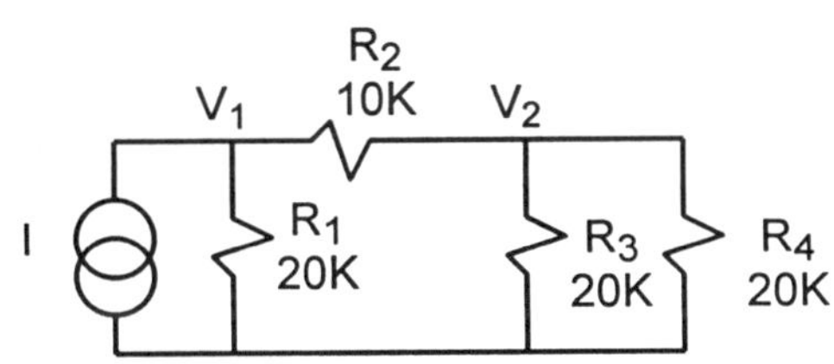

Figure 6.10. **Full Access Equivalency**

When nodal access is limited other component groups, such as device types in series, will form equivalence classes. One example, shown in Figure 6.11, has node V_3 inaccessible. If the voltage at V_3 cannot be measured, the voltage at V_2 is determined by the sum of the component values, making the effect of an individual component indistinguishable.

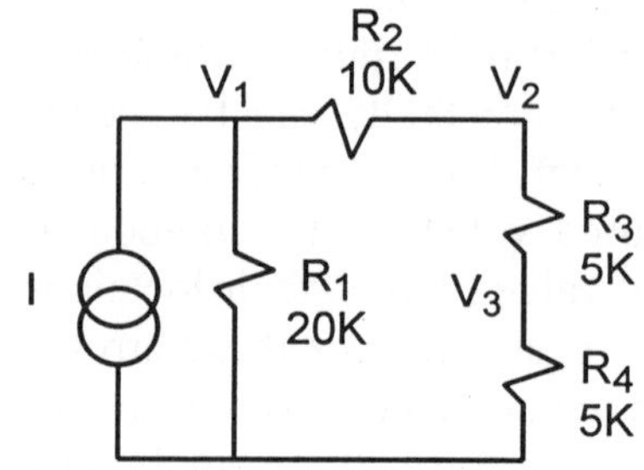

Figure 6.11. **Partial Access Equivalency**

A Monte Carlo simulation of the circuit in Figure 6.11 will not have two edges per component (there are now four components). Rather, there will be two edges per equivalence class, making the simulation appear exactly as in Figure 6.7. The diagnosis for *R1* and *R2* is still unique but the diagnosis for *R3* or *R4* can only be the set {*R3 R4*}. Limited nodal access generally results in a loss of diagnostic resolution due to the creation of equivalence classes. The amount of resolution lost depends both on the number of nodes accessible and their specific location. Other than loss of resolution, the method for limited access does not differ from that of the full access case.

Access can be selected to maximize the diagnostic resolution for a circuit once the schematic of the circuit is available. The choice of access locations depends only on the type and location within the circuit. (See [5] and [6] for a description of an algorithm.)

While the examples shown above have been simple resistive networks, the method is neither limited to resistors or even to passive components. Reactive components generate complex circuit voltages, of which both the real and imaginary parts are bounded. A circuit with reactive components is said to "PASS" when the real part of the measured voltage is within the real hull and the imaginary voltage measured is within the imaginary hull. Active components having small signal AC characteristics may also be included in Figure 6.8. (See [11] for an example of active components.) Measurements must still fall interior to both hulls.

6.2.6 Test Generation

Test generation for structural test requires two steps. First, the circuit must be modeled for simulation by small signal circuit elements. Second, the limits of the convex hull must be determined. The previous example has explored these limits using Monte Carlo simulation. However, this technique is a will become computationally intractable as the circuit grows larger.

Fortunately, there are much more efficient ways to find the edges of the hull. One such method is described in [1]. Other methods are also available in the literature for finding the extreme points from which limits can be derived. See the Directory of Computational Geometry, which, at the time of publication, is available at

http://www.geom.umn.edu/software/cglist/ch.html

The information required for program generation is obtained from the material list, the net list and from the analog BSDL of the IEEE 1149.4 compliant parts. Generally, the material list and net list come from the PC board on which the IEEE 1149.4 parts are mounted. However, test inside the chip is not precluded by this technique since the equivalent description of material and net lists can be made for networks inside a chip. An analog BSDL description will add the characteristics of the compliant device. These characteristics can include devices within the chip connected between the pin and the first switch. Clearly, the development of an analog BSDL for use with the IEEE 1149.4 standard is a high priority task.

6.3 Structural Test Instrumentation

The instrumentation required for successful application of structural testing is not yet available in "off the shelf" parts. Circuits, with other than purely resistive components, require the measurement of complex node voltages. Such instrumentation is presently available in only a few commercial ATE systems. However, quality instrumentation, amenable to large-scale integration, is not difficult to construct and can be obtained from the well-known dual slope conversion process used in high-resolution A/D converters for at least 30 years.

6.3.1 Structural Measurement Instrumentation

Measurement instrumentation for structural methods requires both sensitive and high-resolution measurements. Small signal characteristics are preserved in the IEEE 1149.4 standard by limiting the signal current to a value that ensures the voltage measured is less than 0.1 V. Because active devices are not generally linear, a valid DC operating region must be confirmed prior to measuring small signal characteristics. Measurement instrumentation must support both the determination of a valid operating region and complex AC voltages. There is also another important property for structural measurement instrumentation. The conversion method must have a high degree of linearity.

A dual slope converter is inherently linear. It is important that an input signal, which increases by a small amount, also generates a measured value which increases proportionally. Correct diagnosis depends on the difference in node voltages being meaningful differences and not the by-product of converter bit error. Several dual slope designs have been proven. Figure 6.12 illustrates one such design.

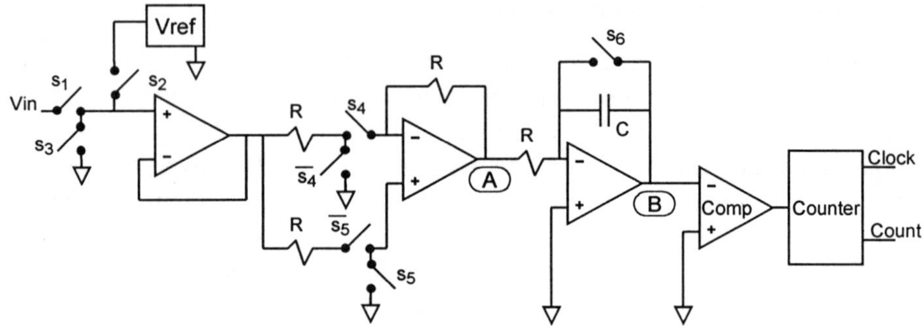

Figure 6.12. **Basic Dual Slope Converter**

The dual slope conversion process conserves the charge on the capacitor *C*. Conservation of charge requires the voltage across the capacitor be zero at the start of conversion and zero when the conversion is complete. Switches s4 and s5 are operated as needed to achieve Equation 6.9.

$$\frac{1}{RC}\int_0^T V_{in}dt - \frac{1}{RC}\int_T^{(T+x)} V_{ref}dt = 0 \qquad \textit{Equation 6.9}$$

If the *RC* product is constant over the interval *T+x,* then this factor can be cancelled from the equation, resulting in

$$V_{in} = \frac{x}{T}V_{ref} \qquad \textit{Equation 6.10}$$

When *T* is a constant, *x* is the time accumulated in the counter at the point where the integrator output crosses zero. Switches s1 and s2 are used to connect to the input signal (*Vin*) to the reference voltage (*Vref*). The choice for the switch settings s4 and s5 are determined at the end of time *T,* thereby causing the integrator output to return to zero. At the end of the conversion cycle switches s3 and s6 are closed and switches s1 and s2 are opened. The integrator output (for a negative input signal) is shown in Figure 6.13. At the end of the conversion, the value of *x* is held in the counter

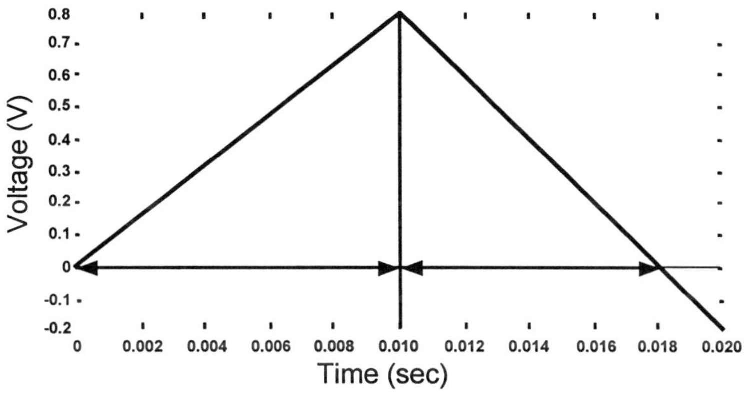

Figure 6.13. **DC Measurement Integrator Waveform at Point B**

6.3.2 Adaptation of the Dual Slope Measurement Technique

The dual slope conversion process can also be used to measure an AC signal with high resolution. Measurement requires the converter (shown in Figure 6.12) to be synchronized to the stimulus source. Switches s4 and s5 are operated synchronously with the source to produce the absolute value of the sine wave during the time interval *T*. The measured value is the average absolute value. An example waveform at the output of the absolute value circuit (for two cycles of the input) is shown in Figure 6.14.

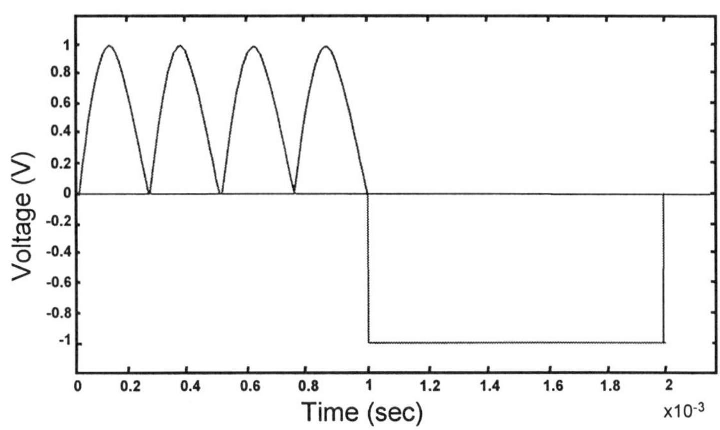

Figure 6.14. **AC Measurement Waveform at Point A**

After integrating *Vin* for time *T*, the remainder of the conversion is completed in the same manner as a DC conversion. The completion can be seen in the integrator waveform shown in Figure 6.15.

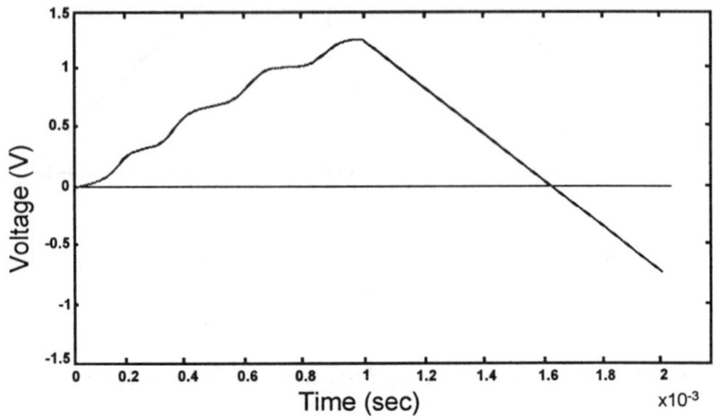

***Figure 6.15.* AC Measurement Waveform at Point B**

For most purposes measurements are calibrated in RMS voltage rather than average value. Equation 6.11 defines the calibration constant needed.

$$\int_0^{\pi} V\,sin(\pi t)dt = 2\frac{V}{\pi} = \frac{2\sqrt{2}}{\pi}V_{RMS} = 0.90031632\ V_{RMS}$$

Equation 6.11

or

$$V_{RMS} = 1.1107207 \int_0^{\pi} V\,sin(\pi t)dt$$

Voltmeters have exploited another property of the dual slope technique for many years. Periodic signals which may sometimes appear in the input signal (perhaps from power line coupling) integrate to zero if their frequency is a multiple of *1/T*. Integration removes the effect of periodic signals on the measurement and allows for accurate measurement of the DC component of the voltage. This property can be used to measure the DC value in the presence of a test signal. Since the IEEE 1149.4 default parameters depend on small signal measurements, the ability to ignore AC signals and measure the DC operating point (which often determine the AC parameters) is very important when testing piecewise linear models. There is a similar property for AC measurement; the removal of a DC component in the AC waveform. Figure 6.16 shows how an 18mv DC signal will appear at the output of the absolute value circuit (point A in Figure 6.12). Clearly, the integral of this waveform is zero when the DC value is constant over the measurement interval. This very important property allows small signal measurements to be made without blocking capacitors.

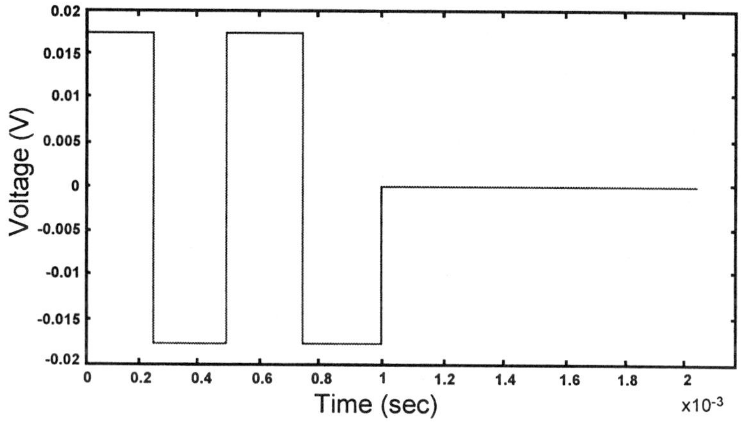

Figure 6.16. **DC Rejection in an AC Measurement at Point A**

One might ask why a blocking capacitor could not provide the same function. Indeed, a blocking capacitor does provide the same function, but at a high cost in the test domain. A long settling delay occurs while the blocking capacitor settles to its final value. To get the capacitor to settle at all, the detector input impedance must also be made finite. Such finite impedance must be modeled in the analysis process. With AC dual slope measurement, the time required to remove the DC signal is exactly the same as for the measurement of the AC signal. Unlike using a blocking capacitor, no transient response is generated.

The method described above measures the magnitude of the sine wave in phase with the source. Therefore, the value returned is the "real" component of the complex voltage.

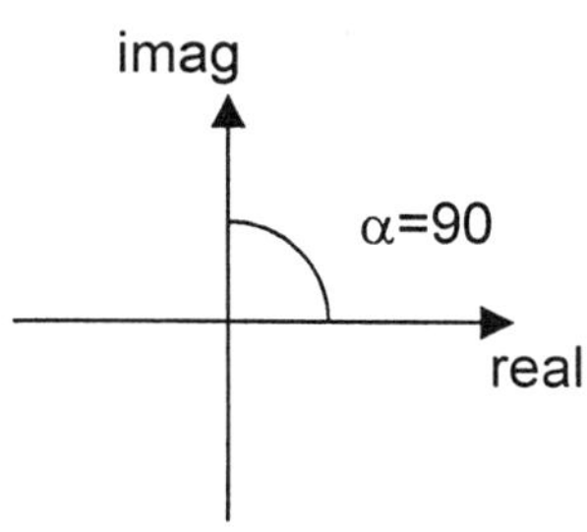

Figure 6.17. **Real vs. Imaginary**

If the operation of s4 and s5 are delayed by 90°, as shown in Figure 6.17, the "imaginary" part of the complex voltage is measured. To make a complete complex node-voltage measurement, the circuit in Figure 6.12 can either be used serially to make consecutive measurements at 0° and then 90°, or implemented as two circuits operating in parallel concurrently obtaining both real and imaginary values. Better performance will be achieved if the parallel model is chosen. Detector calibration is achieved by applying only a DC signal, subtracting the offset, and multiplying by a calibration constant to correct the measured value to the correct value. Note that there are separate offsets for the DC and AC mode since in the AC mode offset in the first stages is eliminated by DC rejection.

In summary, this measurement instrumentation uses a single hardware block to find the DC, real, and imaginary components of a node voltage. Network settling delays are *not needed* for this instrumentation when measuring AC voltages with a DC component. The AC accuracy is very comparable to the DC accuracy and is achieved without an accurate AC source for calibration. This method has produced instrumentation with more than five digits dynamic range, less than 1μV of resolution, and less than a 70msec conversion time. Such converters have been successfully used in some ATE systems for many years [9].

6.3.3 Stimulus Instrumentation

The source needed to realize the measurement must provide synchronization signals to the "real" and "imaginary" detectors. Although there are several ways of providing these signals, the most simple and precise method uses a ROM to construct the digital codes of a sine wave. These codes are then presented to a D/A for conversion to voltage. Digital outputs for the "real" converter (s4-R and s5-R) and "imaginary" converter (s4-I and s5-I) can be generated directly from ROM. Figure 6.18 shows an implementation which provides these synchronization signals.

If we assume that the operational amplifier has a negligible voltage difference between the plus and minus terminals, then a DC value (from the D/A on the left) and an AC value (from the D/A on the right) appear across the resistor *Rs* and determine the current flowing through *Rs*. Since no current can flow into the minus terminal, the same current must flow through a network connected to the current source output. The ability to establish bias conditions in circuits, as shown in Figure 6.18, while concurrently applying a test signal is important for successful application of IEEE 1149.4 test methods.

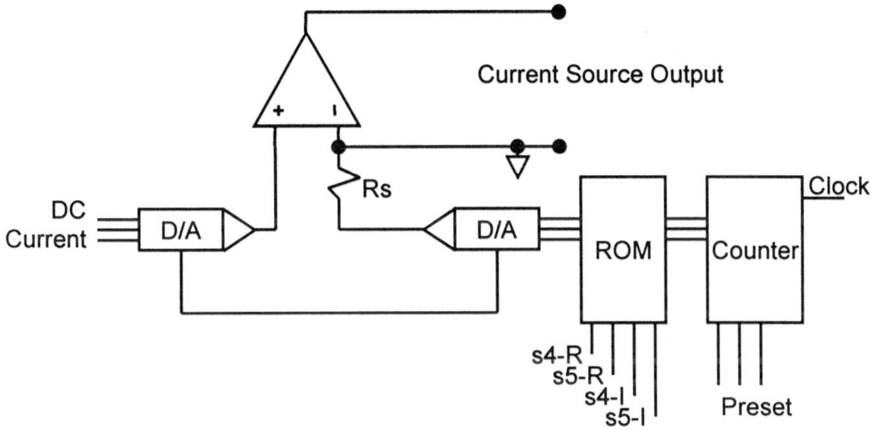

Figure 6.18. **Current Source**

The sine wave generated by the D/A in Figure 6.18 only approximates a true sine wave. Figure 6.19 is an approximation using 20 digital values. The error in this approximation can be found by computing the integral of both waveforms and subtracting the results.

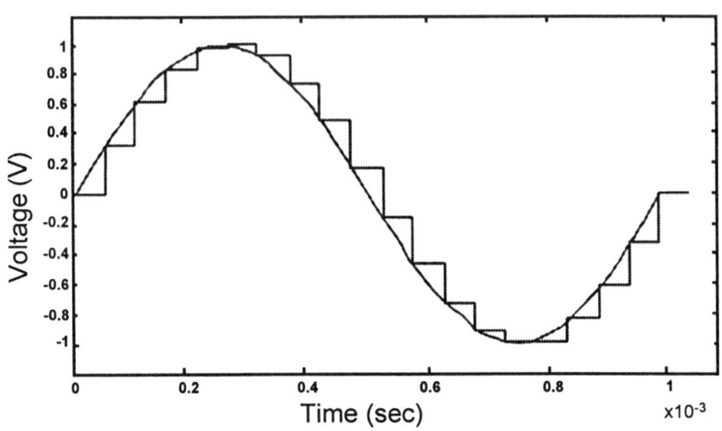

Figure 6.19. **Non-ideal Sine Wave Output**

The integral of the stair-stepped waveform is determined by summing the area of each rectangle having a width of Δt and a height equal to the value of the sine wave at that point. Subtracting the ideal value from this integral yields

$$error = 2kV\,\Delta t \sum_{j=0}^{\frac{N}{2k}} \left| sin(2j\pi f\Delta t) \right| - \int_0^T \left| V\, sin(2\pi ft) \right| dt \qquad \text{Equation 6.12}$$

where $N = \dfrac{T}{2t}$ and k is the number of complete sine wave cycles contained in T. For values of $\dfrac{k}{N} \le 0.1$, the error is well approximated by

$$error \cong -2k\frac{V}{\pi f}\left(\frac{C^2}{3} + \frac{C^4}{45} + \frac{2C^6}{945}\right) \qquad \text{Equation 6.13}$$

where $C = \pi\dfrac{k}{N}$. Note that the error depends only on integer numbers and a well-known constant. This attribute makes correction of this error as simple as multiplying by a constant. Since the source calibration constant K is

$$K = \left(1 - \left(\frac{C^2}{3} + \frac{C^4}{45} + \frac{2C^6}{945}\right)\right)^{-1} \qquad \text{Equation 6.14}$$

This constant can be combined with the other constants used to correct the measured voltage and eliminate the error introduced by the stair step approximation of the D/A.

In summary, the design of the source is straightforward. The resolution of the D/A needs to be sufficient to provide the required resolution, but absolute accuracy is not required. The accuracy is obtained by directly measuring the source with the dual slope converter. Using the metrology about to be described, the source measurement is automatically completed as part of the metrology and does not require a separate calibration step.

SUMMARY

When structural test methods are combined with the IEEE 1149.4 standard, the combination becomes a powerful technique for testing analog parts, both passive and active. Tests are no longer hand crafted, a difficult and time consuming process, but generated automatically like digital boundary scan tests (IEEE 1149.1) are today. When test points are selected to provide optimal access, only some of the network nodes must be connected to an IEEE 1149.4 compliant device. Networks are passed or failed based on whether their nodal voltages could have been created from some combination of parts within tolerance rather than whether the part value itself is within specifications. Diagnoses of failing circuits to a single component (with estimated failing value) or small group of components is also generated as part of the process.

We are seeing an increasing integration of functions, both analog and digital, on a single chip. Functional testing demands that the test engineer understands all the chip functions to be tested. When increasing functionality is combined with "time to market" pressures, it will result in very abbreviated testing or no testing at all. The 1149.4 standard when combined with structural methods, offers the test engineer the benefits of directly testing the circuit design parameters without the need to understand the circuit design, allowing the test engineer to keep pace with this rapidly changing technology.

Structural test methods, however, have two limitations. First, structural testing assumes that *functional specifications can in fact be guaranteed by the design*. When the assumption is true, a structural test will provide a much better fault diagnosis with less test engineering effort. However if the design is not robust, a functional test will *always* be required to differentiate faulty circuits from non-faulty ones. Structural methods also assume that a limited number of faults exist in the circuit block under test. Such an assumption is almost always met in a circuit design released to production.

REFERENCES

[1] *C. Ahrikencheikh and M. Spears*, "Limited Access Testing of Analog Circuits: Handling Tolerances", Proceedings, *International Test Conference*, Atlantic City, NJ, 1999

[2] *R. S. Berkowitz*, "Conditions for Network-Element-Value Solvability", *IRE Trans. Circuit Theory*, pp. 1924-1962, vol. CT-9

[3] *Leon O. Chua, Pen-Min Lin*, "Computer-Aided Analysis of Electronic Circuits", *Prentice Hall*, 1975

[4] *Z. F. Huang, C. S. Lin, and R. Liu*, "Node-FAult Diagnosis and A Design of Testability", *IEEE Trans. On Circuits and Systems*, pp. 257-265, vol. CAS-30, No.5, May 1983

[5] *J.-L. Huang and K.-T. Cheng*, "Analog Fault Diagnosis for Unpowered Circuit Boards", Proceedings, *International Test Conference*, pp. 640-648, Washington DC, 1997

[6] *J.-L. Huang and K.-T. Cheng*, "Test Point Selection for Analog Multi-Frequency Fault Diagnosis", *ECE Technical Report #98-20*, University of California, Santa Barbara, May 1998

[7] *Ruey-wen Liu*, "Testing and Diagnosis of Analog Circuits and Systems", *Van Nostrand Reinhold*, 1991

[8] *K. P. Parker, J. E. McDermid, and S. Oresjo*, "Structure and Metrology for an Analog Testability Bus", Proceedings, *International Test Conference*, pp. 309-322, Baltimore, MD, 1993

[9] *K. P. Parker, J. E. McDermid, R. A. Browen, K. Nuriya, K. Hirayama, and A. Matsuzawa*, "Design, Fabrication and Use of Mixed-Signal IC Testability Structures", Proceedings, *International Test Conference*, pp. 489-498, Washington DC, 1997

[10] *K. P. Parker*, "The Boundary-Scan Handbook, 2nd Edition, Analog & Digital", *Kluwer Academic Publishers*, 1998

[11] *J. E. McDermid*, "Limited Access Testing: Ability And Requirements", *Proceedings of the Technical Program Nepcon West 98*

[12] *Y. Togawa, T. Matsumoto, and H. Aria*, "Linear Algorithm for Branch Fault Diagnosis of Analog Circuits: T F Equivalence-Class Approach", *IEEE Trans. on Circuits and Systems*, pp. 992-1009, vol. CAS-32, N° 10

[13] *T. N. Trick, W. Mayeda, and A. A. Sakla*, "Calculation of Parameter Values from Node Voltage Measurements", *IEEE Trans Circuits and Systems*, pp. 466-475, vol. CAS-26, July 1979

[14] *B. Vinnakota*, "Analog and Mixed Signal Test", *Prentice Hall*, 1998

APPENDIX

This program was used to develop many of the figures in this chapter. The program runs on Matlab Release 11 and perhaps a few earlier versions. The program requires two parameters: a command and the number of points simulated. While only the simple example is coded, it may be modified for larger circuits if the user takes care to ensure the circuit has only three accessible nodes (including ground).

```
function convex_hull_demo(command,n)
% This function is used to demonstrate the convex hull of a very simple circuit
% using the branch admittance model.
%
% Usage:
%       command = " passing" simulates only the passing region
%       command = "individual" shows the individual simulations for a single failing
%       component
%       command = "composite" shows all the single failures with a common set of axes
%       n = number of points simulated
%
% Author:
%       John McDermid -- July 1, 1999

% Define the Yb, A, and J matrices
R1=20000;R2=10000;R3=10000;
I=1e-4;J=[-I; 0; 0; 0];
R=[R1;R2;R3];
Ybn=diag([0;1./R]);
A=[-1 1 1 0;0 0 -1 1];

% Compute the nominal node voltages
Vn=(A*Ybn*A')\(A*J);

% Simulate the passing region
Tol=[10 10 10]'; %Tolerance is entered in percent
for i=1:n
    VN(:,i)=(A*diag([0;1./(rand_value(R,Tol).*R)])*A')\(A*J)-Vn;
end

% Simulate R1 failing
Tol=[99.9 10 10]';
for i=1:n
    VR1(:,i)=(A*diag([0;1./(rand_value(R,Tol).*R)])*A')\(A*J)-Vn;
end

% Simulate R2 failing
Tol=[10 99.9 10]';
for i=1:n
    VR2(:,i)=(A*diag([0;1./(rand_value(R,Tol).*R)])*A')\(A*J)-Vn;
end
```

```
% Simulate R3 failing
Tol=[10 10 99.9]';
for i=1:n
    VR3(:,i)=(A*diag([0;1./(rand_value(R,Tol).*R)])*A')\(A*J)-Vn;
end

switch lower(command)
case 'passing'
    % Plot the passing region
    plot(VN(1,:),VN(2,:),'.g');
case 'individual'
    % Plot the results of each simulation and show the passing region
    subplot(1, 3,1);
    plot(VR1(1,:),VR1(2,:),'.');hold on
    plot(VN(1,:),VN(2,:),'.g');hold off
    title('R1 Faulty');
    subplot(1, 3,2);
    plot(VR2(1,:),VR2(2,:),'.');hold on
    plot(VN(1,:),VN(2,:),'.g');hold off
    title('R2 Faulty');
    subplot(1, 3,3);
    plot(VR3(1,:),VR3(2,:),'.');hold on
    plot(VN(1,:),VN(2,:),'.g');hold off
    title('R3 Faulty');
case 'composite'
    % Plot the composite
    plot(VR1(1,:),VR1(2,:),'.');hold on
    plot(VR2(1,:),VR2(2,:),'.');
    plot(VR3(1,:),VR3(2,:),'.');
    plot(VN(1,:),VN(2,:),'.g');hold off
otherwise
    disp('Command not recognized')
end

% Function to randomize component values
function r=rand_value(R,Tol)
```

INDEX

C

D

E

F

G

H

I

J

L

M

N

P

R

S

T

U

V